Selected Essays on Science and Technology for Securing a Better Philippines

Selected Essays on

Science and Technology for Securing a Better Philippines

Volume 1

Editors
Gisela P. Padilla-Concepcion
Eduardo A. Padlan
Caesar A. Saloma

The University of the Philippines Press
Diliman, Quezon City

THE UNIVERSITY OF THE PHILIPPINES PRESS
E. de los Santos St., UP Campus, Diliman, Quezon City 1101
Tel. Nos.: 9282558, 9253243
E-mail: press@up.edu.ph

The National Library of the Philippines CIP Data

Recommended entry:

Selected essays on science and technology for securing a better Philippines/Editors, Gisela P. Padilla-Concepcion, Eduardo A. Padlan, Caesar A. Saloma.—Quezon City: The University of the Philippines Press, c2008.
p.; cm

1. Science—Social aspects—Miscellanea.
2. Technology—Social aspects—Miscellanea.
3. Science news—Philippines. I. Concepcion, Gisela Padilla, P. II. Padlan, Eduardo A. III. Saloma, Caesar A.

Q175.52P5 303.483 2008 P084000006
ISBN 978-971-542-592-6

UP students on front cover: Marvin Ramirez, Zara Selyn Acaira, Awie Clemente, Roi Christoffer Peralta, May Vergel, Maxine Krizia Dimanlig, Roxanne Roxas, Ervian Ho

Book Design: Nicole Victoria
Printed in the Philippines by Aris Printhaus

Contents

Part 4. On Being A Scientist

Part 5. Immunology, Infectious Diseases, and Public Health Concerns

Part 8. Philippine Biodiversity, Environmental Concerns and Issues

Part 9. Structural Biology, Biophysics, Bioinformatics, and Systems Biology

Preface

THIS BOOK is a collection of essays aimed at providing the Filipino public with viewpoints on a variety of topics in science and technology, scientific research and technological innovation, and national competitiveness in a knowledge-based global economy. The essays are unique—they are written by scientists, academics, and science managers who are based in the Philippines or have spent sufficient time in the country either as a student or researcher at some point in their life.

Advances in science and technology profoundly influence the evolution of human societies and a number of books are already available on the subject matter. To many of us, however, these books suffer from a particular inadequacy—they were written in a general context and without ample consideration of local Philippine conditions, practices, and traditions. In the course of performing my duties and responsibilities as a scientist, teacher, and academic administrator, I have come to recognize the urgent need for multimedia resources that analyze and explain the vital importance of investing seriously and effectively in science education and scientific research and development (R&D) in the Philippines, a country with an economy that is striving to become knowledge-based.

The articles featured in this book have been selected because they promote science in the public mind—a crucial step in building a strong scientific tradition in Philippine society. It is hoped that this book will inspire its young readers to pursue science careers or, at the very least, make them understand what science is about and why scientific excellence is crucial in enabling our society to respond confidently and properly to the socioeconomic pressures that are instigated by a rapidly growing population, an increasingly stressed environment, and a constantly evolving global business climate.

This book came into fruition due to three providential developments. The first was the inception in June 2004 by Dr. Gisela Padilla-Concepcion, Professor of the University of the Philippines (UP) Marine Science Institute, of the weekly column entitled, Star Science, in *The Philippine Star* newspaper. The second was the reexamination of the course offerings of the Science and Society Program that was initiated in June 2006 by the Office of the Dean, College of Science in UP Diliman. The third is the celebration of the centenary of UP which was established in 1908.

The Star Science column is edited by Dr. Padilla-Concepcion and appears every Thursday. It is sustained by the timely contributions of individual stakeholders in the scientific enterprise system of the Philippines. This book is sourced from the articles that first appeared in past Star Science columns.

One tenable outcome of the reexamination was the offering, in the second semester of SY 2006–2007, of a new Science, Technology, and Society (STS) class that was taught with exemplary enthusiasm by invited lecturers from the College of Engineering, College of Science, and the College of Social Sciences and Philosophy. The first offering of the new STS course format was received well by students and additional sections were opened in the following semesters. The effort to revitalize the Science and Society Program also made us aware of the need to develop more multimedia resources that discuss the vital importance of science and technology in Philippine society.

Many activities are being prepared by different sectors of the Filipino community to celebrate the UP Centennial in 2008. This book is a humble gift to UP from the science community in recognition of the vital role that UP has played in the generation of intellectual capital for the Filipino nation.

I ardently wish that the publication of this book would stimulate a period of creative frenzy that will result in the development of more multimedia resources promoting science and technology in our society.

I am truly grateful to Dr. Padilla-Concepcion and Dr. Eduardo Padlan for supporting the project and for generously agreeing to be coeditors of the book. The publication of this book of essays would not have happened without the generosity and goodwill of its many contributors, *The Philippine Star*, Jan-Michael Cayme, and the UP Press. Their vital contributions are duly acknowledged.

Maraming salamat po.

Caesar A. Saloma
Member, National Academy of Science and Technology, Philippines
Professor, National Institute of Physics, College of Science
Dean, College of Science, University of the Philippines Diliman
3 October 2007

Introduction

THE ARTICLES found in this book first appeared in *The Philippine Star*'s Thursday column Star Science from June 2004 to July 2007. Aimed at popularizing science and adding to the *Star*'s regular features on lifestyle, culture, and other areas of human interest, Star Science articles are written by Filipino scientists and science educators on a variety of science and technology (S&T) topics. They are intended for wide readership, for Filipinos—science and non-science professionals, government officials, civic leaders, businessmen and entrepreneurs, students, teachers, laymen, and housewives. With these articles, we aim to improve the science literacy of the general public and contribute to the development of a science culture in our country. We hope to increase awareness of the potential impact of S&T on our daily lives and on socioeconomic conditions in our country.

These articles could stimulate interest, especially among our youth, in socially relevant and important S&T topics. Here we provide them some scientific and intellectual challenges as an alternative, or in addition, to other preoccupations such as the internet and other information communications technology (ICT) activities, music, dance, art, literature, television, movies, entertainment, food, clothing, sports, business, politics, and religion. We feature the work of Filipino scientists and relate them to scientific work being pursued abroad that we hear about more often. We showcase some of our Filipino scientists who are engaged in world-class science with the hope that more of our youth may be inspired to pursue science careers.

With our country's seemingly insurmountable problems, there is a need to project S&T publicly as a potentially significant contributor to increased local productivity and economic growth, the improvement of health and living conditions of Filipinos, the amelioration of our environment, and the protection, conservation, and sustainable use of our natural resources. We hope to influence public opinion and lobby for more government support for scientists and S&T programs in the country. We must impress upon the public the urgent need to improve the state of Philippine science education and S&T research and development, which are sorely lagging behind many of our Southeast Asian neighbors.

Since government cannot provide sufficient support for S&T, there is a need to present to local entrepreneurs and private companies the potential for commercialization of some of the S&T researches being pursued by our scientists. These articles provide an easy and effective means for local scientists to publicize and obtain more support for their work. Scientists in state-funded academic centers will not be perceived as wasting government funds by engaging in useless, irrelevant research, or living in intellectual ivory towers.

Most importantly, we would like to encourage more of our Filipino youth to build careers in science and to contribute in finding constructive and innovative S&T solutions to many of our country's socioeconomic problems. This would be in line with the current direction taken by developing and prosperous countries, which is to build and strengthen their knowledge-based economies. ICT, biotechnology, and nanotechnology are revolutionary technologies of the present and of the future, which will fuel economies toward accelerated productivity, growth, and development. Development of local innovative products in areas where we may have a competitive advantage would boost local commerce, create new jobs, and provide opportunities for import substitution. For those of our society, especially the youth who love nature and the outdoors, proper scientific orientation and training in the conservation of our rich biodiversity and its protection from the destructive effects of environmental pollution will translate to a much better quality of life for us and for future generations.

We thought that a weekly newspaper column would be an effective means of publicizing science, but that it would be difficult to have one writer sustain the column if the articles were to be on a variety of specialized topics. Rather, it would be viable if we could count on the collective effort of many scientists, who would take turns contributing articles in their areas of expertise. The idea was supported by our fellow faculty members at the University of the Philippines (UP) Marine Science Institute (MSI). We then invited MSI Professor Emeritus Dr. Edgardo D. Gomez to cohost the column and he broached the idea to UP Vice-President Dr. Jose "Butch" C. Dalisay, a *Philippine Star* columnist, to explore the possibility of *The Philippine Star* featuring the column. The newspaper was interested and Dr. Dalisay suggested the name "Star Science" for the column. He put us in contact with Mr. Antonio R. Paño, *The Star*'s science editor, who has since been receiving and reviewing the edited articles from us.

We would like to thank the scientists, science educators, administrators, and advocates in UP, La Salle, Ateneo, UST, other local and foreign academic institutions, nongovernment organizations (NGOs), the members of the Philippine-American Academy of Science and Engineering (PAASE), and Her Excellency President Gloria Macapagal-Arroyo, who have contributed articles and allowed us to sustain Star Science for more than three years. They have agreed in writing to the publication of their articles in this book. We would like to qualify that the views expressed by the authors do not necessarily reflect the views of the editors or the publisher.

We thank Dr. Edgardo D. Gomez, Dr. Jose C. Dalisay, Mr. Antonio R. Paño, for making the Star Science column a reality, and Dr. Gomez, Dr. Maria Luisa Virata-Theimer of PAASE, and Dr. Virginia D. Monje of UP for helping us solicit articles for Star Science. With a growing league of young Filipino scientists, we will continue to feature the scientific work of Filipinos in the coming years and to compile the articles in succeeding book volumes.

We hope that our readers find the articles highly informative and useful, as well as enjoyable to read. The book is divided into twelve parts. It begins with articles on the role of science in society (Part 1: Public Understanding and National Importance of Science and Technology), and the importance of science education at all levels (Part 2: Science Education and Creativity; Part 3: Research and Development, and Institution Building; Part 4: On Being a Scientist). Then there are articles on various scientific disciplines and themes and their relevant applications in society (Part 5: Immunology, Infectious Diseases, and Public Health Concerns; Part 6: Herbals and Other Natural Products as Health Remedies; Part 7: Food, Agriculture, Aquaculture, and Biotechnology; Part 8: Philippine Biodiversity, Environmental Concerns and Issues; Part 9: Structural Biology, Biophysics, Bioinformatics, and Systems Biology; Part 10: Mathematics, Physics, and the Engineering Sciences; Part 11: Geobiology, Geology, and Paleontology; and Part 12: Population Issues.)

For readers who may want more detailed information, we have included the current affiliation and email address of the author at the bottom of the first page of each article. The date when the article appeared in the Star Science Column of *The Philippine Star* is indicated at the end of the article.

GISELA P. PADILLA-CONCEPCION
EDUARDO A. PADLAN
31 July 2007

Part 1
Public Understanding and National Importance of Science and Technology

Philippine-American
Academy of Science
and Engineering

An Appeal to President Gloria Macapagal-Arroyo to Increase Government Support for Science and Technology

President Gloria Macapagal-Arroyo and several members of Congress received documents prepared by members of the Philippine-American Academy of Science and Engineering (PAASE) concerning PAASE's appeal to increase the science and technology budget of the Philippine government. PAASE, a nonprofit organization that was founded by a small group of Filipino scientists and engineers in the United States in 1980, has grown in recent years to over 180 members, now with a significant number based in the Philippines.

The campaign is being led by local PAASE members affiliated with leading academic institutions in the country. Several foreign-based PAASE members are coming home until early March 2006 to join local PAASE members in meetings with government officials. Following is PAASE's letter to President Arroyo:

February 15, 2006
Dear President Arroyo:

We are members of the Philippine-American Academy of Science and Engineering (PAASE), an association of scientists and engineers, Filipinos or of Philippine descent, working in the United States, in the Philippines and in other countries. We teach and do research in colleges and universities; we work in industrial laboratories and in governmental and international institutions.

One of the primary objectives of PAASE is to promote contacts and collaborations among Filipino scientists the world over and to help the Philippines in her efforts to establish and maintain a level of activity in science and engineering, in teaching and in research, commensurate to the needs and aspirations of our country. There are now regular contacts between members based in the Philippines and members based abroad, and a number of collaborative projects have been in operation for a number of years. Annual meetings and symposia of PAASE have been held in the Philippines as well as in the United States. Some of us met with you during two of our meetings

Alfonso Albano (2005 PAASE Chairman of the Board of Directors)
aalbano@brynmawr.edu or alfonso.albano@gmail.com

in the Philippines. These annual meetings have served to put Filipino scientists from different parts of the world in contact with each other, and have become opportunities for exploring possible scientific collaborations. Scientists based in the Philippines who have held postdoctoral positions in the laboratories of members based abroad have gone home to establish their own laboratories where they maintain thriving research programs. Some foreign-based members regularly return to the Philippines to teach in Philippine universities and/or work with their Philippine-based colleagues. There are now publications in international research journals coauthored by Filipino scientists based in different countries.

These private initiatives, undertaken by individual scientists, are hardly enough. Much more needs to be done to bring our research climate to a level comparable to those of neighboring countries. A recent study by the Asian Institute of Management (R. E. Escobar, "Philippine Science and Technology: Environmental scanning of domestic and global issues," AIM Policy Center) revealed that Thai scientists publish roughly three times as many scientific articles as their Filipino counterparts; Malaysian scientists, some two-and-a-half times more. Under favorable circumstances, however, Filipino scientists have shown themselves to be extremely productive. The publication records of PAASE members who find themselves in supportive research environments attest to this.

It is for this reason that we appeal to you for help, knowing your stature as a social scientist and an economist yourself. The current level of expenditure for research and development (R&D) in the Philippines is insufficient to support a research enterprise that is anywhere near commensurate to the needs of the country. The 2001 level of R&D spending was 0.078 percent of Gross Domestic Product (GDP), even lower than the 1992 level of 0.2 percent. In contrast, in 2001, Thailand and Malaysia spent 0.27 percent and 0.49 percent of their respective GDPs on R&D.

These differences in R&D funding levels are ultimately reflected in various measures of productivity. In 2000, the number of patents awarded to Philippine residents (per thousand R&D personnel) was 3.5. The corresponding numbers for Malaysia and Thailand are 7.1 and 81.3, respectively. These disparities translate into disparities in the production of goods and services and ultimately in such measures of economic vitality as the GDP growth rate. According to the World Bank, in 2004, the Philippine per capita GDP was US$1,170; the corresponding figures for Thailand and Malaysia are US$2,540 and US$4,650, respectively.

Official data show that while Philippine GDP growth rate per capita was a meager one percent per annum during 1991–2003 (following –0.5 percent GDP per capita growth in the 1980s), the corresponding growth rates in Malaysia and Thailand were about 3.5 percent.

Current Philippine expenditures for R&D are way below the medium-term target of 1.0 percent of GDP set by the Department of Science and Technology (DOST)

National Science and Technology Plan, which expressed the hope of attaining this target in 2010.

We ask that you increase the government support for research in science and technology to 0.5 percent of GDP by 2010. This represents only half of the National Science and Technology Plan target for 2010, but it would be a tremendous start. These additional funds could be coursed through the DOST which would allocate funds through its usual research grant funding processes. An increase in research allocation would also have an immediate impact on science and technology education at the tertiary level as well as on the training of young research scientists, and will certainly provide incentives for trained researchers to stay in the Philippines rather than seek opportunities abroad.

We offer ourselves to help in any way we can. We realize that there are other needs demanding immediate attention, but this is a critically needed investment in the future of the country that we can no longer afford to postpone.

Respectfully yours,
Members of PAASE

February 26, 2006

Philippine-American Academy of Science and Engineering

Science and Technology in Economic Development

This paper was prepared by the Philippine-based members of the Philippine-American Academy of Science and Engineering (PAASE), including Drs. Rhodora V. Azanza, Gisela P. Concepcion, Alvin B. Culaba, Fabian M. Dayrit, Ernesto M. Pernia, and Caesar A. Saloma. It was sent to President Arroyo, several members of Congress and the secretaries of the Department of Science and Technology, Department of Agriculture and Department of Health in February 2006 as part of PAASE's campaign to increase the S&T budget of the government.

Science and Technology and Economic Growth

Increases in capital and labor are important to the growth of an economy. But that is only one part of the real story. Robert M. Solow, Nobel laureate in economic science (1987), showed that capital and labor accounted for only a part of the long-term economic growth of the United States; a substantial part was attributed to advances in the general state of knowledge.

Numerous studies on the economic growth of nations ascribe a significant role to specific factors, such as education, improvements in efficiency, and shifts in the allocation of labor from less productive to more productive activities.

More recent studies—resulting in what is now known as endogenous growth theory—find that technological innovation and economic growth are mutually reinforcing; that is, economic growth can be best sustained through technological innovation, and technological innovation can come about from research and development (R&D).

But who is to lead R&D? In most of the highly developed countries today, the government led in R&D. That is because R&D has generally been regarded as a public good. The private sector in developing countries has neither the resources nor the incentive to generate an optimum level of R&D as it is a costly and risky undertaking. The private sector lacks the incentive to lead in R&D because it cannot fully own or exclude others from the ensuing benefits, and the private rate of return to R&D is typically lower than the social return.

The Philippines' long neglect of science and technology (S&T), besides weak governance, has resulted in its poor economic performance for decades. Figure 1 shows

---

Alvin Culaba (2006 PAASE President), culabaa@dlsu.edu.ph or
Gisela Padilla-Concepcion (2006 PAASE Board of Director), gpconcepcion@yahoo.com

that our Asian neighbors who have had high rates of capital formation, including R&D, have enjoyed higher productivity and economic growth. The clear laggard is the Philippines, where the government's budget for S&T has been on the order of 0.12 percent of GDP on the average over the last several years. By contrast, our Asian neighbors have been investing at least 0.5 percent in S&T.

Clearly, we need a substantial increase in our budget for S&T if we want to achieve rapid and sustained economic growth, increase job creation, dramatically reduce poverty, and significantly improve standards of living for all Filipinos.

We would, therefore, urge this administration and lawmakers to give serious attention to the development of S&T by providing the necessary budgetary allocation. The minimum targets (relative to GDP) should be as follows: 0.2 percent in 2007, 0.3 percent in 2008, 0.4 percent in 2009, and 0.5 percent in 2010. These are very modest targets that would not even bring us at par with our Asian neighbors. By 2020, the aim should be to allocate at least one percent of our GDP for S&T.

We propose that the increased budgetary support be invested in a massive, sustained program of capacity building where world-class S&T centers of excellence linked to regional R&D institutes can provide the right environment to attract the brightest young minds in the country to become PhDs—expert scientists and engineers. The priority R&D programs under the DOST's National Science and Technology Plan would then be implemented by the most competent researchers in the country. We also propose a system of regular external peer review that will ensure optimum efficiency of this S&T investment.

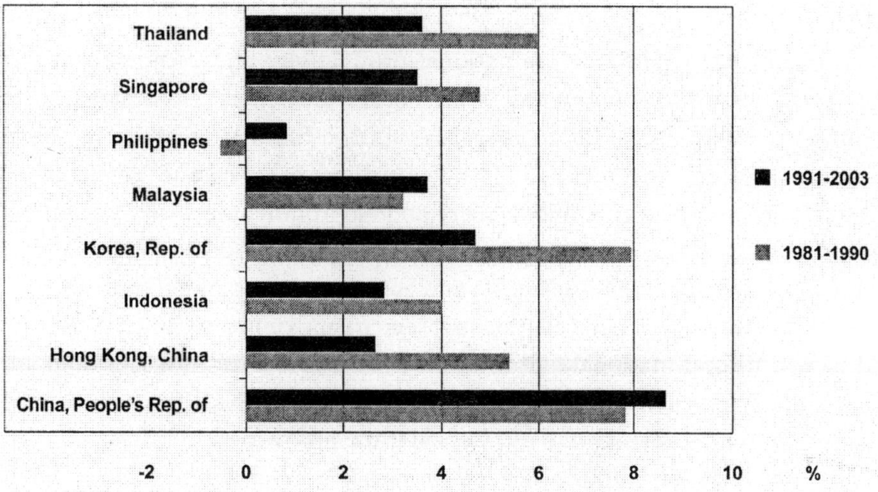

Figure 1. GDP per capita growth rates: Selected Asian countries.

Capacity Building in Science and Technology

The building up of S&T capacity is urgently needed for increased productivity, competitiveness and problem-solving skills. There are many areas for capacity building.

S&T centers of excellence should be established and maintained in the various areas of need. At present, some core areas already exist, but they require completion and additional input of resources.

We need more focused manpower and institutional development programs, such as the Engineering and Science Education Program (ESEP). ESEP had prioritized the training of PhDs in science and engineering and provided needed facilities and equipment.

Regional centers of R&D should be established to support specific industries, but these should be closely linked with the S&T centers of excellence which will lead the country in different areas of R&D.

We need to establish S&T business centers, near the S&T centers of excellence, which will assist, advise and incubate technopreneurship ventures.

We need to invest more resources to strengthen science education at an early stage through the existing Philippine Science High School system.

We need to supplement science education in the primary and secondary levels with up-to-date learning strategies and materials that will encourage our youth to appreciate the value of science to life.

Science and Technology Priorities

The DOST's National Science and Technology Plan (NSTP) has identified priority areas of R&D that would enhance national productivity and competitiveness. This we support fully. PAASE members and other local researchers have prepared proposals for consideration by DOST and other government funding agencies in the following areas: agriculture and livestock, food and nutrition, fisheries and aquaculture, natural products, health and medicine, biodiversity and environment, ecotourism and livelihood, engineering, material science, energy and ICT. These proposals represent only a few of the many projects that Filipino scientists and engineers would like to pursue.

The NSTP further recognizes that in order to maximize its impact, S&T priorities should be tied to both national and local needs to create trust, linkages, and services for the long term. The benefits from increased S&T support should reach the sectors, which the majority of our people depend on. Consider the following:

Over 70 percent of Filipinos depend on agriculture and aquaculture. The land and the sea provide the only sustainable sources of livelihood in the Philippine countryside. More support is needed for agricultural/aquaculture biotechnology which can produce innovative products and services with high technology input and reduced environmental risks.

About 95 percent of business enterprises in the Philippines are small and medium enterprises (SMEs). This sector is virtually bereft of scientific and technological advances because SMEs generally do not have the capability to undertake research and development. There is very little support for S&T that SMEs critically need.

Investments in R&D in agro/aqua-biotechnology and related areas in the various regions of the country have a significant multiplier effect because of their strong local linkages. Unfortunately, there are so many examples of opportunities that we are losing out on because we do not have sufficient S&T capabilities to initiate and support research and development.

Government commitment to fund S&T development can catalyze investments by overseas Filipinos and participation by expatriate Filipino scientists and engineers to help the country. In 2005, OFWs reportedly remitted over $10 billion to their families. Filipino S&T groups with expatriates such as PAASE stand ready to provide their expertise for worthwhile projects. We should intelligently harness these diffused but valuable opportunities into investments for the future. A clear commitment from the government to support key S&T development will serve as the catalyst for this unique opportunity for Filipinos from all over the world to come together to invest in their country's development.

Ensuring Effectiveness of Science and Technology Research and Development

Even while an increase in budgetary allocation for S&T is needed, we must institute a review process that will ensure the effectiveness of the R&D process based on internationally accepted measures of performance. We propose that the review process take the following form:

Regular review of scientific and technical output by a panel of internationally recognized experts. For this we can tap the expertise of expatriate Filipino scientists and engineers of PAASE. This panel can serve as an advisory group under the Legislative Committees on S&T.

The S&T output of our R&D institutes should be up to international standards. While the primary role of the R&D institutes is to support specific sectors, e.g., agriculture, health and industry, they are also expected to publish in reputable ISI-recognized journals. This will help ensure a high level of research, and it is also an important means of establishing the international reputation of Philippine scientists and Philippine products.

R&D targets should be clearly set through a process of consultation which should include the S&T community, national agencies and local governments, and the private sector (industry and local SMEs).

March 2, 2006

HER EXCELLENCY
PRESIDENT GLORIA
MACAPAGAL-ARROYO

Current Government Support for Science and Technology

Below is the speech of President Arroyo during the 27th annual meeting and the 2007 International Science Conference of the Philippine-American Academy of Science and Engineering (PAASE) at the Grand Ballroom of Century Park Hotel in Manila last February 16, 2007.

THANK YOU very much, Secretary (Estrella) Alabastro for your introduction. Dr. Culaba, thank you and congratulations on this very important conference of PAASE, and welcome to the many important Filipino scientists from the different parts of the world, especially America. Congressman (Luis) Villafuerte, other officials of PAASE, other officials of the Philippine government, ladies and gentlemen.

First of all, thank you to PAASE, the Philippine-American Academy of Science and Engineering, for dovetailing this conference with the economy. And this is so fitting. I am proud to say that not only did I talk about science and technology in my last State of the Nation Address, but in my first State of the Nation Address in 2001, I declared that technology is the foundation of future economic development.

Therefore, in our medium-term Philippine development plan, among the strategic measures contained therein is to mobilize and disseminate knowledge. In that same medium-term Philippine development plan, we identified as the key challenges in science and technology the following:

- Make science and technology policies supportive of national development goals and policies. This conference dovetailing S&T in the economy fits into that challenge.
- Enhance competitiveness of the country's human capital.
- Develop a critical mass of scientists and R&D personnel.
- Speed up knowledge creation and dissemination for greater productivity and job creation, especially to the rural areas, and improve the mechanism and programs to promote and encourage technology-based entrepreneurship.

That was in our medium-term plan and it is still there. At first, the Philippines had little of the huge funds required to meet these challenges. The first order of the day for us was to put our economic house in order and find the funds to meet those challenges.

Thanks to congressmen like Louie Villafuerte, we had our fiscal reforms, our peso strengthened, therefore, our interest spreads went down and we saved ₱30 billion last

---

Gisela Padilla-Concepcion (2006 PAASE Board of Director), gpconcepcion@yahoo.com year. Because of our savings and because of our increased funds from revenues and taxes, more funds are now being earmarked to strengthen our educational system, including in the basic grades and years, strengthening math and science in our basic education.

And because of these reforms, the Philippines today basks in glowing reviews of our fiscal and economic reforms. Just today, I came from a groundbreaking ceremony of a pharmaceutical company and nutritional company in Calabarzon (Cavite, Laguna, Batangas, Quezon) worth $80 million, and they said that they did this expansion because of our economic reforms.

The record-breaking high set by the stock market, the peak in overseas remittances, and the raging performance of the peso are clear manifestations that our economic homerun is in play. Not even the intense politicking brought about by the forthcoming elections can spoil the larger gain of the economy.

Our eye is on the ball as we aim our volleys on better-paying jobs, improved infrastructure, including technology, enhanced social services, the alleviation of poverty and hunger, fighting terror, and nurturing the environment—all of which will benefit from technology.

The Filipino people are shrugging off the past and looking forward to the fruits of steady growth, trade, and investment. We must all seize the moment to reach the summit through our unity and hard work, brandishing the Filipino team spirit that has gained renown worldwide.

The Philippines is a team to beat when it comes to excellence, resiliency, and staying power in a competitive world. But there are specific challenges we have to contend with. And if I may digress momentarily from science and technology, I'd like to take this opportunity to express that I'm saddened by the fact that the visa hold has been employed on the deployment of our Filipino nurses as a result of the nursing exams of last year.

I am instructing the Secretary of Labor to immediately look into the findings of the Commission on Graduates of Foreign Nursing Schools (CGFNS) and seek all forms of relief from the visa hold and appeal for reconsideration of the decision. The government shall provide financial assistance to the 2006 nursing board passers for the retake of the exams as called for by the CGFNS with the proper body to oversee the project, including having a common review center. We shall uphold the prestige and excellence of the Philippine nursing profession as we also promote the prestige and excellence of the Filipino scientist all over the world with the plans that we can now have the funds to carry out.

In my State of the Nation Address, in the portion that was drafted by Congressman Louie Villafuerte, I said that the country that wants to be a player in the global economy needs bold and well-funded research and development initiatives of its own for in today's global economy knowledge is the greatest creator of wealth.

And last year, I met with some of you. I met with members of PAASE along with Secretary Alabastro, ICT and education agencies, and Congressman Villafuerte. We agreed on five action plans.

First, a review of policies by a presidential coordinating body. So today, I will issue an executive order creating ... there I have signed it ... the Presidential Coordinating Council on R&D.

Our second agreed action plan was to expand the number of Filipino science and technology experts. This morning, I had a telephone conversation with Secretary of the Budget Nonoy Andaya. I instructed him to release ₱200 million for science and engineering Masters and PhD scholarship programs.

Our third agreement was to establish a science complex. Our supplemental budget for 2006 provides ₱500 million for a UP science complex. And as we were walking in, I asked President Emerlinda Roman, "How's our science complex coming along?" And she said, "It's coming along fine!"

The fourth agreement that we made during that meeting was to strengthen linkages between and among the stakeholders of S&T development to speed up technology transfer to micro, small, and medium enterprises. I am glad that Nina Alabastro and the Department of Science and Technology (DOST) will establish this year four business incubators. I've asked Nina to make one of those four business incubators an incubator for virgin coconut oil. We would like virgin coconut oil to be as famous all over the world as Italian olive oil is. And Nina said, indeed, that's one of her major programs. And one of the incubators will be in Camarines Sur, which is coconut country.

And so indeed I hope to see these four come into fruition this year.

Our fifth agreement was to increase public awareness and advocacy for S&T. I think what you'll be doing here for the next two days is in accordance with that. Aside from this important conference attended by some of our Filipino superstars in the science and technology field in the United States, I count on the PAASE members to regularly provide science and technology information materials to print media.

This is one of the paybacks of our fiscal reforms—that we can now move toward tripling private and public sector spending on R&D. And now that we are closer to the goal to give a bigger share of the budget for science and technology, I am optimistic that together we can fully tap the wonders of science and technology to achieve growth, preserve peace, and improve lives on the longer term.

Mabuhay ang Filipino-American scientists!

Thank you.

March 22, 2007

Caesar Saloma

The Urgent Need to Invest More in Scientific Research and Development

Scientific knowledge is verifiable, reproducible, and impermanent. It is the single most important resource of any advanced society. The late Peter Drucker was able to explain convincingly the vital role that knowledge plays in a modern economy. Drucker is often credited as "the man who invented management (by objectives)." Because the scientific and economic worth of a particular kind of knowledge depreciates rapidly with time, a society needs to develop its own capability to generate new knowledge that enables it to sustain economic growth at a rate that can bring tangible prosperity to its people.

The aim of scientific research is to extend human knowledge of the physical, biological or social world beyond what is already known. In the last hundred years or so, the fruits of research have empowered mankind to defeat ignorance and overcome its innate fear of the unknown. Moreover, scientific research and development has also provided the fuel that drives technological innovation. The continued introduction of novel and better technologies that improve our quality of life amid the intensifying pressures of population growth (in the case of the Philippines) and dwindling natural resources (in the case of the entire world) critically depends on our ability to successfully apply the new knowledge that is gained via scientific research.

This paper argues that increasing the current level of financial support for scientific research makes good economic, social, and even political sense for the Philippine government. At this point in time, the increase in funding will be utilized to develop zones that will gather together scientists and researchers from various areas of specializations and allow them to interact with one another. Zones are places for nurturing scientific excellence where focused research endeavors can be pursued efficiently and where the future generations of scientists, entrepreneurs, and even political leaders of the country are trained competently and effectively.

Only a few zones will be initially developed and sustained to make sure that the increase in financial support for scientific research will produce the maximum desired effect while at the same time, causing only the least amount of perturbation to the overall allocation profile of the national budget. The investments in infrastructure development, equipment acquisition, and manpower recruitment are highly targeted and the measures of progress are easier to formulate and apply.

National Institute of Physics, University of the Philippines, Diliman, Quezon City, 1101, Philippines.
caesar.saloma@gmail.com

The zones will be morphed from existing academic entities that have already satisfied the following conditions. First, they have consistently shown a high level of scientific productivity through publications of research outputs in leading scientific journals of the world. Second, they have existing graduate programs that are attracting and producing technically competent PhD graduates. Although only a few, such entities already exist in the country today.

Filipino researchers are highly productive in foreign laboratories. The reason is the environment and not exceptional talent in science and mathematics as compared with other nationalities. Filipino expatriates blossom and bloom because they are in an environment that fosters creativity and original thinking. More often than not, these same well-meaning scientists cease to produce worthy research outputs when they work in the Philippines. They become distracted by efforts to supplement their grossly inadequate basic salaries through administrative work, consultancy jobs, and heavy teaching loads. Their enthusiasm will be slowly eroded by the apparent lack of urgency and efficiency in Philippine society.

In its human development index (HDI) report for 2005, the United Nations Development Program has ranked the Philippines 84th among the 177 countries that were considered. The HDI is a measure of well-being that takes into account not only the gross domestic product per person but also life expectancy and access to education of the general population. It is employed to assess more effectively the "quality of life" of citizens in different countries. In 2000, the Philippines occupied the 77th position in the HDI ranking. If only income per person is considered, the Philippines would rank even lower at 96th. China is a significant gainer, improving from 96th in 2000 to 86th in 2005. Norway and Sweden as well as Australia and Canada have consistently occupied the top positions in the HDI rankings—the harsh conditions of winter are not an impediment to a contented life.

The Philippines will be unable to improve the quality of life of its people to a level like those of Japan, the United States or Germany if it does not also have their number of PhDs in the natural sciences, mathematics, and engineering. According to a 2004 report in the journal *Nature*, the number of PhDs per 1,000 people is 0.3 (one PhD per 3,316), 0.08 (one PhD per 11,621) and 0.15 (one PhD per 6,533) in Germany, Japan, and US, respectively (King, D. 2004. "The Scientific Impact of Nations." *Nature* 430: 311-17). The average number for countries comprising the European Union is 0.18 per 1,000 people. The Philippines with its (extrapolated) population of about 87.9 million needs to have about 7,500 and 26,500 PhDs if it were to possess the current scientific muscle of Japan and Germany, respectively. The real score for the Philippines is about three orders of magnitude below the mark. There are less than 1,000 PhDs in physics in the country today and the number of PhDs is unlikely to exceed 2,000 in the natural sciences and engineering.

The Philippines does not attract foreign direct investments the way its geographic neighbors do because of its relatively low competitiveness and unfavorable perception of risk and integrity. In its 2003 Global Competitiveness Report (*en.wikipedia.org*), the World Economic Forum has ranked the Philippines 64th, which is below that of Singapore (8th), Taiwan (16th), Hong Kong (19th), Malaysia (26th), Thailand (31st), Vietnam (50th), and Indonesia (60th). The report utilizes the competitiveness index which is a measure that collectively considers the quality of infrastructure, including adequate energy supply, competitive labor, clear and sensible policies and regulations for doing business, and respect for the rule of law. Finland is the most competitive country—it has occupied the number one position in both 2003 and 2005. China, on the other hand, has ranked 46th and 49th in 2003 and 2005, respectively. Thus, a large population of potential consumers alone does not guarantee a rosy climate for doing business. Despite their small populations, the Scandinavian countries are the most competitive in the world because their societies are democratic, well governed, and merit-based. In these countries, goods, services and information are rapidly and efficiently distributed or disseminated. Iceland and Finland are also considered the two least corrupt in the world, according to a 2005 survey of Transparency International.

Filipinos still need to understand more fully the importance of science in national life. The annual budget for scientific research and development has always remained minuscule from one administration to another. Governmental regulations and procedures that have been originally formulated and developed to slow down if not avert graft and corruption, also make the acquisition of research-grade equipment, supplies and materials and even scientific journals and books agonizingly complicated, slow, and unnecessarily expensive. It takes so much energy and effort to start a research laboratory and much more to sustain long-term scientific interest among members of a research group. These obstacles are fatal because scientific research is a "winner-take-all" pursuit. There is no place for the second discoverer or inventor. Urgency and efficiency are the primary concerns of scientists and researchers.

The Philippine society, particularly its academic community, needs to improve continuously its expertise to distinguish and celebrate scientific excellence by adopting standards of excellence that are widely applied in advanced countries. A society without a strong scientific tradition tends to trivialize excellence and rationalize mediocrity in daily life. Many Filipino academics must realize and accept that a professor of a university should not only be good at disseminating textbook knowledge (teaching in the classroom) but must also have a proven track record in scientific research, which requires a different kind of originality and creativity. Being institutions of higher learning, universities should offer graduate programs where a PhD is granted only to a student who has contributed something new and important to the body of knowledge. Graduate degrees are research degrees that are not obtained by merely passing written or oral examinations that only encourage rote learning.

The United Nations Development Program reported that in 2001, the Philippines spent only 0.15 percent of its GDP on scientific R&D and then allocated even less at 0.11 percent the following year. Over an eleven-year period from 1987 to 1997, the Philippines spent an annual average of slightly less than 0.2 percent of GDP to scientific R&D. In comparison, Finland and Sweden each spent more than three percent of their GDPs for the said purpose in 2001. Neighboring Taiwan used 2.1 percent of its GDP for scientific R&D in that same year.

The available information shows that the Philippine government has decreased its annual allocation (as percentage of GDP) in recent years. The reduction was effected against the backdrop of a rapidly increasing national population and a deteriorating foreign exchange rate. The last official census in 2000 revealed a national population of 76.5 million. Based on an average annual growth rate that has hovered around two percent since 1948, the projected 2005 population for the Philippines was 85.23 million, according to the National Statistics and Census Board. Between December 1994 and January 2005, the value of the Philippine peso deteriorated by 129 percent in US dollar terms.

A leading and logical candidate for zone development is the College of Science at the University of the Philippines in Diliman, Quezon City. The college was established in October 1983 and has the largest concentration of PhDs in the natural sciences and mathematics in the country today. It consists of eight institutes (four of which are national in character) and departments and a research institute. More importantly, the college is the primary producer of PhD graduates and research publications in scientific journals that are considered in the Science Citation Index.

The University of the Philippines System had designated 21.9 hectares of land in its Diliman campus for the Science Complex. The area is adjacent to the incubating UP Ayala Technology Park and the bustling C-5 highway. The complex has been envisioned to encourage multidisciplinary interactions among faculty and researchers by consolidating the locations of the various institutes and departments in close proximity to each other.

Today after more than twenty years, the Science Complex is still an unfinished business. Only the buildings for the National Institute of Geological Sciences, the Marine Science Institute and the College of Science Library and Administration could be considered as, arguably, complete. The new building for the National Institute of Physics is less than 50 percent complete and could not yet be fully utilized. The future building for the Institute of Chemistry remains in the drawing board and so are those of Molecular Biology, Biology, Mathematics, and Environmental Science.

An estimated ₱500 million (~$10 million) in capital outlay is needed to complete the buildings for physics and chemistry and to develop the grounds of the Science Complex. The completion of the physics and chemistry buildings is a clear and powerful signal that our political leaders and economic managers truly understand the workings of a national economy in today's highly competitive world. The mere sight of the National Institute of Physics building is certain to inspire the electorate (and their children) and fortify their confidence about our nation's future. Another ₱750 million is required to build the buildings of the institutes of molecular biology, biology, mathematics, and environmental science.

The ₱500 million for capital outlay is not huge. It only represents 0.0476 percent of the proposed 2006 national budget that accounts for 17.6 percent of the Philippine GDP in 2004. It is only half of what the Miami Heat organization pays Shaquille O'Neal to play basketball in a single NBA season.

The intellectual capital that will be created out of the small but steady investment to support and maintain the Science Complex will be enormous and far-reaching. The complex will attract and retain the best young minds of our country. It will appreciably slow down the migration of our young scientists to other countries, which often happens after they obtain their BS diplomas. The lack of a sensible alternative has driven most of our young scientists to go abroad for their PhD degrees. The College of Science will provide them with an opportunity to pursue cutting-edge scientific research under the mentorship of senior scientists.

The prowess of our scientists and researchers will enhance the stature of the Philippines as a responsible, confident, and productive member of the international community of nations. Their expertise would allow our country to respond quickly and effectively against the threats of pandemics and prohibitively high cost of fuel. A country with an established scientific tradition and a proven capacity for technological innovation will be resilient to the vagaries of the world economy that tends to vacillate even more frequently due to the rapid advances in the management, access, and distribution of information. The lessons learned from the development of the Science Complex are vital in the establishment of additional zones corresponding to other fields of study.

The Filipino culture is a unique one that is distinctly different from those of our neighboring countries. It is an amalgam of the rich experiences that Filipinos were able to gain from interactions with other peoples from both East and West. Creativity and innovativeness are natural Filipino traits, especially in music and the arts. In addition, Filipinos have an outstanding facility with the English language, which is the lingua franca of science today. More than technical training and rote learning, great works of science are products of boundless imagination and exceptional creativity. Just read Albert Einstein. Robots can make high-quality cars with zero defect but they cannot imagine. Only people can. I believe that Filipino researchers cannot only do good science—they also have the potential to do it in such a discernible way.

The Philippines needs to invest more in scientific R&D by developing zones of scientific excellence. The returns will greatly outweigh the initial development cost. Let us start auspiciously by completing the Science Complex at the University of the Philippines in Diliman.

March 23, 2006

Gisela Padilla-Concepcion

A Silver Lining for Science and Technology in the Philippines

Being a mother of five, I often think of how best to motivate children to do their best and to succeed. Since they were young, I would stress on my children that they should strive for self-improvement. If their grades improved over a quarter, I was happy. I did not pressure them to be at the top of their class or to be first honors. Their incremental progress was more important. Any accomplishment or award earned on their own was enough. I am patiently waiting to see how my children will turn out. A mother is naturally hopeful and I continue to think that my children hold promise.

Similarly, it is hard to measure success in a society in an absolute way. Technological and scientific accomplishments appear to be one good indicator. Using this measure, the Philippines has not fared well over several decades now. We already know the reasons for this. Political events led to the exodus of some of our brightest minds sometime back. Over a long period, our culture and educational system did not sufficiently instill the discipline required for science and mathematics, and also did not encourage curiosity and creativity enough among our youth. Our leaders did not recognize S&T as the engine of economic development and gave it low funding priority. There was no political will to move S&T forward.

Some of us have identified particular weaknesses in our educational and scientific systems. Several have proposed ways to improve our S&T performance as a country. The proposed benchmarks of success are those measured in technologically successful countries: the directions—those identified globally and which have been taken by our neighboring countries—and the thrusts—those in which we may have an innovative or competitive advantage. Many of our ideas are good. The challenge for us is to have those good ideas implemented.

I think there is a silver lining—an "alignment of the stars"—for S&T in the Philippines today. The scientific community has become united in its determination to move S&T forward and we are making some headway. We should not let up in our campaign—our crusade—to improve government support and our own academic performance in S&T. Since we have gained momentum, we must not lose it. Let us thank the private sector, the technopreneurs and technology brokers who have supported us, and hope that they continue to explore and publicize the potential to make good use of some academe-based research.

The Marine Science Institute, University of the Philippines, Diliman, Quezon City, 1101, Philippines.
gpconcepcion@yahoo.com

As in all endeavors, nothing beats leadership by example in science. The benefits of good science leadership are felt through generations. In the university, an exemplary science teacher, thesis adviser, researcher, research mentor, administrator, and extension worker can inspire colleagues and students by his vision, dedication, performance, and output, and can transform his institute to a world-class organization. Creating a scientific culture in a country requires a continuity of scientific traditions and the upholding of high scientific standards and practices. Continuity is ensured through the constant mentoring of our youth; and mentoring is best done, not just by words, but by example.

Despite the low rankings in S&T of our universities in worldwide surveys today, I believe we can significantly improve our S&T performance soon. There are, in our midst, science leaders who have led by example—as dedicated teacher, productive researcher, and effective, visionary science administrator, all rolled into one. These scientists have been internationally recognized for their achievements, and, quite notably, their leadership has resulted in an upward, accelerated trend in the number of international peer-reviewed publications, patent approvals, and MS and PhD graduates in their institutes. Gradually, their researches are being explored for useful applications by the private sector.

Indeed, the science is only as good as the scientists behind it. Thus, the more science experts we have, the more competent we will be to produce high quality, innovative, and exciting products that enhance our lives and the more prepared we will be to find effective solutions to our food, health, and environmental problems. A critical mass of science experts is what our country needs urgently since our current pool of scientists is critically low. The government should provide enough scholarships and financial incentives to attract our graduates to pursue their PhDs and build careers in science. I think this should be at the forefront of our S&T campaign in the next few months.

Many of us spend a great deal of time teaching and doing research in the university. But realizing the urgency to improve the dire state of S&T in our country, some of us have decided to devote extra time for an advocacy to promote S&T.

In late 2005, PAASE (Philippine-American Academy of Science and Engineering) members initiated efforts that led to what many of us know today as a successful funding outcome. A total of ₱700 million was allocated in the 2006 supplemental budget for S&T—₱500 million for physical infrastructure in the National Science Complex in UP Diliman and ₱200 million for more R&D projects funded through the DOST. More recently, another ₱200 million was allocated for MS and PhD scholarships, in addition to a previous allocation through the CHED.

PAASE members had agreed that the single, worthwhile campaign to mount was to increase government funding support for S&T, and the campaign was to be led by local PAASE members. We knew that it would take more than one letter (or a courtesy call to Malacañang) to catch the attention of our government officials. In February 2006, we put together documents consisting of a PAASE letter to President Gloria Macapagal-Arroyo drafted by our chairman, a position paper written by six local PAASE members on "Science and Technology in Economic Development," individual letters of support and commitment from sixty PAASE members, fifty capsule proposals on R&D projects prepared by PAASE members, and reprints of seventy Star Science articles.

The documents were sent to the President, some Cabinet members, and selected congressmen and senators. We requested appointments to see them. We asked foreign-based PAASE members to join us in these meetings and six of them came home just for this purpose. Congressman Luis Villafuerte responded immediately and became our champion for S&T. Through regular meetings in the succeeding months, we convinced him to include a statement of support for S&T in the President's State of the Nation Address in July 2006, to bring us to Malacañang to see the President in August 2006, and to have the President as our special guest in the PAASE conference in February 2007. We are grateful to Congressman Villafuerte who made all these possible.

At Malacañang, our presentation to the President addressed several aspects of the S&T problem and five recommendations were made: 1) creation of a Presidential Coordinating Council for R&D; 2) revival of the MS and PhD science and engineering scholarships program; 3) completion of the National Science Complex in UP Diliman; 4) support for R&D programs that would benefit rural communities, particularly the agriculture and fisheries sectors; and 5) increase in public awareness of science.

The President and our government officials have responded positively to the recent appeal of the Philippine science community. PAASE, being a nongovernment, multi-institutional, nonprofit, and purely voluntary organization, was effective in leading the campaign. At the PAASE conference last month, the President updated us on the five recommendations we had made to her. She announced the release of an additional ₱200 million for PhD and MS scholarships. She signed Executive Order 604 on the creation of a Presidential Coordinating Council for R&D which will prioritize and coordinate R&D programs, and monitor and evaluate R&D output. With two slots for representatives from the academe, hopefully, this new body will help ensure the relevance and high quality of the country's R&D output.

As a teacher and researcher, I can now more confidently tell our students and researchers that finally our country is on the right path in S&T. Soon our government will be able to provide them scholarships for their PhD studies. Soon we will have a good working environment with better infrastructure and equipment. With high-quality scientific output and better publicity for our research, soon the private sector may be enlightened to explore applications of our research.

Gradually, S&T in our country is improving. Like our children, many of our young scientists appreciate our mentoring and the supportive environment that we are striving hard to provide them. Many say they will rise up to the challenge to become productive and self-reliant. Soon our efforts at improving ourselves individually and collectively may bear fruit and we may join the ranks of the S&T achievers of the world.

March 15, 2007

Flor Lacanilao

Public Understanding of Science

"Development goals that do not recognize the importance of science and technology in economic transformation are likely to fail, especially those aimed at reducing poverty and raising income levels" (*Harvard Report: Meeting the needs of developing countries,* 2001). Science in the Philippines, however, has hardly progressed to enable us to produce or adapt useful technologies for sustained development, and research is the culprit.

Research and Development Process

One way to improve the situation is to know first the R&D process. This consists of information production (research), information dissemination (extension work), and the use of information (development). Development depends on the quality of disseminated information, which relies on the way research is done. To do research properly, study results must be published in a research journal adequately peer-reviewed and accessible for international verification.

In science, the normal way of publishing results is the international refereed journal. Such published study is known as a scientific paper or valid publication. This is the scientist's primary output, and is the source of information for the growth of scientific knowledge or science. The paper has to pass through adequate peer review before publication. It has to be accessible through indexes and normal library channels to enable international experts to do follow-up studies and verify the results, a process the scientific method requires. But we have limited experts in various fields to do adequate peer review and verification studies. Still, most of our research papers are published locally. Therefore, they have doubtful validity and hardly contribute to the growth of science. These are the basic problems of research in the Philippines, the reasons we cannot sustain development.

The next phase of R&D is information dissemination that usually starts with review articles authored by respected scientists. Scientists are researchers who have published their studies in international journals. Those who have published enough papers on a given subject in such journals write review articles by gathering reliable information

Formerly with The Marine Science Institute, University of Philippines, Diliman, Quezon City, 1101, Philippines; and the Southeast Asian Fisheries Development Center.
flor_lacanilao@yahoo.com

from important journals. From review articles, information is disseminated to the general public and other users through various means—newspapers and magazines, newsletters, textbooks and manuals, radio and television, and the Internet.

Whatever the means, the success of extension work depends on information quality. This is the reason why doing research properly—including publication in adequately peer-reviewed and accessible journals—is important.

Writing review articles, and to some extent extension materials, is the scientist's second role in R&D. They are addressed to the general public and other users of information. This function differs from the scientist's primary role of producing scientific papers that are addressed mainly to other scientists for verification. Many fail to see the distinction between these two roles of scientists. They say, for instance, that research papers should be published locally to make them available to users. Since we do not have enough scientists in a given field to adequately review manuscripts, most locally published data are of poor quality, and using them for development programs is the common cause of R&D problems in the country.

Another use of information is for generating technologies. Some review articles are on technology production using scientific papers. As noted earlier, these contain useful information that forms the raw material for the growth of science. Hence, science is the source of information for generating technologies. To be useful, technologies should also be made by scientists, who should not be mistaken for those who have published only in newsletters, institutional publications, and most conference proceedings. Papers in these publications are not taken seriously, and are not considered when assessing the science and technology (S&T) performance of nations.

If desired development continues to elude the Philippines, blame the technologies we claim to have. They used largely poor quality information from publications other than peer-reviewed international journals. This information did not contribute to the growth of science and no amount of extension effort will fully satisfy users of technologies made from information of doubtful validity.

The relation between R&D and S&T can then be illustrated as follows: RESEARCH to SCIENCE to TECHNOLOGY to DEVELOPMENT.

Development depends on technology, which depends on science, and ultimately on research. Thus, the basic component is research. If the country has development problems, one can identify a major cause if he or she knows the role of each component in the series. It can also be seen that development will hardly follow even if the funding or number of PhDs is increased without the correct research output. This is evident in the Philippines.

Funding is a necessary but not sufficient precondition for research and development. It will only improve research if every funded study results in a scientific paper. The prescribed increase in R&D budget for developing nations assumes that, like in developed countries, research is done properly. This is what leads to national progress, which can support more research for more progress, and so on to industrialization.

For more on the R&D process, read "Essentials of development" posted at the UP Diliman website (*http://www.ovcrd.upd.edu.ph/index.php?option=com_content&task=view&id=462&Itemid=81*).

Research and Development problems in the Philippines

As noted above, problems holding back the growth of science in the Philippines are rooted in wrong research practices, not funding as is commonly believed. Poor graduate training is a major cause. Or is it an effect?

An important requirement for an advanced degree in science is the thesis, which is meant to be training in research. And research is not completed until results are published. But in the country's graduate schools, except some in UP, the end of graduate training is the bound thesis, rather than its publication. Hence, most holders of graduate degrees in the country do not think of proper publication as part of research.

Many in the graduate faculty who avoid international publication failed to develop the needed research capability. Hence, they are unable to exploit scientific advances and to equip the country's future researchers and technologists with useful skills. Unpublished faculty members produce unpublished PhDs and become graduate faculty members and repeat the process.

Many of our PhDs in science who got their degrees abroad did not publish their theses, knowing their home institutions would recognize the degrees even without their publication. Whereas new PhDs without publication experience in international journals do not get faculty positions or research grants in developed countries, in the Philippines they are given automatic promotions. They include those in the graduate faculty, industry, and science administration.

Promotions, even to full professor, are given without justifiable indication of contributions to knowledge or consideration of valid publications. The common practice is to give more importance to promise (graduate degree) than performance (useful publications).

A research paper published without adequate peer review and not accessible through indexes and normal library channels, is gray literature. Examples are research papers in newsletters, institutional publications, and most conference proceedings. Production of papers continues for such publications because they entitle the authors to promotions, honoraria, or even awards. But as has been shown elsewhere, such publications hardly contribute to the growth of science and technology and do not count in ranking nations based on S&T performance. Most of our science administrators and researchers forget that publication in international journals can improve the country's capability to advance local science and technology.

We publish research journals without enough qualified researchers to manage them and adequately review manuscripts. We should review the intentions of Philippine journals because obviously in their present state, they are not promoting Philippine science. Science organizations should also stop giving awards to papers published in these journals (except the few that are now ISI-indexed journals, defined as journals covered in the indexes of the Institute for Scientific Information).

A worse practice is to use unpublished data for policymaking, development programs and other purposes. This is common in the local implementation of projects. The practice is prevalent because of contractual demands from the government and international funding agencies, totally ignoring the established procedures of scientific research. It is an outcome of failure to make publication as the purpose of data gathering. Even those seeking high positions include in their achievements a long list of "unpublished research."

There were large increases in the budget of the Department of Science and Technology, particularly in 1991–1996 (*The S&T Post*, October 1995). Increases in the DOST budget during this period had a yearly mean of 36 percent or from ₱854 million to ₱3.4 billion. But research output, extension materials, and other publications consisted largely of project reports, institutional publications, and other gray literature, which did not count in S&T assessments. They made up more than 95 percent of the total number of papers produced. Only 249 papers were published in international journals in 1995, half of which was from the International Rice Research Institute in Los Baños and other international R&D organizations in the country, and about a fourth was from the University of the Philippines (*Science Citation Index* 1995).

Further, among seven nations in the region, including Thailand, Malaysia, and Indonesia, the Philippines had the least progress in S&T between 1981 and 1995. Whereas its Asian neighbors had 37 to 300 percent increases in the number of international journal publications (the standard measure of S&T performance), that of the Philippines increased by only seven percent, and this was contributed largely by international R&D organizations in the country (*Science Citation Index* and previously published in my paper "Science in the University of the Philippines" *Diliman Review* 47(2):11-18, 1999).

Nations with more productive scientists were ahead of the Philippines in economic growth. Smaller Taiwan had twenty-three times more and tiny Singapore, six times more accredited publications in science than the Philippines in 1994–1995.

By 2000, the Philippines was still behind the seven countries. It has, however, made significant progress. This is expected to continue since a few leading universities have been giving cash incentives for articles published in international journals. For instance, since 1999, the University of the Philippines has been giving an award of ₱50,000 (later raised to ₱55,000) for every paper published in an ISI-indexed journal. By 2003, the number of such publications from UP went up to 182, a two-fold increase (*Current Contents*, all versions; thanks to Dr. Gil Jacinto). Although we still have a long way to go (China's total publications were 58 times that of the Philippines in 2000; Japan's, 160 times), we are on the right track.

Contrary to common claims and practices in the country, we have hardly generated, fully adapted, or transferred useful technologies. We should have realized this by now. When our agricultural products are refused by importing countries for contaminants or diseases, we are unable to take remedial action because of scientific inability. As a former director of graduate studies in physics at Princeton said, "You need to know how to do research properly before you can begin to think about commercializing discoveries" (Kaiser J. 1995. A business blueprint: how to build a better Ph.D. *Science* 270:133-34).

In local extension work, the quality of information is often ignored. Most extension materials contain information from project reports, newsletters, institutional publications, and conference proceedings. They are prepared largely by well-meaning but unqualified people forced to do a scientist's job. Hence, the materials are of doubtful integrity.

Most users of information, like decision makers, teachers, or the general public, rely on the information reaching them. They may not be able to judge the quality of the information. If most of what they get is substandard, as what often happens in the Philippines and other poor countries, errors are propagated and people hardly learn something useful. As a former chairperson of the US Atomic Energy Commission observed, "The public will remain uninformed and uneducated in science until the media professionals decide otherwise, until they stop quoting charlatans and quacks, and until respected scientists speak up" (Ray DL. 1990. Who is to blame when the public misunderstands science? *The Scientist*, 16 April 1990).

For more on R&D problems in the Philippines and related references, read "Celebrating the UP Centennial" posted at the UP Diliman website (*http://www.ovcrd.upd.edu.ph/index.php?option=com_content&task=view&id=461&Itemid=81*).

Needed Changes

The above problems are perpetuated by the indifference to reform so entrenched at all levels of research administration in the country. The vicious cause-and-effect cycle must be broken. It has wasted much time and resources. R&D funds have been spent to produce unpublished papers (graduate theses, project reports, lectures, etc.) and gray literature with the thought that the job has been done. While the practice has benefited the authors in the form of promotions, honoraria, and awards, it has hardly contributed to institutional growth and national progress in science and technology.

There is too much talk about technology and development, when attention should focus on changing our research practices. Evaluation should be in accordance with the established tradition of scientific publication and accepted measures of performance. International indicators of science and technology development, not publicity and claims of achievements, should be the standard measure of S&T performance in the country.

To continue the steps taken by some universities in improving research, nationwide changes are needed in graduate training, research practices, and performance evaluation. A simple way to speed up the development of science is to publish study results in ISI-indexed journals (see ISI website: *www.isinet.com*). Publications in such journals should be required in (a) the granting of a PhD degree, (b) appointing members of the graduate faculty and journal editorial boards, (c) giving research grants, promotions, and awards, and (d) appointing research managers and science administrators. We must build up the competence at all levels of our research endeavor. Doing so will enable us to develop the needed competitive ability.

Researchers should publish their studies, even if the funding agency wants only a final report. This applies also to the graduate thesis, even if the program requires only a bound thesis. To serve as training for writing and publishing research papers, the graduate thesis should be written as a publishable manuscript, ready for submission to an ISI-indexed journal. A master's degree may require only a publishable manuscript, but a PhD degree should require at least one publication from the thesis. The researcher's contribution to the development of science and education will come from such publications.

Three guides in writing a manuscript are: (a) a good book on scientific writing and publishing, (b) a guide for authors of an ISI journal, and (c) sample articles from the journal. Day and Gastel's *How to Write and Publish a Scientific Paper*, 6th edition (2006), is an excellent book. The three writing guides can improve the chances of a manuscript in getting accepted in a good journal. Select journals that do not have page charges (there are enough of them in your field) if you don't have funds for publication.

Changing the criteria for appointments, promotions, and awards will not be easy. The usual problems with incumbents, politics, and mediocrity have defied past attempts. For instance, graduate faculty often voted against changes because members are already enjoying benefits with only progress reports, and most of them don't have publications in international journals. Eventually, however, only respected scientists will dominate positions in all ranks of our research enterprise. This is the situation that had facilitated advancement of science and technology in nations that achieved sustained economic growth and improved human condition.

Note that most of the proposed changes will not need added cost. If a part of the DOST budget is used as cash incentives for research publications like what the UP is doing, we will see the dawn of scientific revolution in the country. Meanwhile, the National Academy of Science and Technology must be more active in policy debates related to science-based initiatives. It should play a major role in our economic reform and social transformation. Among its members are literate scientists who can promote science literacy, especially among government and industry leaders, and ensure that scientific research is incorporated into all of the country's development strategies. There is no better alternative for industrialization than advanced science and technology.

Printed in two parts on May 19 and May 26, 2005

Flor Lacanilao

Problems with Media and Scientists

The public will remain uninformed and uneducated in science until the media professionals decide otherwise, until they stop quoting charlatans and quacks, and until respected scientists speak up.—The Scientist, 16 April 1990.

There is trivia overload but little science reporting, says *Philippine Daily Inquirer* columnist Juan Mercado. Citing UP development communication professor Felix Librero, he asks, why journalists can't do for science and technology what the Philippine Center for Investigative Journalism does very well for politics?

This observation reflects the public's poor understanding of science in the Philippines. The problem is not too little science stories being reported, but too much of them from nonscientists. The information that the public gets is largely propagated errors, but taken as information in science. Hence, there is widespread public ignorance of basic scientific concepts and procedures, which only true scientists can explain.

When they try, however, they have to compete for media space and time with non-scientists, who write and talk better and greatly outnumber scientists. Scientists, on the other hand, are trained and practiced in scientific writing, the style of which is adapted for a distinct purpose and audience. They are then generally poor storytellers. It is still their duty, however, to communicate science to the public.

The media can improve their role in spreading science. It is important for media people to know who the scientists are since they produce the useful information. Scientists should be given more access to media and editorial aid, when scientists have science stories to tell. Media and scientists are the two promoters of science literacy and only by working together can they be effective in giving the public useful science information.

RP's Information Production

As discussed in previous papers in this column, scientists are researchers who have published papers in international refereed journals, which are indexed in Current Contents (CC) or Science Citation Index (SCI). These indexes are produced by the Institute for Scientific Information (*www.isinet.com*). Scientific papers or valid

Formerly with The Marine Science Institute, University of Philippines, Diliman, Quezon City, 1101, Philippines; and the Southeast Asian Fisheries Development Center.
flor_lacanilao@yahoo.com

publications are research papers appearing in CC- or SCI-indexed journals, referred to as international journals. A research paper published elsewhere is gray literature (information produced by nonscientists without adequate peer review and verification). The most common output of research in the Philippines, gray literature is the main source of information disseminated to the public for education, development programs, and decision making.

Because of its doubtful validity, gray literature does not count in established assessments of S&T performance. It is the main reason why our country and universities always lag in international rankings of S&T capability and economic progress. Our science policymakers are yet to be aware of this, which is long past due.

The country produced only 249 international journal papers in S&T in 1995. The rest were unpublished reports and gray literature. The international journal papers made up only five percent of the country's output. Nearly half of the international journal papers were produced by the Los Baños-based International Rice Research Institute, which had only about three percent of the country's PhDs in S&T. Worse, except for the University of the Philippines, the national production of scientific information has hardly improved. At least there is hope as the UP leads.

Examples of Poor Science Reporting

If journalists aim to do for S&T what the Philippine Center for Investigative Journalism does very well for politics, as earlier suggested, but will acquire information from nonscientists and the DOST, they will just continue propagating errors in science. Saying this in a nice way will not help the overall science program. It will just hide the truth.

An example is a recent cover story in *Newsbreak* (January 30, 2006) on alternative medicine. Not one among those interviewed has any published papers in an international medical journal on the efficacy of herbal and coconut oil products. The boxed statement in the lead article says, "Conventional and alternative medicine can go hand in hand to provide integrative or holistic healthcare." This is dangerous if interpreted as meaning consumers can take herbal drugs while taking prescribed medicines.

Numerous studies have shown the adverse interactions between herbal drugs and prescription medicines. Our medical doctors must know this as there are also interactions between certain prescription medicines. Thus, unknowingly the authors and editor are endangering consumers' safety when taking herbal drugs. Could this potential danger to consumers be offset by benefits that readers get from the report?

One would hardly fault some companies promoting herbal products for profit, their main aim. Nor would hosts of talk shows and documentaries be blamed for advertising them. But why would some of our medical doctors and the government agencies in health and S&T promote the use of herbal products and virgin coconut oil? And how can we trust the Bureau of Food and Drugs, the agency that approved a product of virgin coconut oil being advertised in newspapers?

There are not enough studies to guarantee these products' efficacy and protect consumers' safety, as my study on medicinal plant research has shown. No scientific review article has confirmed the efficacy and safety claimed by any of the herbal or coconut oil products. On the other hand, scientific reviews of the literature on the use of herbal products worldwide have documented numerous examples of liver, kidney or other organ damage.

Meanwhile, all forms of media, including talk shows and documentaries, continue to report claims of beneficial effects of herbal and coconut oil products promoted by nonscientists.

Why Don't Scientists Speak Up?

Some of them do or try. But they are overwhelmed by non-scientists. Here are some reasons:

First, unpublished PhDs (no publications in international journal) outnumber published PhDs in S&T many times over. Reasons include the following: (a) With few exceptions at UP, PhD programs in science require only the bound thesis instead of valid publications, (b) most academic and R&D institutions recognize the doctoral degree even without publications in international journal, a practice not seen in any developed country, and (c) unpublished PhDs are given research grants by our funding agencies, and they continue to produce gray literature that entitles the author to honorarium, promotion, or even award.

Second, academic and R&D people in the humanities and social sciences publish more popular articles in science than scientists in the natural sciences and technology. They also appear more on television and radio programs. They participate actively in activities of community organizations (e.g., religious, environmental, health) that support the public's science education.

Third, unlike most nonscientists, scientists are largely poor writers for the general audience. They are trained and experienced in scientific writing which has different rules. Research reports that they write are addressed to fellow scientists for verification, not for general public consumption. Whereas the popular article provides information to the curious reader, the scientific paper "seeks to argue persuasively a new claim" to peers with similar training. The scientist's style clearly differs from that of popular communication. Getting used to the journalist's style is not easy.

Improving Media's Role

Reliable sources of scientific information are available. Media people have only to learn how to get them. For instance, they should start by distinguishing a scientific paper from gray literature, a scientist from a charlatan. There are indicators for each of these differences. The scientific paper, for example, separates the scientist from the charlatan. Find out from the Google Scholar (list of publications) and the CC or SCI index (list of international journals).

One suggestion for editors is not to routinely refuse papers from scientists (once these have been verified) for reasons of poor style. Instead, why not edit the language to suit the general audience? Further, why reject science stories previously submitted or published elsewhere? The popular science article should be treated like a news story that can be published in different newspapers for public information or entertainment. It is not a publication of original data that is barred dual publication in scientific journals. Science stories are intended for the information of maximum audience. What then is the reason for the desired exclusive coverage of science stories by our newspapers and magazines? A lot of useful information from scientists does not reach as many readers as intended because of such editorial policies.

Whereas the scientist's primary role is to produce scientific information (research), dissemination of verified information (extension work) is a scientist's social responsibility. As discussed above, however, they are not trained to write for the community. But some try; a few struggle. This is where editors of science, health, or technology sections can help. For instance, section editors can invite scientists to submit papers on their research projects or interests, which can then be edited to appeal to the general audience. This way, editors are assured of getting useful scientific information from scientists rather than waiting for papers of doubtful quality from people of unknown skill.

Journalists can write about a scientist's work or get information from scientists' review papers in international journals. Scientific review articles not only show which subjects have been adequately studied but also tell which studies have been adequately verified. The quality of cited literature reflects the quality and integrity of a review article, whether or not this is from an international journal. Then journalists can do for S&T what the Philippine Center for Investigative Journalism does very well for politics.

Finally, the key to expand the public's understanding of science is for media people and scientists to recognize their respective roles and to work together and for our scientists to ponder the sad state of science literacy in the country while saying, "We have met the enemy, and he is us!"

July 27, 2006

Part 2

Science Education and Creativity

Literature, Science, and the Imagination

Gemino Abad

LITERATURE AND science, when either one is undertaken and achieved, have in common one great human faculty or power of the mind—the imagination. All great scholars, scientists, and artists are men and women of vibrant imagination. For me, the imagination is simply the finest form of intelligence: it comprises reason and intuition, our physical senses, feeling, and memory, all alive at all points of contact with reality.

Another bond that holds science and literature together, where both serve humanity in equal measure, lies in their quest for reality. But what we call reality is nothing less than the continuing mystery of life and creation, and we alone of all living creatures are the servants of its one living Word—one, because all life and creation is one web. We are its indefatigable initiates because it is a mystery in which we have all our faith and in which our mind finds its profoundest delight and fulfillment unraveling it.

We are all on this quest of reality—to understand, and so, master reality only in order to serve it well. This is what it means to be real to ourselves: we assume the responsibility for life and nature—both human nature and the natural world.

There is no quest where there are no questions. Both science and literature are committed only to such a quest. Neither one respects any politics or ideology except where they, too, join the same quest. Both science and literature move and have their being on the universal plane. The universal plane is not the site of eternal verities but a clearing of everlasting questioning.

Here then are two big words—imagination and reality; these words hold an infinity of human endeavor where language is the crucible of understanding. For the works of literature and the works of science are essentially works of language. Such works always show a brilliant interplay of reality and imagination—on the one hand, the workaday dragons of fact that we would tame, and on the other, the treasure that every dragon guards.

Language is the supreme human achievement. Both literature and science weave a language by which a perception or insight into reality is endowed with a meaningful form. Such weaving of a language, such expression through letters and figures, only realizes, or makes real to our mind, the phenomena—the appearances or manifestations—of nature: again, both human nature and the natural world; yet, and still, their reality

Department of English and Comparative Literature, College of Arts and Letters, University of the Philippines, Diliman, Quezon City, 1101, Philippines.
gabad@yahoo.com

lies deeper beneath our language, is beyond mere nomenclature. What we can know of reality is only a possible text of what we call "our reality"—how we perceive it, from our own standpoint; we are then always at risk.

The works of literature and the works of science, are, as I said, works of language. Yet language itself is already work of imagination. "When the imagination sleeps," says the philosopher Albert Camus, "words are emptied of their meaning." Every name or word, every theorem or way of looking, is only a dwelling-place for that item or aspect of reality that we seek to understand. But to understand is only to stand under: "a green thought in a green shade," says the poet Andrew Marvell.

In science and in literature, language is essentially translation, a word from the Latin *transferre*, *translatus*, meaning "to ferry across." By whatever language, mathematics or a given natural language like English, we only translate the immutable silences of nature, we only ferry across our words, our vocabula and mathema, or learning, our own soul's cargo: our insights and conceivings. What is requisite in science is to have a sharp sense of what may be lacking in our knowledge or conception of anything in reality; likewise, what is requisite in literature is a lively imagination of what may be ideal or possible for humanity.

We are nature's ferryman; we give her our speech.

There is only one difference, as I see it, between literature and science, but it is a difference that enriches us all. Science abstracts from human experience or the observation of phenomena, but literature renders the experience itself as imagined as lived. Science establishes the facts of the reality it observes in order to see it again whole—the idea of it: a theory, a way of looking, by which a specific understanding of it is reached. In this way is a meaning fixed for the reality in the observer's ken. In literature, on the other hand, the imagination's field is the individual human experience where it affirms a person's essential freedom and dignity. As Jung puts it, "the only reality is the individual," to which Aristotle would say, "There is no science of the individual." Literature then has no theory and establishes nothing. It only mimics or simulates a human experience, in one's own historical and cultural scene, in order to see it again whole, as imagined as lived; the form of it in story, poem, or play is the matter of art, and its content is the matter of interpretation. In this way what is achieved is not meaning but meaningfulness: the meaningfulness of the living of a human experience as imagined; what is offered is not a theory of the living of it, but only the sensation of being wholly alive to oneself and a sense of what it is to be human.

In all this, language is the crucible of understanding—and the work that is accomplished is veritable work of imagination.

April 26, 2007

Gisela Padilla-Concepcion

Developing Creativity in Science among Our Youth

Science as a Field of Human Knowledge

To develop creativity in science, first we must understand what is unique about science as a field of human knowledge. Science involves the study of the physical or material world. In science, we study all aspects of Nature, which consists of all material things, including man and the environment. The scientific method, which is considered a refinement of the human cognitive process, is traditionally described as a chronology of activities which begins with a hypothesis to solve a problem or explain a phenomenon in the physical world. This is followed by observations and experiments to test the hypothesis, and it ends with the hypothesis being proven right or wrong. Deductive and inductive reasoning, analytical thinking and synthesis, are exercised throughout this process.

Today science is conducted in a less structured way and generally involves the continuous interplay of empirical, experimental, and rational methods. This is aimed at understanding the underlying principle, mechanism, or function of a physical, chemical, or biological phenomenon. The scientific facts or "truths" discovered and propounded by scientists must be shown to be verifiable and reproducible, being drawn from conclusions obtained using methods which are as objective and quantitative as possible. Indeed, through the years, certain scientific "truths" have been overturned or disproved by new ones. We say that science is not infallible or that it is self-correcting.

Depending on the field of interest or specialization in science, and there are countless today, one method may be used predominantly over the others. Thus, physicists can either be experimental physicists or theoretical physicists. The field of biology, generally considered an empirical science in the past, now benefits from a great deal of experimentation, manipulation, and mathematical modeling, due to the development of new techniques in cell biology and molecular engineering and the availability of very powerful instruments and computers. The trend in science seems to be toward understanding the complexity of Nature at greater depth and breadth, from the systems biology level down to the molecular level. Moreover, scientists in specialized fields are coming together under multidisciplinary research programs, and mathematicians are interacting more with scientists. Thus, we see the growth of fields such as biophysics,

---

The Marine Science Institute, University of the Philippines, Diliman, Quezon City, 1101, Philippines.
gpconcepcion@yahoo.com

molecular medicine, and computational biology, to name a few. These new combined perspectives and "multitasking" efforts have led to deeper insights and greater creativity in solving problems in science. For example, today, biological principles and biological structures are being used as models to solve engineering problems in a more environment-friendly way. (Spider silk fibers, mussel glues, and deep-sea sponge fibers are being used as models to produce ropes stronger than steel, glues that can stick to nonstick Teflon, and better optic fibers, respectively.)

Creativity in Science

Human knowledge, the foundation of human culture and civilization, builds on old or existing knowledge, particularly in science. The term "research" underscores the idea that "new" science builds on a huge existing body of scientific knowledge that is well-documented. Age-old scientific principles, as well as newer scientific findings, are found in books and journals and are available to young students for them to understand and analyze, and subsequently to test, contest, disprove, expand, refine, and modify. We build on what we already know. Basic education in mathematics and the sciences equips our youth with the ability to be creative in science, to change it and move it forward, to make new discoveries and add these to basic scientific knowledge, and to develop innovative technologies or applications that could benefit mankind. Advancements in science have occurred at an accelerated or geometric pace in the last 100 years, far exceeding those from the time discoveries by early man were first recorded.

In school, we would like our youth to be provided with the basic foundation to make discoveries possible. Serendipity, as it is often said, favors the prepared mind. Many important discoveries were made, seemingly, by chance or luck. A classic example is the discovery of the antibiotic penicillin by the microbiologist Alexander Fleming, who had found a mold or fungal overgrowth inhibiting a *Staphylococcal* (bacterial) growth among a pile of Petri dishes that had been left for days near the sink in his laboratory. It took Fleming's trained eye and scientific mind to realize that what he had observed could be very useful in treating bacterial infections.

Being creative in science is tough and requires brain power. Creativity involves seeing new patterns and connections, putting together parts to make a whole and breaking the whole to make new parts. It is viewing something from a different perspective, considering possibilities or alternatives, or thinking unconventionally—thinking straight when the pattern is fuzzy, thinking nonlinearly when the pattern seems straight! It is looking at the forest beyond the trees, and also at the trees that make up the forest. It is seeing as essential what is invisible, but also seeing what is visible as essential. Science, described by Robert Pirsig in his book *Zen and the Art of Motorcycle Maintenance*, is the "high country of the mind," and being creative in science means seeing and doing new things from this vantage point.

Persons Who Can Be Creative in Science

Can anyone equipped with the proper tools and background to do science be creative in science? It would seem that there are scientists who are naturally creative and those who are not very creative. If one had a big brain, with lots of convexities (folding and refolding) in the cerebrum, implying that one has many interconnected neurons, then a high level of creative activity could occur in one's brain. And with a highly developed frontal lobe, which is the seat of emotions such as strong motivation and passion, then one may have been born with the "brains" or the genes to become creative. In short, the creative trait is, first and foremost, anatomic, genetic, or inborn.

The creative process in science begins with the mental processing of a lot of information gathered from the physical world and ends with a tangible result—new knowledge, a new creation, product or discovery. For all these, one needs a superior mind or a powerful brain with a dense circuit of highly interconnected neurons that can process numerous stimuli and generate equally numerous responses quickly, almost instantaneously. Mental processing is not done "in series" or "in parallel," but through a complex circuit or network of neurons.

And what role does motivation, passion, enthusiasm, or excitement play in the creative pursuit of science? Creative scientists seem to possess these strong emotions which serve as the driving force for the sustained, sometimes lifetime, obsessive pursuit of a field or subject of interest. This excitement or passion is evident during talks or seminars given by eminent scientists. James Watson, famous for solving the structure of DNA in the mid-50s, writes about his passion for science in a book entitled *A Passion for DNA*.

Can creativity in science be learned or nurtured? Creativity as a trait may be largely genetic, but to a certain degree, stimulating and training a person to think creatively, especially someone young, can make that person more creative. Because of neural or brain plasticity, constant stimulation in animals results in long-term potentiation, the equivalent of memory and learning in humans, and we may presume that something similar happens in humans. Thus, we may be able to nurture creativity among our youth by continuously providing them with the necessary stimulation and training to think creatively. The first five years of life are the most formative years, when the learning curve is steepest, and when a child's skills or talents for language, the arts, and even perhaps for scientific observation, can be awakened. Children can become "naturalists" or Nature lovers, and may grow up to become scientists, by being taught to observe the ways, cycles, and balance of nature, early in life.

Can creativity in science be stifled or discouraged? There are societies or cultures where convention and tradition can stifle creativity in the young. Creativity or the lack of it is partly cultural. We know of countries which have consistently led the world in technological inventions and revolutions because individuals in these countries have enjoyed the freedom to think innovatively and defy established rules and procedures. On the other hand, there are countries where a rigid, hierarchical, educational, or academic system tends to discourage creativity and promotes adherence to tried-and-tested solutions to problems. Students who are inclined toward the physical sciences and mathematics become engineers rather than scientists and their countries produce excellent engineers and engineering products. In some instances, individuals who lack innovative skills resort to copying inventions or "pirating" the ideas of others.

Can an untrained, unschooled individual, someone who is not science-literate, be innately creative and make significant discoveries in science, just as an unschooled or nonformally trained artist can produce good art work? Can a person "create something out of nothing" in science? The answer seems to be a qualified "no" because scientific discovery requires a proper background in mathematics and the basic sciences, although technical or vocational education can provide the background for simple inventions. One can also think of special cases such as "herbal specialists" in ethnic communities possessing "traditional knowledge," who formulate cures for various ailments based on simple observations and experiments made with medicinal plants that grow around them. We could consider these unschooled "herbolarios" or "naturalists" as possessing some degree of scientific creativity.

People Who Can Inspire and Guide our Youth to Become Creative in Science

Parents can provide a stimulating home environment for young children to develop their natural curiosity and inquisitiveness in the early years. Children can be encouraged to observe Nature—to ask the "what, when, where, how, and why" of anything. Children are given toys to tinker with to develop their experimental and manual skills. They should be allowed to break apart their toys and put them together again. Formal guidance or training in science of our youth is then passed on to the teachers in the pre-school, grade school, intermediate school and college levels of education.

Science teachers can inspire students with excellent lectures and well-planned experiments designed to make science interesting and discovery exciting. Behind every successful scientist is a high school teacher who has influenced the student to pursue a science degree in college. College is a turning point or a decision time in the life of a young person. A student who is inspired because of the stimulating and persuasive ways a high school teacher teaches astronomy, physics, geology, chemistry, or biology in the classroom, will choose to pursue a science course in college because of sheer fascination with science, despite the prospect of a financially unrewarding scientific career or an uncertain future after college.

Science students and researchers should seek out outstanding research mentors to work with. Research professors in college and graduate school and leaders of research projects are the sources of inspiration of young researchers and also serve as their most important role models for creativity in scientific research. Research mentors play the most influential role in the way science is conducted by their young associates. It should be emphasized that the quality of research depends principally on the persons involved, and the research leader is mainly responsible for the quality of work in his laboratory. It should be the wish of every research mentor to have his students and researchers exceed his accomplishments or outdo him. A truly creative researcher is one who will deviate from or expand the work of his mentor and pursue his own interests, while an uncreative researcher will continue to do his mentor's work without variation or innovation. To move science forward, it will be good to encourage and train our young researchers to think and to learn to research independently.

A country's political leaders, science administrators, and science agencies should provide the vision and mission that will guide the general direction that S&T (science and technology) research will take to become useful and relevant to society. This provides a general framework by which great leaps of creativity and productivity in science could be made. A national S&T program with ambitious and far-reaching goals can capture the imagination of a nation or of the world, unite many people, and produce a lot of new scientific information as well as new useful technologies. The Cassini probe circling Saturn brought images of Saturn's rings back to Earth at the speed of light. The US interplanetary probe and space shuttle exploration programs, initiated in the 1960s, have provided us a great wealth of information about the Earth and information within and outside our solar system. (We recall that it was only a hundred years ago that the Wright brothers flew the first plane.) The "Star Wars" antimissile defense program of the US, initiated in the 1980s, is credited with having ended the Cold War with the USSR and causing the demise of the Soviet Empire and communism. Both the space exploration and anti-missile defense programs are also credited with having spawned many new commercial technologies, e.g., bigger capacity IT chips and computers, and miniaturized instruments.

We also recall that science fiction and futuristic books have long fired up the dreams and imagination of many generations of humanity. Scientists have been inspired to make the fantasy of scientific gadgets and robots described in these books a reality. Today, writers continue to capture our imagination because so much of what was imagined then has actually come true! We can think of creative books such as H.G. Well's *The Time Machine*, Jules Verne's *Voyage to the Center of the Earth*, Lewis Carroll's *Alice in Wonderland*, and TV shows like *Dick Tracy* and Gene Rodenberry's *Star Trek*, which all provided important ideas for scientific development.

Ways to Develop Creativity, Independence, and Responsibility in Science among Our Youth

Thus, to develop creativity in science among our youth, we must equip them early on with the tools of mathematics and the basic sciences. We must stimulate their curiosity of the natural world and encourage them to become inquisitive. We can guide them toward new ways of looking at things, at the same time allowing them to do their own thinking that would be different from the perspective of their mentors. It is imagination and creativity tempered with discipline and analytical prowess that translate to significant strides in science. Our youth should be encouraged to read the accounts of the discoveries and the biographies of outstanding scientists such as Nobel Prize winners who exhibited these traits and could serve as their models. Young researchers should seek out good and kind mentors. These mentors will guide them well by their examples and encourage them to pursue higher academic degrees and more advanced technical training. Mentors can help our young researchers appreciate the value of science to human life, society, and the environment, and encourage and prepare them to become creative, independent-minded, and responsible scientists.

Printed in two parts on May 5 and May 12, 2005

Maria Victoria Carpio-Bernido and Christopher Bernido

Essentials versus Peripherals: Our Experience in Basic Education

In 2002, we introduced a dynamic learning program (DLP) of strategies in the high school of the Central Visayan Institute Foundation (CVIF), three years after full immersion and observation that revealed the weaknesses of prevailing educational programs and strategies. With so much effort and backbreaking work, results were nil or negligible. We then designed the CVIF DLP as a package of solutions to partially solve the problem of a complex system subject to multiple constraints. In particular, we aimed at designing a low-budget yet effective educational program that:

- is suitable for large classes (>25 students);
- is less dependent on textbooks;
- requires less equipment;
- addresses the English proficiency problem without sacrificing deep understanding of science and math;
- addresses the spectrum of students from slow to brilliant, from poor to rich;
- is suited to physical and psychological stages of the youth;
- reduces teaching personnel requirements;
- is less dependent on the personalities and abilities of teachers; and
- has built-in checks of dysfunctional behavior observed in Filipinos while boosting their natural creativity and ingenuity.

Typical of theoretical physicists, we went "back to first principles." We started with setting up targets and the standards by which to measure success. Clearly, we could not define these in terms of the national situation considering the country's very poor performance in both national and international achievement tests (generally <50 percent proficiency marks). On the other hand, even in advanced countries, there is some alarm over the deteriorating performance of schoolchildren and declining interest in the basic and applied sciences—the driving factors for the technological advances now enjoyed by mankind. To define our goals we thus had to think deeply, distinguishing between the essential versus the peripheral, invariable versus fashionable, primary principles versus

Research Center for Theoretical Physics, Central Visayan Institute Foundation, Jagna, Bohol, Philippines.
cbernido@mozcom.com

secondary, universal versus local and parochial. Schools of thought as to the ends and methodology of education were examined simply as "schools of thought" and not as binding mandates.

At present, our long-term goal is to develop our young people up to the apex of a pyramidal hierarchy of learning. Visual-kinesthetic exploratory learning forms the base, followed by qualitative-conceptual/verbal (explanatory) learning. Of higher order is quantitative/mathematical explanatory learning. The apex—quantitative-mathematical synthesis—is difficult to achieve but is the fountainhead for high-impact creativity and accelerated technological advancement.

Although we thought of these levels of learning to properly train students of physics, these may be applied to other fields such as sports and the performing arts. Athletes who undergo scientific training programs generally do better. Talent and physical endowments are not enough to be competitive when shaving off hundredths of seconds from world records.

After setting up targets and clarifying standards, we devised strategies and procedures to accomplish our mission in the quickest time, with minimal change in physical and organizational infrastructure, despite inferior resources. These strategies are enveloped in the CVIF DLP, a program mixing traditional methods, individual instruction, distance learning, the Montessori philosophy, and reviving classical approaches such as the no-homework policy. The components of the program are Parallel Classes, Activity-based Multi-domain Learning, In-school Comprehensive Student Portfolios (instead of notebooks), Comprehensive Teachers Portfolio (instead of Lesson Plan Notebooks), Strategic Study and Rest Periods, and Integrated Cultural and Values Formation. (Synchronized classes allow a scenario of "Learning as One Nation" with lectures of a national expert teacher aired live on TV to benefit more students. This would bypass the country's perennial lack of qualified science teachers. The project is being considered for piloting. On January 16, 2006, the Fund for Assistance to Private Education hosted a meeting of the directors and staff of the Department of Science and Technology-Science Education Institute, the Philippine Science Centrum, and the presidents of the Center for Educational Measurement, the Samahang Pisika ng Pilipinas, and the CVIF, to brainstorm implementation schemes for the project.)

We implemented the CVIF DLP with the Basic Education Curriculum in 2002 and then started the slow deliberate scientific process of continuing improvement and decreasing margins of error (though not precluding quantum jumps with resonance of insights). Focus on targets made us resilient against the misgivings expressed by those who found our methods slow and strange. We knew that before positive results of a major enterprise can be enjoyed, a slowdown may first occur. After the dip, a slow rise begins, and then exponential progression can be anticipated. By 2003, 2004, and 2005, out of about a dozen students who took the University of the Philippines (UP) College Admission Test annually, four, five, and six students passed, respectively, after eight consecutive years without any passers. (UP's reasonable fee for quality education is helpful for our students who are mostly children of farmers, fishermen, laborers, and tricycle drivers, among others.) Students now do better in our open-ended evaluation activities. However, we are still far from our target number of graduates admitted to good colleges and universities, and who do not find difficulty in college math and science courses. Our other targets (within five to seven years) include competitive performance in international scholastic aptitude tests, and internationally publishable research work of teachers and advanced students. From low baselines in 1999, we are slowly approaching normal conditions for this school year (SY). Starting SY 2006-2007, we shall move more aggressively toward our long-range goals.

Academic performance is not enough. We desire the development of the whole person. Scientific culture is not bred in isolation but catalyzed even as we inculcate civic culture. For example, rules on punctuality in our school demand synchronization of clocks with international time checks. Synchronization demands the availability of quality clocks, thus increasing sensitivity to precision and accuracy—valued by scientists. This has also taught the discipline of timekeeping and the value of a single minute. For a meeting scheduled at 3:45 to 4:45 p.m., teachers generally come two minutes before, assured that the meeting starts at exactly 3:45 p.m. and will end at 4:45 p.m. Following the example, student council officers schedule and start meetings on time.

In teaching math and science, we underscore the stringent rules of logic and the discipline of the scientific method. This value integration seems to have diminished the propensity for gossip, hearsay, and unchecked opinions. There is also better appreciation of rules and laws of the community and the nation at large as important for order, progress, and the development of all members of the community.

Echoing the Socratic Method, we have also evolved a more scholarly school culture counter to the more materialistic popular culture. (While respecting "parallel cultures," we note that some cultures foster excellence and progress while others foster squalor, poverty, and mediocrity. Note also that there are athletes who are intellectual, and university professors who are not.) In our most recent three-act play, "St. Thomas More," Sir More is portrayed as a loving and dedicated family man who had his daughters educated in a time when women were relegated to less intellectual ventures. One scene has Sir More's children discussing theories of the earth and the solar system (an opportunity to recap lessons on Copernicus, Kepler, Columbus, and Magellan). Also, the Machiavellian tendencies of King Henry VIII and his advisers were contrasted with the selflessness of Sir More in his perception of the hierarchy of family, king, country, and God. In such plays, our students are exposed to real dilemmas faced by great men and their choices which determined their role in history.

The excitement and challenge of educating our students continue to propel us in our work. We do not deny that there are many difficult moments when our fortitude, stamina, patience, and perseverance are sorely tested. However, these are nothing compared to the many moments of profound joy and reverence as we observe our students in the throes of purest creativity and exploration of truths invisible to the physical eye. Moreover, it was when we immersed ourselves fully in the school that we began to see answers to many questions. Why do many children, all the way up to adulthood, fear or hate math and physics? Are there ways of making these subjects enjoyable even while being rigorously mathematical and not just conceptual? How do we transcend linguistic and cultural barriers to uncover universal truths? What would drive our nation to genuine progress and development? (For us, genuine development is not the high-rise condominiums, computers, fancy cars, and high-tech gadgets. These are but peripherals. True development means that the people of a nation have achieved high levels of civilization—they are honest, industrious, dutiful, and can maintain clean, peaceful, productive towns and cities in all parts of the country.) We do not have all the answers yet but, clearly, education of the youth remains a fruitful source of insights.

April 13, 2006

Science Lessons Forum for Elementary and Secondary Schools

ANGEL DE DIOS

THE COMPUTER classroom, like any classroom, is a place, first and foremost, for learning and not teaching. This is a simple but nonetheless an important point. When we focus on learning, we provide an environment that is rich in opportunities for discovery and inquiry. On the other hand, an emphasis on teaching may carry the detrimental scars of negative experiences adults may have acquired in the past as well as their own limitations. Learning requires facilitation. It begins with resources and continues with guidance. Students, especially children, are generally visual, sensing, active, and sequential learners. It is within this premise that the Science Lessons Forum has been assembled.

The Science Lessons Forum contains learning resources for teachers, students, and parents. Each lesson starts with a news article. The forum can therefore be regarded as learning science as a current events-based subject. The news article (obtained from press releases from universities and articles provided by *eurekalert.org*, *livescience.com*, *sciencedaily.com*, and other science news sources) is usually a text explaining in layman or popular language a recent scientific discovery. These articles are generally short as they are only meant to arouse one's quest for knowledge. The bulk of the science lessons lies in the exploratory questions found at the end of the news article. These questions are asked and answered by links to websites that have been chosen for their clarity, correctness, and style. Efforts are made to arrange the questions in some useful pedagogical order. Most of these websites have been chosen for their visual content. Some of them are dynamic; some are interactive; some carry audio material. Each lesson ends with interactive games relevant to the topic to further attract the interest of the learners.

For example, in one of the topics, the leading news article relates the story of the recent Guimaras oil spill. This article is then followed by the basic question as to why one should be concerned with oil spills. The answer to this question is provided by a link to a site called "oil and water don't mix" made by the Australian Maritime Safety Authority, which provides a highly interactive cartoon-based presentation on the immiscibility of oil and water and the harmful environmental consequences of oil spills.

The Science Lessons Forum is based on a philosophy that incorporates what we know about learning into teaching. The forum does not recommend that we abandon the

Department of Chemistry, Georgetown University, Washington DC, USA.
dediosa@georgetown.edu

traditional lecture-based classroom. Instead, the suggestion is to modify the lectures with a style that takes into account the preferences of young learners. With these in mind, one can see the rationale behind the structure of each lesson.

As a summary, the Science Lessons Forum provides resources that hopefully will facilitate learning in the classrooms. These, unfortunately, are just resources and simply embody a starting point. To make the forum work, it requires participation. The forums will only facilitate learning if there are facilitators. And this is the call for all parents and teachers. Reading and examining the contents of the forum is the first important step.

One of the important features of the Science Lessons Forum lies in its outline, in which learners are guided on how to explore concepts, ideas, and phenomena. It is expected that the time spent on these lessons will translate into developments in the areas of reading, mathematics, and reasoning. Thus, it is projected to have an impact on almost every aspect of education. The placement of the science lessons in a forum creates an atmosphere of feedback. Messages and queries can be posted and conversations on various topics can be cultivated. These pages are dynamic and they could become interactive with our participation. Although the webmasters could access the statistics of the forum and determine how frequent the science lessons are being used, what is more significant is the fact that the readers—students, teachers, and parents—can post. It is this aspect that makes the Science Lessons Forum unique. At this early stage of the forum, we hope that this interactive capability, which is built on the framework of the Science Lessons Forum, will grow with time. The Science Lessons Forum currently has 300 topics with thousands of links, and can be found in: *http://www.paete.org/forums/viewforum.php?f=12*

The Science Lessons Forum is part of the "Alay Computer" project of Paete, Laguna. Some of the topics have already inspired the local leaders of Paete, Laguna while they search for better ways of protecting their environment and for more sustainable means of livelihood. In this year's National Achievement Exam, the schools of Paete obtained mean percentage scores in the upper 60s and mid-70s, a marked improvement from a couple of years ago when their scores were in the 25-35 range.

November 23, 2006

Angel de Dios

"Alay Computer" in Paete, Laguna

I spent three weeks of this past summer in the small town of Paete in Laguna, conducting a series of lectures to parents and teachers in the town's three public elementary schools. With so many professionals leaving the Philippines, it is time for those who have left to reach back to their homeland. With financial help from overseas Paetenians, a computer classroom has been built in each of the elementary schools of Paete. And this past summer signaled the next step beyond financial aid. The project now requires one's time and knowledge and more importantly, a more direct involvement. Born in the Internet, the "Alay Computer" project of Paete is a collaboration among the town's schools, the municipal government, and individual Paetenians. The project was launched in 2002 with the original objective of training the students of Paete in information technology and computer office skills. Three years later, the computer classrooms have become an integral part of synergistic activities that will address not only the education but also the livelihood, welfare, and environment of Paete.

Paete was recently declared as the "Carving Capital of the Philippines." It is a town that lies right at the foot of the Sierra Madre mountains, and on its west is the lake of Laguna, one of the largest freshwater lakes in Asia. Thus, the opportunity of launching programs that will benefit art, culture, and Nature conservation is genuine in the town of Paete. The role of information technology in programs that will address concerns in the areas of livelihood, environment, education, and social awareness is, indeed, extremely important. Outlined here are possible ways of using technology to support and enhance programs that are already in place. These objectives comprise the main focus of the local government of Paete, thus these proposed activities are expected to receive full support from the local officials and citizens of Paete.

The computer classrooms are currently housed inside the public elementary schools of Paete. Therefore, one of their main uses is the enhancement of both teaching and learning in the schools. Teachers, as well as parents, have been introduced to the possible uses of the Internet. The three weeks spent in seminars highlighted the use of the Internet in enhancing the students' basic skills in reading, writing, logic, and arithmetic. We have compiled a list of resources on the Internet that will provide discovery-based learning activities, multimedia presentations, lesson plans, quizzes, and puzzles in one page: *http://bouman.chem.georgetown.edu/batangpaete.html*

In a training session that lasted one week, almost 100 teachers and the same number of parents were introduced to examples taken from these Internet resources. Hence,

Department of Chemistry, Georgetown University, Washington DC, USA.
dediosa@georgetown.edu

slowly but surely, we can begin to incorporate the Internet as a learning and teaching tool in the schools. The schools are now equipped with multimedia projectors and screens, bringing the Internet into classroom instruction and learning.

Webcams have also been installed in the classrooms to support a Web-based mentorship. These are expected to assist the continuing involvement and support of overseas Paetenians. Discussion and updates on the projects continue in a forum maintained on the Internet: *http://.paete.org/forums/index.php*

The educational component of the computer classrooms is, indeed, a very important component, but we realize that we can expand the use of these classrooms to support, in a synergistic way, other programs of the town that will address other areas outside of education. Here we present some concrete examples.

The town of Paete is blessed with several nongovernmental organizations (NGOs). The computer classrooms can be used during weekends and evenings to help train all the NGOs in Paete so that these groups can take advantage of information technology in their planning, presentation, discussion, and solicitation. It is then proposed that the computer schools hold training sessions for these groups that will include training in e-mail, forum participation, word processing, slide presentation, and Web authoring.

The town of Paete will coordinate with the Office of Social Concern and Involvement of the Ateneo de Manila University to launch an immersion program for computer science, environmental science, natural science, and economics juniors and seniors of the Ateneo. Since Paete is only about 100 kilometers from Metro Manila, students from the Ateneo can easily travel to Paete on weekends to help train the NGOs in the use of information technology.

The town of Paete is likewise blessed with overseas professionals who can participate in the training and mentoring, either in person or over the Internet. There are several webmasters based in Canada, the US, and Japan, who can easily lend their expertise in training the students, teachers, parents, and NGOs in the use of information technology. One important aspect of the Internet is that it provides an excellent medium of communication, in which brainstorming sessions can be held as well as one-on-one correspondence, as in distance learning.

The town of Paete, with its culture and natural landscape, provides an excellent destination for grade school and high school students. In fact, during the months of September and October, buses of grade school students from Metro Manila visit the town. The computer classrooms can easily serve as launching stations for these trips where students will be provided a slide presentation of the town's history and culture. In this manner, the children will be better prepared for their sightseeing of Paete. Since the classrooms are located inside the elementary schools of Paete, students from the elite schools in Manila will get the chance to meet the local students.

Small business enterprises can also benefit from the use of the Internet. Since Paete's products cater to visitors and tourists, they can be promoted and marketed on the

Internet. Therefore, these enterprises should also be provided time for training in the computer classrooms.

These groups can be assisted by "immersion" students from the management and economics departments of the Ateneo. To succeed in Internet commerce, effective advertising is necessary. The schools of Paete will thus engage themselves in projects involving the use of computers aimed at promoting the products of the town. Computer products such as calendars, bookmarks, Web pages, and screen savers that highlight Paete's art are excellent projects for the students and teachers of Paete. These products can then be sold to overseas Paetenians who could use them as tokens or small gifts to their friends and coworkers in the US, Europe, Japan, and other countries. The proceeds will be used for the maintenance and upgrade of the computer classrooms. The mayor of Paete has already set an example when he used the computer classrooms to publish a book of his poems that has already generated thousands of dollars in revenue. The synergistic aspect is obvious. At the same time that funds were being obtained to support the classrooms, Paete's cultural heritage in the form of poems was also being promoted.

Paete is unique in the overwhelming presence of its overseas citizens on the Internet. The mailing list on Yahoo! currently has more than 400 members, and with the computer classrooms in Paete, the much-needed link between overseas and local Paetenians has been established. Part of the three-week series of seminars was, in fact, devoted to training the citizens of Paete in the use of the forum and the mailing list.

The computer project in Paete is an example of a community moving as one. Very recently, the World Bank donated eighteen desktop computers to the "Alay Computer" project. Five of these computers have been placed inside the Sangguniang Bayan of Paete to provide access to out-of-school youth and families of overseas workers. There are likewise plans of installing a wireless Internet infrastructure in the town, hopefully through the support of the Cisco Foundation Inc. With support from other institutions, the objectives that have been briefly outlined here are, indeed, within reach. Paete, however, is not merely on the receiving end. For example, in the immersion program, Ateneo students will be exposed to real-life situations, providing them with excellent practical training. Science students will see their course work in action and in a real setting. Better education will provide better employees and customers to companies in the future. With the projects described here, Paete will likewise serve as a great beacon for other Filipinos to follow.

These programs can be easily monitored for assessment. The participation of the NGOs on the Internet and information technology can easily be quantified in terms of presentations made and webpages published. The schools have already benefitted from the computers as manifested by the recent success of their school papers. Since the beginning of the computer classrooms in schoolyear 2003, the Paete Central Elementary School has won thirty-six trophies in journalism, including the Catholic Mass Media Award (as best organ), in 2003 and 2004, the two years the computer classrooms have been in operation. Of course, with the incorporation of the Internet into classroom instruction, we will begin to use the national reading and arithmetic proficiency tests as measures of success. We will also pay attention to the retention rate of students in the schools as we hope that the provision of computers and multimedia tools inside the classrooms will encourage children to stay in school.

It is the community spirit that enabled the "Alay Computer" project to reach its current status. With the involvement of the community in the design and implementation of the project, funds donated truly reached their maximum potential. All the people behind the project worked as volunteers and received no salaries or commissions. The local population also contributed to the project by launching its own fundraising events. The participation is always overwhelming. The community involvement provides the necessary framework for this project to be regarded by the people of Paete as truly their own. The project goes far beyond simply providing computer classrooms to elementary schoolchildren in Paete. These classrooms enhance the teaching of the instructors via access to Internet resources and teaching tools. But since the project involves the entire municipality, the town is able to exercise, for the first time, an opportunity to work together for the future of their children. The impact of the project on how the adults view education is enormous as we project an increased awareness of the importance of education, which hopefully will yield lower dropout rates and lead to a decrease in child labor in the woodcarving shops of Paete.

The "Alay Computer" project is supported, run, and administered by the Paetenian Alliance for Education and Technology Inc., where I currently serve as honorary director. It is a private, nonprofit, nonstock, nongovernmental organization registered with the Securities and Exchange Commission on November 5, 2002 with Company Reg. No. CN202-59015. Its nickname is PAETEch Foundation and is based in Paete, Laguna, Philippines.

December 1, 2005

Caesar Saloma

On Being a Teacher

We yearn to learn from a teacher. In an ideal world, that is possible, or at least in the movies, it is not difficult to sympathize with or even adore teachers. Think of Mr. Miyagi (played by Pat Morita) in the *Karate Kid* or Henry Jones (played by Harrison Ford) in the Indiana Jones series. Many of us grew up admiring these heroic characters and only a demented mind would disagree that both Miyagi-sensei and Professor Jones represent something that is noble and right in civil society. We want to emulate these characters and it is not because they symbolize immense wealth or near absolute political power—Miyagi and Jones had none of these. The reason seems to be our innate recognition that teachers are invaluable in human societies—they impart knowledge that allows us to understand ourselves better. Civilizations are shaped and enriched by teachers.

Teaching is a vocation that requires technical skill, selflessness, and commitment. Every discipline, from the basic sciences and mathematics to medicine and engineering, needs its own coterie of talented teachers in order to avoid stagnation and decline. Teachers ensure that skills are continuously refined and passed on to the next generation. They guarantee that specialized knowledge is shared and tested incessantly through time. To accomplish these difficult tasks, teachers often put the interest of their students first above their own even in the absence of a tenable material reward.

The need for competent and dedicated teachers is most dire in our graduate programs in the natural and applied sciences, mathematics, and engineering. The Philippines is unable to produce a sufficient number of PhD graduates due to the lack of qualified mentors to supervise graduate students. A 2004 study published in the journal *Nature* reported that there is one PhD for every 3,316, 11,621 and 6,533 heads of population in Germany, Japan, and US, respectively. In the Philippines where the 2007 population is estimated to be around 88 million, the total number of PhDs is likely to be below 2,000. To achieve the PhD density profile of Japan, our country needs more than 7,500 PhDs.

A PhD degree is a research degree. It is the last formal degree that could be earned in the conventional system of matriculated training. The PhD degree is awarded to a student who has contributed something original, novel, and valuable to the body of scientific knowledge. The constant generation of new knowledge is strategically vital to our country because it is the fuel that drives technological innovation, which in turn allows industries to prosper in a ruthless global economy. More than 440 years after the

National Institute of Physics, University of the Philippines, Diliman, Quezon City, 1101, Philippines.
caesar.saloma@gmail.com

birth of Galileo, scientific knowledge has attained such an advanced state that it is now impossible for a student to satisfy the stringent requirement of a PhD degree without the guidance of an experienced mentor.

The incapacity of our current higher education system to train technically competent scientists and engineers in significant numbers hampers the efficacy of our country in attracting a substantial amount of foreign direct investment which is crucial in creating well-paying jobs for our people. The scarcity of viable graduate programs is a major cause of the persistent migration of talented BS graduates in the sciences and engineering. The brain drain that is happening is more due to the absence of a sensible option than to the lack of love of country.

Even in advanced countries, schools and universities are not known to be among the most generous of employers in terms of salaries and benefits. The situation is worse in an emerging economy like the Philippines. A survey published in the journal *Science* in November 2006 reported a US academic median salary for 2006 of $78,382 and $60,809 for male and female, respectively. The corresponding salary in the private sector is 1.34 and 1.29 times higher. In the University of the Philippines, the maximum salary of a full professor (rank 12) is ₱361,356 or $7,374.61 (one dollar = ₱49) which is one order of magnitude less that his/her US counterpart. A young PhD normally in his late 20s or early 30s who is holding the rank of an assistant professor receives an annual salary of ₱245,184—not enough to decently support a housewife and an offspring.

It is heartening to note that both the government and the private sector are investing more seriously to build more school classrooms and to hire more teachers. Many of our countrymen believe that the Philippines must improve the quality of the science and mathematics instruction that is delivered in most elementary and high schools. And they are right. However, our national investment in science education would be futile if our country is unable to provide excellent graduate programs that will retain the best and the brightest young Filipino minds. The skillful graduates that are produced by our science high schools and colleges will only end up in the graduate programs of foreign universities. Neighboring countries like Japan, South Korea, Singapore, and Taiwan ROC have spent heavily sprucing up their science and engineering departments. They are now on the lookout for talented graduate students. If the current situation with its graduate programs is not addressed effectively, investment by the Philippines in basic science education will be an unintended subsidy to the higher education efforts of its wealthi er neighbors.

A university is a place for higher learning. Its reputation is derived from the creative prowess of its faculty. The US has the best higher education system in the world and its universities set up endowment funds to attract and retain the best available faculty. Great professors attract talented students (and not the other way around). Together they organize into research teams that draw the serious funding attention of the private sector and government agencies. Individuals who are part of these powerhouse teams are the most likely to make excellent contributions to science, win the Nobel Prize, and bring pride to a nation.

Companies and universities around the world are engaged in a relentless search for talent. Both the government and the private sector should recognize the profound implications of this global trend to the self-esteem and future prosperity of the Filipino nation. They should initiate sensible ways of celebrating and recognizing the contributions of mentors and as a result, encourage young scientists and researchers to follow their path.

March 8, 2007

Caesar Saloma

On Mentoring

A mentor is someone who advises, supports, monitors, and nurtures the progress of a younger, less experienced individual. He or she is usually older and more experienced. The continued success of the scientific enterprise to shape human civilization depends greatly on the capacity of mentors to train the next generation of productive scientists in society. A well-qualified mentor of a graduate student possesses the noble qualities of a good teacher and the proven track record of a successful researcher because a PhD is a research degree that is granted after the contribution of a work that is original, novel, and consequential to the body of scientific knowledge.

Principally through the dedicated efforts of the local science community and its advocates in both the government and the private sector, interest in pursuing an advanced science career is rising among the Filipino youth, especially among the ones with more talent and better training during high school and college. The trend is not only propitious for science but also for the future socioeconomic well-being of Philippine society. Technological innovation is not possible without the application of new knowledge that is generated by scientific research and development.

Innovation allows industries to develop and offer better and more exciting products and services. A vibrant economy that is fueled by the timely introduction and efficient use of (new) technologies will enable the Philippines to withstand or absorb the attendant pressures of a rapidly increasing population, dwindling natural resources and a calamity-prone geographic location. In 2004, an article in the journal *Nature* pointed out a strong correlation between scientific productivity and the economic wealth of countries around the world. Taken together, the various surveys that have appeared in the weekly newspaper *The Economist* reveal that countries, which are strong in scientific research and development (R&D) are also perceived to be less corrupt and less risky to foreign direct investment. Doing business is also less costly, a climate that is more likely to benefit consumers in such countries. Earnest investment in R&D does not only strengthen a national economy, it also builds the self-esteem of a nation.

The Philippines needs more mentors to mitigate its dire shortage of advanced scientific manpower that prevents it from thriving in a knowledge-based global economy. The need to address the scarcity is urgent because the science community is the primary source of intellectual capital in an advanced society. A mentor provides a young graduate student the valuable opportunity to efficiently acquire essential technical skills and precious laboratory experience that are crucial to his future productivity as an

---

National Institute of Physics, University of the Philippines, Diliman, Quezon City, 1101, Philippines.
caesar.saloma@gmail.com

independent scientist. Mentoring is the lifeblood that sustains scientific tradition and the bequest of knowledge and experience that is achieved by mentoring, giving science the power to formulate an increasingly accurate explanation of the physical world.

The qualities of a mentor are honed through years of scientific productivity and self-rectification. Filipino mentors, in particular, need to develop a sound understanding of their working environment to enable them to rise above the unfavorable socioeconomic conditions of doing R&D in a society where scientific culture is still in its infancy. Even productive researchers or dedicated teachers have to toil to become successful mentors who are able to strike a delicate balance between personal engagement and detachment in the course of supervising a PhD student. Mentoring requires a considerable amount of selflessness, generosity, and endurance to contend with the ups and downs of scientific research and the exuberance of youth.

The effect of the brain drain is lethal particularly among young BS graduates who depart because they could not find a nurturing research laboratory for their PhD training in the Philippines. In my observation, their failure is not due to the lack of trying or love of country. There are simply too few research laboratories around and their number will increase only if additional mentors are found. Mentors build laboratories and guide students, ultimately inspiring them to achieve the remarkable joy of making a scientific discovery. They play a key role in slowing, if not reversing, the needless emigration of Filipino scientific talent.

Becoming an esteemed mentor is a worthy challenge that all researchers and teachers should take in the course of their professional life.

November 9, 2006

Eduardo Padlan

On Scientific Publications

I collect cartoons. There is one by Nick Downes that I clipped a few years ago that shows a professor being told (as a thug is getting ready to shoot him): "Surely you were aware when you accepted the position, Professor, that it was publish or perish." You see, "publish or perish" is used in academic circles to mean that if you do not publish (papers) you do not get tenure or you do not get promoted; in other words, you do not get recognition. These days, the number and quality of one's publications are taken as a measure of one's productivity, of one's worth. It even matters in which journals one publishes. There is an insistence, for example, that publications be "peer-reviewed," i.e., judged worthy of publication by others in the field; non-"peer-reviewed" publications do not count as much. In the Philippines, it is a big plus to publish in international journals. In fact, some institutions even give monetary rewards for publishing in international journals. (In marked contrast, the National Academy of Science and Technology (NAST) Philippines, bless its collective soul, yearly gives awards to outstanding scientific papers published in local journals.)

Not all scientific results are of international interest. For example, the association of gastrointestinal disorders with the amount of rainfall (probably a fictitious association) in the town of Puling (a fictitious town) in the province of Bugat (a fictitious province) in the Philippines, would be of little interest to the rest of the world and will most probably not be accepted for publication in an international journal. However, that finding would be of great interest to the inhabitants of Puling and other towns in the Philippines and will almost surely be a feature article in the Puling Bugat Journal of Science (a fictitious journal). How do we decide what is a worthy publication and what is not?

(One publication is often cited as one that has made a tremendous impact on the lives of billions and that has been reprinted and quoted innumerable times, but which was never "peer-reviewed" nor published in an international journal—the Bible.)

Scientists publish their experiments and experiences for a purpose, although that purpose has changed over the years. Originally, scientists published their data, the analysis of those data, and their theories based on their results, so as to share with their colleagues (the other scientists in the world) the knowledge that they had gained from their experiments. Like other endeavors, science builds on previous work—the more work had been done and the more findings are made available to the public, the easier it

Formerly with the US National Institutes of Health and currently with The Marine Science Institute, University of the Philippines, Diliman, Quezon City, 1101, Philippines.
edpadlan@yahoo.com

would be for others in the field to know which new path to take, what else needs to be done, what mistakes to avoid, etc. Further, if we knew what others are doing, we would not waste our time and resources by duplicating their work. Collaborations may even get started by the sharing of ideas and experiences. The sharing of knowledge is not only through publications. Results and hypotheses are often shared during personal visits to other labs, at meetings, through letters, even in phone conversations. It is all in the spirit of sharing—of cooperation.

That was the norm when I started doing science in the mid-1960s, although I soon learned that not everyone played fair—there was some competition and results were kept secret. But by and large, sharing was the prevailing attitude.

It is very different now. With the start of biotech companies in the mid-1980s, secrecy became more and more prevalent. You cannot blame the companies. They exist for corporate profits. How could they possibly compete if they disclose their trade secrets?

(There was a meeting held in the late 1980s in which a representative from a biotech company discussed at length their results on a molecule they were working on, but would not name the molecule! The moderator, a scientist of the old school, led the audience in a round of hissing. That was how the scientist community felt about secrecy at the time. That has changed. Nowadays, many (most?) scientists keep their findings secret until their commercial value has been assessed and secured (through patents). Scientists have gone corporate! As Howard Schachman wrote in a recent article, scientists nowadays have a new creed: "patent and prosper."

More and more, the spirit of sharing, of cooperation, is being replaced by competition. Even "peer review" of publications sometimes suffers. Remember that your "peers," those who would be passing judgment on your papers, are the other scientists in your field, i.e., your competitors! So, one hears stories of papers whose publication had been "hindered." And, of course, your competitors get to find out from your manuscripts what you are up to and what your latest findings are. One can only hope that those instances are rare.

Personally, I prefer a system in which all papers are published. Science is self-correcting. If you keep publishing results that are wrong or experiments that were sloppily done, people will stop paying attention to your work. And there are few things worse that could happen to a scientist, or to anyone for that matter, than to be ignored by one's peers. All self-respecting scientists will publish only "good" papers.

In my opinion, every scientific experience should be made known to everyone, including—and especially—experiments that went wrong or that were extremely difficult to perform (but not sloppy work). We learn from our mistakes; we can learn from the mistakes of others, too.

In a dissenting opinion, Dr. Giselle Concepcion, who read an early version of this article, points out that there still has to be some form of review. "While expert scientists can discern what is good or bad science, that may not be true of students. Bad science could be damaging to our youth," she correctly states. "Self-correcting science, learning from one's or others' mistakes, like natural selection and evolution, will be wasteful and will take too much time," she adds. I guess I have to agree. But I still think that every scientific experiment, data, interpretation, theory, hypothesis, etc., provided it is not fraudulent or incompetently done, should be made available to others.

Every piece of knowledge is important. There is a university in the US whose library plans to get a copy of every written piece of work in history. There is a church denomination that is currently compiling the genealogy of every man and woman on Earth. There is a continuing effort to catalog every plant and animal species in the world. On the occasion of his 250th birthday, musicians in his native Austria played every note that Mozart had ever written. The key word in all of these efforts is "every." And these are almost surely not the only endeavors in which every bit of data on a given subject is being collected. I am certain that soon every scientific paper, even the ones published in the Puling Bugat Journal of Science, will be accessible through the Internet.

When that happens, the original purpose of publishing scientific findings will be achieved. When that happens, there will be no distinction between an international publication and a local one. But until then, here in our country, a measure of a scientist's worth, his productivity, is the number of his (or her) international publications.

To be productive in science, one has to be creative. But being creative is often not enough. One may be very creative, but he (or she) would have a hard time producing publishable results if he (or she) is saddled with loads of teaching and administrative responsibilities—a situation in which most scientists in the Philippines find themselves. Further, it would be difficult to produce if one does not have the funds, the equipment and other necessities to do the work.

Clearly, most scientists in the Philippines are at a great disadvantage. We are not lacking in creativity, but we usually do not have the resources to accomplish what we are capable of doing. We may have ideas that are of international importance and which can compare with those of the best scientists in the world (just look at how well many Filipinos abroad are faring), but we here (with some notable exceptions) simply cannot work on those ideas and compete with the rest of the world because of lack of resources. But there are lots of important problems—local problems—for which we do not have to compete with the rest of the world. And we can publish our results in local journals. Those of us who work on local problems are doing the country a lot of good. Of course, we do not get much recognition from those who insist that we publish in international journals.

Maybe we can make our local journals internationally "visible" and thus be included in international indices (the sign of international recognition). That may not be too difficult to achieve. If we could convince our more productive local scientists and our compatriots abroad, who are able to produce internationally competitive results, to publish seminal papers or review papers in local journals and afterwards cite those (local) papers in their other (international) publications, then the international community will become aware of our local science and our local journals.

Let's do it!

June 7, 2007

Fabian Dayrit

A Strategy for Developing Advanced Human Resources in Science and Engineering

March and April comprise the graduation season, a time of celebration and expectation. But over the past years, it has also become a time of great anxiety as the graduates and their families ask: *Are there enough jobs out there? Will the available jobs pay a decent enough salary?*

For the country as a whole, there is an equally important question that must be answered: *Are we producing the graduates we need to meet our country's needs?*

The Philippine Education Policy

Historically, the education policy of the Philippines has focused mainly on primary and secondary education. This was a rational policy for the 1950s when a good high school education would have been sufficient to get one a decent job. Today, however, because a high school education is no longer enough, the Philippines must consider a more proactive policy toward university education, something which our more successful Asian neighbors have been doing.

If the country is to successfully compete in the global economy, it must produce graduates in higher education who are at par, if not better, than our peer countries, particularly in areas of science and technology. Unfortunately, we have not been doing very well at all.

Based on the latest data available from CHED (year 2003), the Philippines had a total of 401,787 graduates in higher education (college and graduate degrees). Of this number, only 4,872 graduates were in the natural sciences (or only about 12 science graduates for every 1,000 university graduates!) and 53,487 in engineering and technology.

The combined total of natural science, engineering, and technology graduates in 2003 accounted for less than 15 percent of all graduates. This pales in comparison with countries such as Singapore, Japan, and China where science and engineering (S&E) graduates account for over 70 percent of university graduates.

Department of Chemistry, School of Science and Engineering, Ateneo de Manila University, Loyola Heights, Quezon City, Philippines.
fdayrit@ateneo.edu

However, even more indicative of the poor state of our S&E in higher education is the number of PhDs that the entire country produces: in 2003, there were only twelve PhDs in the natural sciences and six in engineering!

Graduates by discipline and level, 2002-03

Discipline Group	Baccalaureate	Pre-and Post-Baccalaureate	Master's	Doctoral	Total
Engineering & Technology	42,187	10,989	305	6	53,487
Natural Science	4,692	2	166	12	4,872
Other Disciplines	287,428	39,527	14,743	1,730	343,428
Total	334,307	50,518	15,214	1,748	401,787

(Source: Commision on Higher Education)

The focus of basic education is to teach basic values and to enable our students to read and write, add and multiply. A bachelor's education is viewed as a requirement to get a job and earn a living. On the other hand, investment in MS and PhD is for global competitiveness. How can we expect to be globally competitive if we produce only 12 PhDs in science and six PhDs in engineering per year, while our Asian neighbors train them in the hundreds and even thousands?

Data on S&E Professionals

The statistics on the number of our S&T professionals pales in comparison even with UNESCO guidelines, much less with those of the leading economies in the region. UNESCO recommends a target of about 380 scientists and engineers per million population. The Department of Science and Technology reported that the Philippines only had 157 scientists and engineers per million population (1992 data). Singapore had 2,318, while Japan led all countries with 4,909.

Today, the Philippines is competing mainly through its comparative strengths in English and the export of labor and domestic helpers. But this cannot continue to be our long-term national aspiration. We must start to invest more in the training and productivity of our advanced human resources in science and engineering. Our pride in the Filipino's "abilidad" can only take us so far; it is not a substitute for real abilities.

Certainly it cannot be said that the country has not invested in the training of PhDs in science and engineering. Indeed, there have been successful programs in the past, but they have not gone far enough, and they have not been sustained. I wish to point out some of the characteristics of these programs because they reveal our approach to advanced manpower development that must be changed. In order to put together an effective strategy of developing advanced human resources in science and engineering, we must do three things.

Budgetary Outlay

First, we must support the development of human resources in S&E as part of our regular annual budgetary outlay in order to ensure its continuity. Most of our advanced human resources development programs have been either supported largely by foreign grants or loans, or have been special projects with a limited time frame.

At the end of World War II, our scholars were the American *pensionados*. During the 1960s and 70s, many of our PhDs were trained in the US, supported by the Ford Foundation and the Fulbright Program. European scholarships were offered by the British Council and Germany's DAAD (Deutscher Akademischer Austausch Dienst or German Academic Exchange Service).

There were modest attempts of the Philippine government to fund PhD programs overseas, but many scholars chose not to return after their studies, or returned briefly, only to leave again for greener pastures.

During the 1980s, the UP–Ateneo–La Salle PhD Consortium obtained assistance from Australia and Japan (JSPS and Monbusho). The Consortium pioneered the "sandwich" PhD program, wherein the students took most of their PhD courses and, whenever possible, started their research locally. They then went to a carefully selected foreign laboratory for periods of six months up to two years to carry out their research. They then returned to the Philippines to complete their write-up and defend their dissertation. The sandwich PhD program enjoyed very good success as evidenced by high rates of completion and retention.

In 1992, the DOST spearheaded the Engineering and Science Education Project (ESEP), a World Bank-funded program which sought to upgrade our human resources and laboratories in universities. Unfortunately, this project ended in 1997 and has not been followed up.

Meanwhile, much of the equipment, which was obtained under the ESEP, is deteriorating from inadequate maintenance or is approaching the end of useful life.

Today, the DOST and CHED continue to provide MS and PhD scholarships in S&E, but it is apparent from the data that a scholarship program alone is not enough. A much wider project must be put in place.

Because most of our advanced S&T manpower efforts have depended on foreign funding, programs have been sporadic and erratic. This also indicates that the development of advanced S&T human resources is not yet a key concern of government and so we do not have a policy for the development of such capabilities. However, in today's highly scientific and technological world, such a policy for continuous support is what is urgently needed.

Advanced Education and Productivity

Second, we should widen the scope of support and coordination to include provisions for the enhancement of productivity. Our programs to develop advanced human resources in S&T all end after they have attained their PhDs. The new PhD then returns to an ill-equipped laboratory and is given an administrative position to better advertise his degree. In no time, this PhD becomes frustrated and is reduced to obsolescence.

Advanced S&T human resources development today includes PhD training, and the development of a productive environment. We need to expand the scope of support that we provide our PhDs, especially in S&E where the provision of good laboratories and productive environment is important.

National Development Objectives

Third, we should link our program of manpower development with national development objectives. Most of our concerns now require highly sophisticated capabilities in science and engineering, including areas such as environment and ecology, agriculture and fisheries, food production, health and medicine. While we have already identified and have started focusing on these areas, the scale, depth, and coordination need to be improved.

Because such a policy to develop science and technology for national development demands a big investment and a radical change in our educational objectives, it requires political will. At its core, it is a commitment to invest in excellence. Indeed, our investment in the development of our scientists and engineers reflects our national aspirations and the way we view our national destiny.

March 30, 2006

Fabian Dayrit

Enabling Competitiveness with Science and Technology Human Resources

Many people equate the possession of high-tech gadgets with being in an advanced state of development. For example, if we can buy the most advanced cellphones or laptops or enjoy the comfort of fancy malls, then we must be technologically advanced.

Unfortunately, many people confuse the ability to use technology, with the ability to create technology. The former is mere consumption, whereas the latter requires science and technology (S&T) capabilities. Sadly, the Philippines has largely been only a technology consumer.

So, what does it take to create successful technologies? While there are many factors that need to be put in place, the single most important indicator is the presence of a productive and competitive S&T community. This single characteristic is a reflection of the state of the system.

S&T funding is obviously important, but it would be a mistake if we think that simply providing the funds will improve our situation. In fact, this type of thinking is part of the problem. We need to spend our money wisely and attend to the needs of developing a real human resource in the same way that successful companies husband their human resources.

The most competitive globalized companies compete to recruit the best people, give competitive salaries, provide the already smart graduates with more competitive skills, build a conducive environment, and keep them productive in the company. Beyond simply providing the financial resources, we must create an attractive S&T environment in the Philippines that the young will want to build their careers in. Clearly, this will only work with unwavering political will, unity of purpose, and well-thought out measures.

Engineering and Science Education Project I

From 1992 to1998, the Philippine government, through the Department of Science and Technology, implemented the Engineering and Science Education Project (ESEP) with financing from the World Bank. ESEP sought to upgrade engineering and science

Department of Chemistry, School of Science and Engineering, Ateneo de Manila University, Loyola Heights, Quezon City, Philippines.
fdayrit@ateneo.edu

through a multipronged approach which included the following components: local MS scholarships, local and international PhD scholarships, postdoctoral fellowships, equipment upgrading, civil works, library upgrading, and networking.

ESEP had a number of good features. First, ESEP went beyond just giving scholarships. ESEP employed a more holistic approach to manpower development principally by aiming also to improve the research environment; for example, by providing the necessary equipment, building laboratories, upgrading library resources, and very importantly, by encouraging networking. Second, although ESEP did not achieve the high numbers that it initially projected, it was nevertheless very successful in terms of achieving its main goals. The completion rate improved significantly from baseline values; the return rate of the graduates was very high; and scientific productivity improved due to the sandwich linkages and new facilities. But most importantly, if one surveys local universities today, one will find that many of the leaders were ESEP graduates. Third, despite its relatively high administrative cost, which is typical of a World Bank project, the WB system imposed transparency and discipline with regard to the program planning and implementation.

What were the difficulties with ESEP? First, ESEP ran for a limited period only—six to eight years—and by the time things were starting to function more smoothly—schools were starting to cooperate, PhD students and projects were getting started locally, etc.—it was time to wrap up and assess the program. It was barely enough time to go through one cycle of PhD students. Clearly, a longer period is necessary if we are to introduce a new paradigm in science and engineering education. By the time ESEP finished, we were back to our baseline level of manpower development.

Second, ESEP did not provide for maintenance of equipment, most likely because WB expected this to be taken up by the Philippine government. Unfortunately, our system does not seem to recognize the importance and the cost of maintaining high tech equipment.

A Holistic Strategy

While we recognize the urgent need to improve our S&T human resources, we must do it smartly and devise a holistic strategy:

- We should provide attractive scholarships.
- We should improve the research environment by providing the necessary equipment, laboratories, library resources, and other facilities essential for productivity. Without these, the students whom we train will simply seek more satisfying jobs elsewhere and continue to succumb to the brain drain.
- Our program should be a long-term commitment.

- We should encourage innovation, interdisciplinary collaboration, and partnerships between academe and industry and academe and government, both national and local. Local governments and members of Congress can support components of this program by upgrading their local institutions to benefit their own development priorities.

Globalization has also meant increased mobility, especially for the most talented. If we are to succeed, we should start to think of S&T human resource development as a competition for the best minds. We should realize that what will attract students is not so much the scholarships, but their career after they finish their degree. We can succeed only if we can make a scientific career in the Philippines attractive.

March 1, 2007

Maria Corazon de Ungria

Role of the Filipino Leader in the Development of the Asia-Pacific Community

The role of a leader is often shaped by the characteristics of the people being led. How do we define a Filipino? In an archipelagic country strategically situated at the entrance to Asia from the Pacific and composed of 7,107 islands with more than 110 ethnolinguistic groups, each with its own cultural history, is there even such a thing as a unique Philippine identity? Based on initial research conducted by the UP DNA Analysis Laboratory that aims to study the relationships of different indigenous groups such as the Ivatans of Batanes; Bugkalots, Kalangoyas, and migrant Ibalois of Nueva Vizcaya; Kalingas of Kalinga province; Aetas of Bataan; Atis of Panay Island and Guimaras; six groups of Mangyans in Oriental Mindoro; Maranaos in Lanao del Norte; and the urban groups such as the Tagalogs, Bisayans, Ilocanos, and Ilonggos, there are certain traits that are common to all groups that may help identify and define the role of the Filipino leader.

First is inherent multipluralism; that is, the ability of different groups to coexist peacefully in one geographic area. For instance, the seven tribes of Mangyans of Oriental Mindoro—the Buhids, Tawbuhids, Hanunoos, Irayas, Tadyawans, Batangans, and Alangans, possess their own language, history, beliefs and cultural traditions, and yet live together without dispute. On a larger scale, the general populations of Tagalogs have coexisted with Bisayans, Ilocanos, Ilonggos, and Dabawenos without conflict.

Multipluralism at work is best exemplified by the EDSA revolution, where Filipinos from all walks of life and different religious, political, and ideological beliefs worked together as one to oust a dictatorial regime without bloodshed.

Second, is the capacity to adapt to new environments and the "can do" attitude, even under challenging conditions. Integral to adaptability is creativity and ingenuity. Take, for example, the Ivatans of Batanes, which lies at the junction of the South China Sea and the Pacific Ocean and therefore experiences the worst typhoons. Yet, the Ivatans have managed to live in Batanes by building their houses out of limestone, planting mainly root crops with *camote* as their staple food rather than rice, and developing numerous means to preserve food. How about the Ifugaos who had built the rice terraces which is not only a work of art and considered the eighth wonder of the world, but is also a highly effective system of rice irrigation given the available mountainous terrain?

DNA Analysis Laboratory of the Natural Sciences Research Institute, University of the Philippines, Diliman, Quezon City, Philippines.
mcadu@uplink.com.ph

On a larger scale, the adaptability of Filipinos is exemplified by the numerous overseas Filipinos who are living in different parts of the world. Unlike other cultures which tend to keep to themselves and form their own communities, Filipinos are known to integrate themselves into the "host country." On the local level, the Filipinos' ability to cope with adversity is remarkable, e.g. vendors selling umbrellas when it starts to rain; children selling slippers when it starts to flood; people living in flooded areas taking a *banca* to work; students lining up in photocopying shops to obtain cheap copies of their textbooks; mothers cooking instant *mami* noodles instead of rice to make ends meet; and women setting up small variety stores in almost every corner of the neighborhood.

Third is the capacity to work together as a team as shown by the inherent *bayanihan* spirit. For example, this author has witnessed the community of Ivatans help out in preparing the burial of a member, the Kalangoyas assist in the wedding preparations of a couple, and the Atis of Aklan help a family build their home. There is a prevailing spirit of camaraderie and teamwork in the communities which is crucial for nation-building and regional development.

Given these characteristics, it is evident that the Filipino leader can contribute significantly to the development of the Asia-Pacific region through his multipluralistic perspective, respect for democracy, adaptability to current and future challenges using his skills and creativity, and his capacity to work with a team of diverse players striving for the common good.

January 25, 2007

Flor Lacanilao

Training Graduate Students

We tried to catch up with our neighboring countries by rightly building up our S&T capabilities. Programs included increasing R&D funding and technical manpower. After a decade, however, established indicators showed no progress in S&T performance. Obviously, the S&T programs missed a basic objective—to improve research.

For instance, in 1991-1996, the DOST budget increased four-fold or from ₱854 million to ₱3.4 billion. It doubled to ₱1.7 billion in 1992. Furthermore, holders of graduate degrees in science and engineering also significantly increased. The UP College of Science alone granted 509 advanced degrees from 1985 to 1994, with 133 PhDs. Our S&T performance, however, did not improve. Total yearly output of publications in international journals in 1981-1995 by national R&D institutions remained the same.

Scientists are again calling on the government to allot more money to improve S&T. It's about time. I hope now programs would give more weight on the proper way of doing research. One needs to know how to do research properly before he can begin to think about commercializing discoveries.

If we hope to achieve our objectives this time around, we can no longer ignore the established norm of adequate peer review and verification of research results. Peer review and verification have more than three centuries of developing tradition. They have served science well. We can apply them in training graduate students. This can focus on thesis work, which is meant to be training in research. Since research is not completed until results are properly published and the best indicator of a good graduate training is publication of the thesis, we should require publication in the graduate program.

Below are some pointers to improve the training of graduate students in science. They are also useful for graduate faculty and those involved in R&D.

Choosing a Research Problem

Funding agencies usually dictate the areas of research to support and often leave grant applicants no choice but to work on problems outside of their main interest. As a result, the quality of the output suffers. The interest level of the researchers would have been higher had they been allowed to research on a problem of their choice. Also, grant

Formerly with The Marine Science Institute, University of Philippines, Diliman, Quezon City, 1101, Philippines; and the Southeast Asian Fisheries Development Center.
flor_lacanilao@yahoo.com

recipients lose the chance to sustain their creativity and develop competitive ability. In our present state of S&T, we are yet aiming for a critical mass of competitive scientists.

Another problem is the type of research, whether to go basic or applied. Poor countries are more on applied research, whereas basic studies are common in developed countries. In applied research, time for benefit is short, predictability is high, payoff is small, and scope of applicability is narrow. But in basic research, time for benefit is long, predictability is low, payoff is big, and scope of applicability is wide. These differences perhaps explain why rich nations get richer, whereas poor countries find it hard to develop.

Many of our researchers often ignore basic knowledge when doing applied research. Although relevant studies are available in the journal literature, they are seldom cited in most grant proposals and published papers. Hence, results from applied research often fail in field applications. Developed nations have shown that the greater the need for applied research, the more important basic studies become.

The big payoff and wide scope of applicability of basic research deserve more comment. The great scientific discoveries that changed our ways of life were not intended. This serendipitous nature of basic research has given us wonder drugs, lasers, computers, biotechnology, and many others. Every scientist who contributed to the development of biotechnology, for example, did not have biotechnology in mind. This stresses the point of leaving the choice of problems to scientists. As Nobel laureate Joshua Lederberg once said in connection with the development of biotechnology, "It would have been tragic were any industry to have had a veto in deciding what would truly be of greatest industrial consequence."

Literature Search

A major cause of poor research performance is the failure to search the journal literature. A common complaint is the lack of journals in the library. No library in a poor country can afford to subscribe to the journal needs of researchers. But references can be accessed through indexes that cover important journals. Recently, Google Scholar has added a speedy way of accessing literature. Whereas literature review should be exhaustive, literature citation should be selective.

Libraries should therefore subscribe to indexes and limit other subscriptions to important review journals. Indexes vary in number and quality of covered journals. The most widely used indexes in the natural and social sciences are issued by the Institute for Scientific Information (*www.isinet.com*).

Current Contents (CC)

Choosing journals and articles for reference will improve the quality of the data gathering and the publication arising from it. An indicator of article quality is the reference list. The integrity and quality of an article will depend on the quality of the bibliography added to it. If the reference list of an article is more than 75 percent gray literature and if you are a serious researcher, then don't waste time reading the article.

Data Gathering

Data gathering is the first main part of research; the second is proper publication. It is important to think of publication as an objective when gathering data. This improves data gathering as one thinks of peer review, references and methods, adequacy of data, controls, replicates, etc.

The best time to develop proper work habits is during the data gathering part of training. Much of the quality and usefulness of results depends on work habits. There is no substitute in observing the professor in his laboratory, or in the field, to learn the scientific habit. Unfortunately for our country, we rarely find a scientist whose only absence in his laboratory is holding class.

One work habit that should be corrected early in graduate training is sloppiness. Errors are caused by careless handling of data. If they leak through the peer review, errors are propagated and can result in serious damage to the scientific literature. The other cause of errors is fraud. Sloppiness gives rise to unintentional errors; fraud, to intentional errors. Their damaging effects on science are the same. But we are often more concerned about fraud than sloppiness, when sloppiness is much more prevalent than fraud. Sloppiness is easier to correct during graduate training, but the tendency to commit fraud can be deeply rooted.

Manuscript and Publication

Normally, the first experience of publication comes from the thesis. The master's thesis or a chapter of the doctoral dissertation should therefore be a scientific manuscript, written following a "guide for authors" of a CC journal. A master's program may require only such publishable manuscript. But a PhD degree should require at least a published chapter of the thesis. There is no sense in writing the thesis differently then rewriting it for publication.

Three guides will be needed in preparing a manuscript: (a) a good book on scientific writing and publishing, (b) a guide for authors of a chosen journal, and (c) sample articles from the journal. A good book on scientific writing and publishing will show how to prepare the manuscript—stating the title, writing the abstract, introduction, and materials and methods, designing tables, preparing illustrations, etc. A guide for authors and sample articles will show the details that should be followed strictly.

Presenting a Paper

The purpose of oral presentation is to inform the audience and to improve the manuscript before publication. Scientists in the audience, through their questions and comments, can improve the paper. Some are helpful to young researchers, particularly to those serious in their work, which is seen in the manner of presentation. It is then important that one learns the basic rules of presenting a paper orally. These are described in a good book on writing and publishing scientific papers.

Presentation should cover only the main points. Tables and figures should show results clearly and briefly. Prepare summary tables and figures selected from those in the manuscript. For example, reducing the number of columns and rows of tables will allow bigger prints. Avoid presenting tables and figures as they appear in the manuscript. Unlike in the written paper, projected images are seen only briefly. Well-prepared and properly sequenced slides will allow smooth presentation, with the presenter talking less and letting the slides convey the message. Some presenters read everything printed on the slide, forgetting that everyone in the audience can also read.

Mechanisms of quality control

A manuscript usually undergoes a preliminary or informal peer review. The reading of a manuscript by a professor or colleague is an example. Another is presenting a paper in a seminar or scientific meeting as described above.

Then the manuscript goes through the standard device for quality control—the peer review. For adequate peer review of a manuscript, good journals have normally three referees active in research on the same specific subject. We do not have enough such reviewers in the country to adequately review manuscripts; this is one reason why researchers should publish in international journals.

Once the paper is published, it goes through the second and final stage of quality control—peer verification. This occurs if the paper has appeared in a journal covered in widely used indexes. International peers can then conduct verification studies by repeating experiments to disprove or confirm published results. This explains why improperly published results have largely failed in our technology generation.

Research proposal

This topic is discussed last to show that knowledge of research is needed in making a research proposal. If one knows the research process as outlined above, then preparing a proposal should be easy. Problems may still be met, however, until we have enough scientists to review proposals.

The three elements of a research proposal are (a) the funding agency and peer review, (b) scientific merit, and (c) proponent's capability. I will only reason for the proponent's capability. This is most crucial for advancing S&T, and it should be the main concern of funding agencies to avoid wasting funds. Evaluation of research proposals is simplified if publication experience is made the criterion in screening applicants. For example, only those with publications in CC-indexed journals should qualify for research grants as practised in developed countries.

With this requirement and the availability of funds, we can expect every proponent to produce a scientific paper, which is what we expect from research, and what we need to advance science, so that we can have more useful technologies, and finally move on to real development. Whereas science alone cannot save the Philippines, the Philippines without science cannot be saved.

March 16, 2006

Part 3

Research and Development and Institution Building

Edgardo Gomez

From Third World Backwaters to First World Institution

Few readers may realize that some pioneering research on Philippine coral reefs led the way from the backwaters of the Third World to the status of a First World institution. This article will briefly trace this development while serving to introduce a series of papers on marine science and technology in the Philippines written by scientists working at the UP Marine Science Institute. I will use as the main theme one of our great marine natural resources, the highest diversity coral reefs of the world.

In today's environmental jargon, one often hears of "hotspots" of biological diversity or "biodiversity" for short. For the layman, a hotspot is a geographical entity that has high or unique assemblages of species (like orchids, birds or fish), but which organisms are under serious threat for their very existence! The Philippines is now considered a hotspot for several groups, including trees, birds, and corals. This article will be about the last group. I have chosen to focus on this, for several reasons, among which is the fact that the Philippines lies at the apex of a region of the world known as "The Coral Triangle," because of its roughly three-sided geographical spread, with the Philippines on the north, the Sunda islands of Indonesia to the southwest, and the Papua New Guinea/northern Great Barrier Reef to the southeast. This geographical entity was not yet recognized as such three decades ago when the story I am about to tell had its uncertain beginning.

A new research center was approved for establishment by the University of the Philippines in 1974 with virtually no funds. Almost at its inception, the author conceptualized a research proposal to investigate the coral resources of the country; this project gradually highlighted our signal advantage of having the largest number of known species of corals in the world. This fact attracted a few foreign scientists to look into what we were doing and to slowly establish communications with our handful of young researchers. The contacts were easily established because the local scientists could communicate easily in English. More importantly, the small team that was newly formed was driven to achieve results and eager to seek help wherever they could. This openness and eagerness sent positive signals abroad that, indeed, there was some potential for doing collaborative science in a Third World country.

Let me step back a little and make mention of the fact that a then retired Filipino scientist, who had worked single-handedly, was already recognized for his pioneering

The Marine Science Institute, University of the Philippines, Diliman, Quezon City, 1101, Philippines.
edgomez@upmsi.ph

studies on the taxonomy of corals. This was the late Prof. Francisco Nemenzo, the father of the incumbent president of the University of the Philippines. While he had made a name for himself, his influence was limited as he had not been able to establish a cadre of competent Filipinos to carry on and expand his work. (But I will come back to Prof. Nemenzo later in this article.)

The "new" angle to coral reef studies involved a team approach. Realizing that the national scope of the new investigation would be daunting for a single group to tackle, I enlisted the collaboration of former Secretary Angel Alcala, then the head of the Silliman University Marine Laboratory in Dumaguete City, to form the second team, while I set up a third team in Cebu City, involving UP Cebu and a graduate of the University of San Carlos. To telescope the results of this project, we arrived at a status report on the condition of Philippine coral reefs. The figures arrived at 25 years ago are now household knowledge, 5 percent excellent, 25 percent good, and the rest either in fair or poor condition. Besides sounding the alarm bell on the serious condition of our natural resources, this survey was a landmark effort. It was the first national inventory of the coral reefs of any country in the world. Today, there is a global effort generally under the name of Global Coral Reef Monitoring Network or GCRMN that regularly assesses or evaluates the condition of coral reefs in virtually all the countries that have these resources. It may also be worth noting that we were represented in the Scientific and Technical Advisory Committee of the GCRMN from the start, taking on the chairmanship subsequently, and are now represented in the Management Committee.

Partly on account of this project, the Philippines was invited to host the 4th International Coral Reef Symposium (ICRS) in 1981, when there was just a nucleus of coral reef scientists in the country. The ICRS is the premier international scientific meeting on coral reefs. The small nucleus based at the former Marine Sciences Center (now the Marine Science Institute or the UPMSI), in spite of its inexperience, rose to the challenge in two ways: by presenting a credible number of scientific papers and by organizing an international symposium for the first time in such a professional manner that many still remember that conference as one of the best organized, even after more than two decades. One highlight of the 1981 symposium was the conferment by the UP of an honorary doctoral degree, honoris causa, on the late Prof. Francisco Nemenzo in recognition of his outstanding work on the taxonomy of corals.

Recently, Andy Hooten, a consultant to the World Bank Environment Division, did a tally by country of the primary authors of the papers published in the proceedings of the last two ICRS meetings, the 8th in Panama in 1996 and the 9th in Bali, Indonesia in 2000. The Philippines was second to Brazil among all the developing countries in the 8th ICRS, being ninth among all countries, with fourteen presentations, the same number as Canada. However, in the 9th ICRS, it rose to third overall, tied with the United Kingdom and the first among the developing countries. On the latter occasion, it followed the United States and Australia and tied with the UK with a total of twenty-six papers. The 10th ICRS was recently held in Okinawa, Japan with some three dozen Filipinos in attendance, the majority being staff, students, and graduates of the UPMSI. It was heartening to hear unsolicited comments from foreign scientists, expressing their gratification in seeing so many young Filipino scientists deliver first class papers on their research work.

Hence, it is not surprising that in the first truly global research program on coral reefs, the newly approved Global Environment Facility or GEF Targeted Research on Coral Reefs and Capacity Building for Management which has an initial budget of $22 million, Filipinos are represented in five of the six Working Groups and a Philippine resident American is a member of the sixth. We have the largest developing country representation in this worldwide program of high-level coral reef research which addresses issues on coral bleaching, connectivity, diseases, modeling, reef restoration, and remote sensing. Further, the UPMSI, with its Bolinao Marine Laboratory, has been designated one of four Centers of Excellence (CoEs) where the bulk of the research work will be undertaken. It may be noteworthy to mention that these centers are distributed in four seas of the world, with the Philippines representing the North Pacific. This choice will have widespread benefits as the major aims of the CoEs are capacity-building and the linkage to management in addition to the facilitation of the research investigations.

The details stated above bear witness to the fact that a small, active group of Filipino scientists working in a state-supported institution is held in high esteem and treated with genuine respect and admiration by their international peers. In the field of coral reef research, many supporters from the developed world are willing to go the extra mile to be supportive of our efforts. This is further evidence that it is possible to attain world-class status through professional endeavors and hard work, even if local funding is very limited. The only natural asset in abundance were the coral reefs which, unfortunately, are now much abused principally by both desperate and greedy fishers. It is hoped that the insults on these productive ecosystems be mitigated, lest the research of a First World institution in a developing country be for naught.

August 5, 2004

Caesar Saloma

Physics in the Philippines Today

The Philippines is a country with a population of 82.3 million in 2003 that is growing at an annual growth rate of about 2.34 percent. Its unemployment rate was 11 percent in January 2004. Of the total population, 12.98 percent is concentrated in the National Capital Region. The literacy rate is 92.28 percent among household population 10 years and above, and the GDP per person in 2001 was $1,050. In terms of the human development index, the country occupied the 77th position in 2000 (rank in GDP per person: 97).

Registered earnings from merchandise exports for the year ending February 2003 reached $2.72 billion. Receipts from electronic components accounted for 53.7 percent of the aggregate export revenue, representing an increase of 4.5 percent to $1.464 billion from $1.4 billion in the previous year. In 2003, the industry sector (manufacturing and construction) contributed 31.5 percent to the Gross National Product.

College life begins at age sixteen or seventeen after four years of high school and six to seven years of elementary education. Schooling at any level may be obtained from a state or private institution; there are at least sixty-five universities in the Philippines today. The medium of university instruction in the natural sciences and engineering is English and the academic year starts in June of each year and ends in May of the following year.

An undergraduate degree in physics or applied physics may be obtained from one of eighteen schools, ten of which are state institutions. Four physics departments are classified by the Commission on Higher Education as centers of excellence (CoEs): Ateneo de Manila University, De La Salle University, Iligan Institute of Technology in the Mindanao State University, and National Institute of Physics in UP Diliman. Another four are classified as centers of development (CoDs): Silliman University, University of San Carlos, Institute of Mathematical Sciences and Physics in UP Los Baños, and UP Baguio). Regular undergraduate programs in physics or applied physics take four years to accomplish, except at the National Institute of Physics (NIP) in UP Diliman where it is a five-year course.

In the second semester of the academic year 2002-2003, the undergraduate physics/ applied physics population in the CoEs and CoDs was 876 which is distributed as follows: Ateneo (120), De La Salle (100), Iligan Institute (72), NIP (350), Silliman (19), San Carlos (69), UP Los Baños (120) and UP Baguio (26).

National Institute of Physics, University of the Philippines, Diliman, Quezon City, 1101, Philippines.
caesar.saloma@gmail.com

The gender ratio in the undergraduate population is close to unity. A good source of promising physics students is the Philippine Science High School System which has seven campuses all over the country. The system is operated by the Department of Science and Technology which also provides college scholarships for physics and applied physics students. The University of the Philippines administers a number of academic scholarships to physics students. Recently, Intel Philippines has also provided a number of scholarships to NIP graduate students.

There are less than eighty PhD degree holders in physics in the Philippines, more than half of whom are with the academe—twenty-six are with the NIP. The physics departments of the privately owned Ateneo and De La Salle have about ten PhD faculty members each. Each center of excellence offers an MS degree program in physics, with Ateneo, De La Salle, and the NIP also offering PhD degree programs. The average graduate population is 127 [Ateneo (twenty), De La Salle (twenty), Iligan Institute (twenty-five) and the NIP (sixty-two)]. About 90 percent of NIP graduate students study full-time. A publication in an ISI-abstracted science journal is a minimum requirement for graduation in the NIP's PhD program.

The number of publications by Philippine-based physicists in journals abstracted by the Institute of Scientific Information (ISI) had grown steadily from sixteen in 2000 to thirty-three in 2002. In 2003, the number of ISI publications dropped to fifteen. Between 1995 and 2003, about 83 percent of the ISI publications were authored by researchers who were or had been affiliated with the NIP. Of the ISI publications of the NIP, 53 percent were in optics, signal processing, and complex systems. It is worth mentioning that researchers from the privately owned Research Center for Theoretical Physics in Jagna, Bohol, have also produced a number of ISI publications in mathematical physics over the years. The number of ISI publications generated by non-NIP researchers has increased notably since 2001.

Most physicists who have remained active in research are members of the Samahang Pisika ng Pilipinas (SPP) which was founded in 1979. The SPP, which had 250 members in 2002, holds a scientific conference each year where researchers from all over the country meet to present their latest findings. All papers for presentation are peer-reviewed before acceptance in the conference. All approved presentations are published as full papers in the conference proceedings.

The Society has experienced continuous growth in the number of conference presentations from 101 in 2000, 119 in 2001, to 133 in 2002. The research areas covered in past conferences were in condensed matter physics, superconductivity, lasers and its applications, optics and photonics, image and signal processing, plasma physics, liquid crystals, statistical physics, field theory, complex systems, and physics education. The number of presentations by NIP researchers has remained at 60 percent of the total in each of the conferences in the last three years. This implies an increasing number of technical contributions from other Philippine schools.

The basic salary range of a professor at the University of the Philippines (UP) is $5,438 to $6,752 (US$ = ₱53.52) per annum before tax which is at 30 percent of gross income. A regular faculty member has a workload of twelve units—one unit is equivalent to one lecture-hour a week. A PhD faculty member at the NIP can also apply for a three-unit research load credit. The basic salary of a professor at the privately owned Ateneo de Manila is higher with a regular teaching load of 30 units per academic year. The student tuition in Ateneo is almost five times higher than UP's.

UP also awards research grants to qualified faculty members in the amount equivalent to their annual salary. Since 1998, a cash award of $1,000 is given to a publication in an ISI-abstracted journal, which is divided equally among the authors—only those who use the affiliation of UP in their papers qualify for the awards. Similar incentive schemes for ISI publications are also implemented at the Ateneo de Manila and De La Salle. Recent PhD graduates of the NIP can also avail themselves of postdoctoral fellowships from the UP System that allow them to do research at foreign laboratories for a period of one year.

For researchers in the physics community, the most important source of research grants for personnel and equipment is the Philippine Council for Advanced Science and Technology Research and Development (PCASTRD) of the Department of Science and Technology. The PCASTRD provides research funding in the areas of biotechnology, electronics, instrumentation and controls, information technology, materials science, photonics, and space technology application.

Within a fourteen-year period (1989-2002), the PCASTRD has released a total amount of ₱164,909,776.42 in research grants, of which ₱67,340,254.11 (40.83 percent) and ₱15,524,343.04 (9.41 percent) went to the materials science and photonics sectors, respectively. The biotechnology sector received 27.45 percent of the total financial disbursements for research and development. In 2002, Intel Philippines gave an equipment grant to NIP in the amount of 155,000 euros.

Physics research in the Philippines has never been in a better state than it is today. ISI Essential Science Indicators revealed that after the 10.17-year period, from January 1992 to February 2002, the Philippines had been able to enter into the rank of countries with the highest citations in physics.

Filipino physicists are acquiring the competence and confidence to recognize more interesting research projects and publish in high-impact scientific journals. This is critical if the number of available PhD supervisors is to increase in the country. Noticeable improvements in research infrastructure and facilities and in the number of publications in ISI journals have turned the NIP graduate school into a viable option where young talents can pursue advanced careers in physics. As a consequence, a slowdown is happening in the migration of valuable BS Physics graduates leaving for abroad.

The quality of college freshmen who are entering NIP has been improving steadily. The number of BS graduates is also increasing—more than forty were produced in the academic year ending May 2004. A steady stream of BS graduates is crucial for the continued growth of the NIP graduate school. The physics community is also relatively young—about two-thirds of the current SPP members are below forty years old.

The semiconductor and electronics industry in the Philippines recognizes the indispensable element of R&D for survival in a fiercely competitive environment. The high cost of conducting R&D, in terms of equipment and manpower requirement, leads to the conclusion that it is best carried out in the universities and there is now a conscious effort in both industry and academe to interface with each other. The primal need to thrive and prosper is forcing the industry to financially support R&D efforts in the academe. On the other hand, the private sector's obsession with the bottom line is certain to reduce, if not eliminate, boondoggling in the academe.

The Philippine physics community must capitalize on its current gains to bring its scientific capability to greater heights and enable a majority of Filipinos to enjoy the benefits of science and technology. Success in R&D depends not only on talent, technical capability, and determination on the part of researchers, but also on infrastructure and financial support mechanisms in the environment.

There is a pressing need to improve state auditing and customs regulations governing R&D expenditures. Regulations should be designed to promote urgency and efficiency in R&D activities (rather than to curb corruption in government). Conscious effort is also needed to heighten public understanding of the true nature of the scientific inquiry. Among those who are literate, many still harbor the misconceived notion that scientific knowledge is static and that all information is already available (for free) on the Internet. To them, investment in science is not essential.

An overhaul of the system of incentives in the academe is overdue to make successful scientists abroad to even consider working in the Philippines—the basic salary of a UP professor is lower than the US federal minimum wage of US$10,712 per annum. To make ends meet, a typical UP faculty member has to take extra workload that inevitably jeopardizes his performance as a teacher and researcher.

At least in the near future, the state will remain to be the main sponsor of scientific R&D. There is a clear need for the state to increase its level of R&D funding if only to match deleterious effects of inflation (six percent in July 2004). Between January 1990 (US$ = ₱24.46) and June 2004 (US$ = ₱55.98), the Philippine peso had depreciated by about 129 percent against the US dollar. The twelve-year period (January 1990-December 2003), the R&D budget (in US$ terms) of PCASTRD had increased by only 7.31 percent—an average rate of less than one percent a year. To put things in perspective, the 2003 budgets of the National Science Foundation and the National Institutes of Health in the United States increased by 5 percent and 17 percent (year-to-year), respectively. In 2001, the Philippines spent 0.15 percent of its Gross National Product on R&D, while the US, Japan, South Korea, Taiwan, and Malaysia devoted 2.58 percent, 3.09 percent,

2.96 percent, 2.16 percent, and 0.49 percent of their respective GNPs. In 1992, the Philippines spent 0.22 percent of its GNP on R&D.

In the Philippines, the odds of encountering a Filipino with a PhD in physics are one in a million (literally). However, there are signs that this rarity is going to change for the better.

Printed in two parts on October 21 and October 28, 2004

Raul Kamantigue Suarez

International Science: Function, Dysfunction, and Flowers in a Grassy Field

I READ a series of valuable articles by Dr. Flor Lacanilao published in the column Star Science which, I am told, sparked negative and defensive reactions because they discussed how science should be done and how the Filipino scientific community has largely failed in its mission to serve the people. As a Filipino scientist who received his doctoral education in Canada and now works as a professor in the US, I have had the opportunity to get a sense of how science is done internationally through many years of research expeditions, conferences and direct contacts with international scientists. The purpose of this article is not to provide a scholarly analysis of how well various countries are doing in science and how they managed to do it. Neither is it meant to be a frontal attack upon the way science is administered or conducted in the Philippines. I am sensitive to the sometimes valid question of whether I have any right to be critical, given my abandonment of *mi patria adorada*. I would like readers to regard this article as more of a series of anecdotes that will hopefully provoke introspection and discussion among Filipino scientists, science administrators, and the tax-paying public. It is the latter, after all, who pay scientists to do science and who are owed their money's worth.

There is no perfect system in any country, so this is why, as I suggest in the title, what one sees internationally is a broad range or spectrum—as in the colors of the rainbow—a continuous gradation from "function," albeit imperfect, to complete "dysfunction." By functionality, I mean that the scientific community plays a positive role by generating new knowledge, by subjecting findings to peer review (submission of scientific manuscripts to refereed journals), through success in having these submissions accepted and published (then read and cited by others in the international scientific community), by accountability for the way tax money is spent, and by serving as the engine of intellectual advancement and economic development.

US Science—"Lean and Mean"

The US is considered to be the world leader in science, so I shall start with some facts and anecdotes about US science. Perhaps a common notion among Filipinos might

Department of Ecology, Evolution and Marine Biology, University of California, Santa Barbara, California, USA.
suarez@lifesci.ucsb.edu

be that there is so much money for and so many jobs in science, as well as so many great scientists to interact with and learn from that the US must be "science-heaven on earth." Well, yes, there is a lot of money, but this has to be thought of in relation to the number of scientists, the number of research universities and institutions, and the needs of society. Taking all these into account, there is little doubt that US funding for science is insufficient and that this is at least partly because too many tax dollars go into such things as foreign invasions and supporting what Eisenhower referred to as the "military industrial complex." But I should digress no further.

The consequence of the above is that the lives of university scientists are now very different compared to what they were like in the 1960s, and very different from what most readers of this article may envision. Let me illustrate. When scientists apply for research funding from the US National Science Foundation (NSF), a federal agency, each application is reviewed by six to eight anonymous referees. Their written reviews are evaluated by a panel of fifteen or so scientists under the supervision of a program director. Each panel member is assigned to present several proposals to the group. Then, everyone is invited to comment on the strengths and weaknesses of each proposal. The panel ranks the proposals and the program director makes the final decision concerning which ones to fund and how much money to give. The typical funding rate is between 10 and 15 percent—meaning 10 to 15 applications get funded out of every 100 submitted nationwide each time the panel meets (panels meet twice a year). So the very low success rate reflects how little money there is. Having been a panel member myself, I can say that a funding rate of 30 percent would be more fair and would still fund the most highly deserving scientists. Nevertheless, the system is such that a proposal of poor quality, submitted by a mediocre scientist, could not possibly be funded. Panels award grants to the "top 10-15 percent" based on the quality of the proposals and the track records of the applicants.

Professors in major universities typically spend more time writing grant applications, doing research, and writing research papers than teaching. This is as it should be—universities do not just teach; they exist for the advancement of knowledge. But the enterprise costs money and virtually all US universities survive financially by charging research grants "indirect costs." What this means is that at the University of California at Santa Barbara, about 45 percent of my NSF grant money goes to the university. In addition, universities pay faculty nine-month salaries, so many of them add two months of summer salary into their grant budgets. This system greatly increases the amounts individual scientists must ask for and contributes to the low funding rates. All these characteristics lead to my description of the system as "lean"—referring to its lack of fat.

Now, I'll get to how it is "mean." The young US scientist with a recent PhD degree and a few years of postdoctoral research training faces a formidable challenge. Landing a faculty position involves successfully competing with fifty (or maybe even a hundred or more) highly qualified applicants. To become tenured (which means to be made a permanent faculty member) requires research and the publication of peer-reviewed research papers *of sufficient quantity and quality*. Obviously, this requires that sufficient funds be raised to support the research. Therefore, this young person has to be, at the very least, one of the top 10 to 15 percent of US scientists from the perspective of funding agencies. I say "at the very least" because tenure decisions are based on peer review and not on grant dollars alone. The quality and quantity of research publications, teaching and service are scrutinized within one's department, at higher levels within the university, as well as externally by several external referees. Those who don't measure up are denied tenure and asked to leave; in a few elite universities, only 10 percent of new faculty makes it to tenure. I should add, however, that it would be a mistake to think that the unfortunates never move on to live happy, productive lives. There are jobs in industry, small colleges, government agencies, etc. But for those who survive, performing successfully in a system with extraordinarily high expectations, among ambitious, opinionated, ego-inflated colleagues (each at or near the top of their respective fields) can be an enjoyable life—in a somewhat perverted way. However, failure to land a faculty job, denial of tenure, and the demands of the profession even among those who succeed have also led to embittered lives, failed marriages and, in a few instances, shots fired and the SWAT team in the coffee room. Long after tenure, internationally recognized professors who may have already authored leading textbooks and 100 or more scientific papers in their fields still compete in a system that funds 10 to 15 percent of applicants. US science is a meritocracy, geared to support only the best work done by the best people.

Various Degrees of Leanness and Meanness

Other Western, industrialized countries show variations on the theme of lean and mean. Much of Canadian science operates under a more benign funding system. The Natural Sciences and Engineerng Research Council (NSERC), the Canadian equivalent of the American NSF, tends to spread research money around more than the Americans do. It is a more "socialist" approach where most, if not all, good scientists obtain some research funding, and only superstars get a lot. To stay in the game still requires ideas and research productivity, measured in the quality and quantity of research publications. The Canadian system is envied by many American scientists, especially those in my field. A downside is that expensive research, e.g., molecular biology, is difficult to conduct on NSERC funding alone, and scientists who perform such research must constantly scramble for money from additional sources. There are those who complain that mediocre research gets funded and that this money would be better spent by giving larger grants to better researchers. But Canadian universities do not charge their professors indirect costs and pay them twelve-month salaries. This makes possible higher grant funding rates and lessens the tendency to value faculty in terms of how much money they bring in, a practice considered by some to be both capitalist and American. Not so, because British scientists now struggle with funding rates as low as those in the US and lack of state support for universities—something that began with Margaret Thatcher, I am told. In at least one of the best British universities, they have become obsessed with trying to measure (and increase) quality and productivity by demanding that scientists publish a minimum number of scientific papers per year, and only in the "best" scientific journals with the highest impact factors (a way of quantifying a scientific journal's quality, previously discussed by Lacanilao).

Policies may be misguided or could be improved in the preceding examples, but they should be enough to convince readers of a number of things: being a scientist is not easy; rather, it is a tough, fast-paced, fiercely competitive life, not too dissimilar from the life of a professional athlete or concert musician who must constantly perform at the highest level. This is because the scientific systems in these examples of "imperfect functionality" tend to be lean and mean, and are structured to support only the "best." The resulting quality control produces what is arguably the best science these countries are capable of. Charlatans tend not to be hired by respectable institutions, are unable to secure government funding to support their work, and are deprived of the opportunity to waste taxpayers' money. The *positive* outcomes are seen in how research and development fuel economic growth (there are studies that actually show the extent to which this is so), in how universities have evolved to become "playgrounds of the mind" (to borrow a phrase from Jose Rizal)—where independent scholars concern themselves with global climate change, the current biodiversity crisis, alternative energy sources, stem cell research, or the design of ancient Roman toilets, regardless of what George W. Bush says. And we know from history that from healthy, vibrant, and productive scientific communities sprout discoveries that lead to new world-views and such things as light bulbs, penicillin, and computers—like flowers sprouting in a grassy field.

Lessons from Latin America

So far, my stories have only provided readers with images of First World science. Moving on to examples of greater imperfection, I would like to offer some observations concerning a few developing countries, starting with Brazil. In the Brazilian state of Sao Paulo, I am told that one percent of tax revenues go directly into science. Like Filipinos, the Brazilians do not have a perfect system. They have their share of power-hungry and/or incompetent bureaucrats, unpublished scientist-administrators, destructive professional jealousies or rivalries among scientists, and instances of outright corruption. As a society, they are also recovering from years of political dictatorship and face social problems such as poverty, crime, and lack of education. One can walk on a downtown street in Sao Paulo and imagine being in Europe, turn a corner and find oneself again in the Third World amid garbage, beggars, potential robbers, and kidnap rings. But democracy is at work in Brazil. They have a populist president who is committed to change (and, by the way, has a good relationship with Washington arch-enemy Hugo Chavez). Despite their problems, they have a system enlightened enough to use tax money to properly equip laboratories, adequately fund research projects, finance attendance at scientific meetings abroad, send their young scientists for research training (for example, to my laboratory in California), fly foreign scientists (like me) to the University of Sao Paulo to give lectures. As an editor of a leading international scientific journal, I get to oversee the rigorous peer-review process that manuscripts from my Brazilian friends are subjected to. Clearly, Brazilians can do world-class science, and they do it in Brazil. It is not so surprising when I go from my place of work in Santa Barbara to Los Angeles (on my way to visit loved ones in Canada), that I often fly in a plane called the Brasilia, designed and manufactured in Brazil, or that my former car was a Brazilian Volkswagen. An anecdote worth including involves a Brazilian student who spent a year in my laboratory: he says he was once reported to the police when he arrived at a small Brazilian town to catch hummingbirds for research. His permit had been approved but was not in his possession, so they threw him in jail. Even my ex-convict visiting student thought this was an admirable demonstration of Brazilian environmental consciousness.

I am writing this article in Colima, Mexico, where my graduate student and I are conducting research in collaboration with a Mexican scientist and two of his students. My collaborator says two-thirds of his salary is based on his productivity—if he did not publish scientific papers, his salary would be one-third of what he currently receives. On two occasions, I conducted research at the Smithsonian Tropical Research Institution's station on Barro Colorado Island, in the Panama Canal. There, we got to know scientists and students from various Central and South American countries who were conducting research alongside Americans, Canadians, Australians, and Europeans of various nationalities. The Hispanic scientists had international reputations, were well–published in peer-reviewed journals, and were excellent role models for their students. I also had a visiting Chilean student in my California laboratory who got her own grant from the Chilean government and brought samples to my laboratory for biochemical analysis. While she was washing cuvettes (sort of a rectangular test-tube that fits in an instrument called a spectrophotometer) in the laboratory, I asked her what the word for cuvette was in Spanish. She said *cubeta. Lavandera de cubetas* is now a PhD student funded by the Fulbright Foundation at the University of California at Berkeley. Her home department at the Universidad Catolica, in Santiago, Chile, is staffed by a number of internationally recognized scientists (I know two of them personally) who are great role models to their students, assets to their country, and active contributors—through their peer-reviewed publications in international journals—to the advancement of scientific knowledge.

It is interesting to think of how many of these countries were also former colonies of European countries, subject to the same (or at least very similar) influences as those seen in the Philippines, i.e., former dictatorships, anti-communism used to justify oppression, government corruption, inequities between the rich and poor, Spanish or Portuguese and Church influences on culture and way of life, history of covert or overt US meddling in national affairs. They should have as many excuses as anyone or any country for having a completely dysfunctional scientific system.

Avoiding the Road to Dysfunction

In my youth in the 1950s and early 1960s, the Philippines was full of talent, nationalism, and promise; in many ways, it somehow still "worked" as a country and its progress seemed unstoppable. It is very common for Filipinos my age and older to lament the fact that then, the Philippines was ahead of many developing countries but now has been overtaken by them. I promised not to be critical. But we are in the twenty-first century, and the dearth of peer-reviewed articles originating from the Philippines leads me to fear that the country of my birth may be on the road heading toward complete dysfunction. It may be considered improper for an outsider to offer unsolicited "advice" to a scientific community not lacking in intelligence, expertise or nationalism. Despite the current sad state of Philippine science, within this community are "flowers in bloom in the grassy field." As we all learned in grade school, these need sunshine, fresh air, clean water, and good soil. We learned from Rizal's life that it should not be considered a bad thing to write of darkness, foul air, and dirty water. Filipinos shot the messenger in 1896, and his message was censored for decades afterwards by the Church and banned in some of the best universities in the country. It is in this light that I view negative reactions to objective analyses of the state of Philippine science. How sadly counter-productive! How contrary such reactions are to the interests of the Filipino people! But for every Filipino in Rizal's firing squad in 1896, there were thousands of others who joined the revolution. Today, for every person who does not want to read or hear of dysfunction in Philippine science, there are many more who realize that it is time for change. May the flowers bloom.

Printed in two parts on April 5 and April 12, 2007

Lawrence L. Ilag

Patenting Academic Research

Intellectual property has become a defining force in the scientific and business worlds. To be globally competitive, developing nations are establishing intellectual property laws consistent with global norms. As part of my legal education at Harvard, I sought to explore the impact of intellectual property on developing nations. With Harvard Law School's approval and financial support, I spent several weeks in the Philippines—at the University of the Philippines in Los Baños (UPLB), one of the country's top research universities—to get a firsthand glimpse of the challenges that confront a developing nation in its efforts to grapple with intellectual property protection, specifically patent protection of biotech research. This article discusses some of my observations there.

From the start of my externship, it was apparent that the importance of intellectual property was only beginning to appear on the radar screen of the university. The university's legal counsel seemed to have only a basic familiarity with patent doctrines. The legal office did not appear alarmed that the university lacked an efficient system to resolve patent-related issues and patent in-house scientific research. There were also no strong linkages between the university's legal department and the Office of the Vice-Chancellor for Research and Extension (OVCRE), which oversees the development and implementation of research and extension programs of the university.

Recently, the OVCRE established the Intellectual Property Rights (IPR) Office, the first of its kind in the entire University of the Philippines system, to oversee UPLB's patent applications. When I was there, the IPR Office was managed by a nontechnical person. Similarly, a nontechnical person handled the patent applications of the National Institute of Molecular Biology and Biotechnology, a UPLB-based research institute at the forefront of biotech research in the Philippines. Most likely, these personnel lack the necessary understanding of the technologies disclosed in the patent applications and the ability to engage the scientist-inventors in substantive discussions about the adequacy of the claims or the specifications of the patent applications. In this regard, I believe that a science PhD or a patent lawyer would be best suited to head the IPR Office and would lend more credibility to the office's initiatives. The current situation may be symptomatic of the relative unimportance of, or unfamiliarity with, intellectual property rights locally. This contrasts with the situation in industrialized nations where intellectual property protection is vigorously pursued and enforced.

In the university, patent applications may be written by the scientist-inventor with minimal substantive input from the patent counsel. This practice fails to recognize that

Finnegan Henderson Farabow Garrett & Dunner, LLP, Washington, D.C. USA.
larry.ilag@post.harvard.edu

a patent application, albeit a technical disclosure, is also a legal instrument that has to satisfy particular requirements of law. If written without the help of a patent expert, a patent application will likely fall short of satisfying the legal requirements. Consequently, a patent will not be granted, or if granted, it will likely be invalidated or restricted in scope when challenged in court. Cutting corners by not employing the services of a patent expert may thus prove costly in the long run.

Although the university has endorsed acquiring intellectual property protection of its research as a matter of official policy, recent efforts will not suffice to achieve that policy. Many factors might account for this. Heading the list is the lack of financial resources. Research programs and facilities already suffer from constrained budgets and obsolescent equipment. It may seem inappropriate to shrink the research budget further by channeling some of it to the acquiring and enforcing of intellectual property rights.

A robust patent portfolio, however, can yield returns that far outstrip the costs. The potential may be illustrated by the following example. In the 1970s and 1980s, a university faculty member, with the help of graduate students, developed an environmentally friendly method of controlling plant parasitic nematodes. This invention generated a lot of interest abroad. Its market was even estimated to reach $87 billion. Owing in part to the university's unfamiliarity with patent laws, the university transferred its interests in the invention to outside groups. The technology was refined elsewhere and is now patented, with the university having no patent rights in the invention. Had the university developed and patented the technology on its own, it would have had more leverage in, and more revenues from, licensing and other partnership deals. Of course, not every patent will generate net income. For every one that does, many more will not. Patent protection can therefore be an expensive proposition, especially if the expenses of enforcing patents in litigation are added in. A patenting strategy that is well thought-out thus plays a crucial role in maximizing returns and containing costs.

Another benefit of patenting research is control. A patent confers a right on the patentee to exclude others from unlicensed practice of the patented invention. If the university's research remains unpatented, business-savvy enterprises might improve on the research and patent the improvement, ironically depriving the university of the full benefits of its research. Patenting research allows the university to continue unfettered in pursuing its lines of scientific inquiry, refining its inventions, and generating new ones.

The university's patenting strategy remains in the rudimentary stages. For it to evolve into a viable platform, efforts to integrate it into the university's research initiatives need to be ramped up. Invigorating patent awareness and application may be accomplished in a number of ways such as instituting a patent-based requirement for academic advancement, streamlining the university's internal procedures for patent procurement, engaging patent counsel, and organizing colloquia on intellectual property. With patents gaining even more currency, the university having endorsed the pursuit of patent protection, and the Philippines being a signatory to the Trade-Related Aspects of Intellectual Property Rights (TRIPs) agreement (which commits it to conform to international norms relating to intellectual property), now is the time that patents come to the forefront of the university's—and for that matter, the country's—consciousness.

March 24, 2005

Ellery Jose

Transferring Technology, Creating Wealth

As a graduate student of technology management (a relatively new graduate program of UP), I am often asked by my acquaintances to differentiate my degree from an MBA or MPA. I explain it in terms of how they differ in content and emphasis, a very academic definition. My classmate gives a more illustrative definition and a better idea of technology management. For him, it is simply "making money out of technology." I have since adopted his definition and I was glad that this idea resonated in a recent article that I read. It was about several Filipino entrepreneurial teams that won in local and international business plan competitions. Featured were a graduate student from Silliman University who invented a technology for waterless live fish transport, a team from FEU with their business plan for construction boards made out of aluminum packaging, and engineering undergraduates from UP Diliman who prepared a business plan for affordable, high-quality medical equipment. Congratulating them was Presidential Consultant Jose Concepcion III who said, "It's time for Filipinos to make money."

However, this idea of "technology for wealth-creation" still has to be cultivated in the Filipino consciousness. In the Philippines, a science or engineering career is not looked at as a pathway to becoming rich. When I was an undergraduate, students knew that academic research, especially if they generate inventions, had potential commercial value. But somehow, it was uncommon to think of using our scientific or technical know-how to build a business enterprise. We know that the University of the Philippines has not yet realized the potential value of its technologies or benefited greatly from them, unlike Stanford University, which has already earned $552 million in royalties from its patents since 1970. Based on a survey, for the year 2000, American universities have received a total of $1.2 billion in royalties from their licensed technologies, which have generated $60 billion in product sales and 400,000 new jobs. And this does not yet include the multiplier effects of the increased economic activity on the entire national economy.

Universities enable wealth-creation by transferring the knowledge they create primarily through published work and through their graduates. Knowledge can also be embedded in inventions that arise from university research and are the subject of technology transfer. A university will then apply intellectual property (IP) protection in the form of a patent or utility model and then seek potential investors who may want

Technology and Supply Chain Management Professional
efjefj2001@yahoo.com

to license the invention. It is actually like a software license that gives a user a right to use MS Word but not own it, which means that it cannot be reproduced and distributed to other PCs unless permission is granted because the copyright for it remains with the owner, in this case, Microsoft.

A technology license will always include provisions on payments (royalties, up-front fees, milestone payments), exclusivity (MS Word is nonexclusive because it is licensed to anyone who wants to use it and pay the required fee), length of effectivity and obligations of the investor or licensee. I was saddened to hear from an Ateneo professor that one of his colleagues invented a very useful device but sold its patent to an investor rather than licensing it, thereby losing all his rights to the technology. Yes, this is an option, but not preferred, for obvious reasons. Licensing allows maximization of revenues from the technology because it gives the patent (or utility model) owner a longer (and ultimately greater) stream of income, unlike selling, which means just a one-off payment based on the present (and potentially lower) valuation of the invention. If it were licensed, then depending on the terms of the license, the value of the payments may be renegotiated.

Biological materials, like cell lines or tissues, may be transferred through a material transfer agreement (MTA). Some of the MTAs I have seen from UP microbial culture collections limit the use of the microbes for research work. An MTA is an excellent way of controlling the distribution of reproducible materials like cells even if these are not patented and its scope can extend even to the progeny material. If it is not protected by IP, then other entities may use the same material freely as long as it was from another source not governed by an MTA.

The next level of technology transfer would be in the form of spin-off firms and equity participation. A university may want to receive equity in an enterprise that will commercialize a certain technology, instead of royalties. Faculty members would most likely be the founders of these firms themselves and the university may provide business incubation services to them. This is a very different setup that is not yet widely used.

It will take time before UP evolves into a full-fledged "entrepreneurial university" or for the Katipunan area, with UP, Miriam College, and Ateneo University as research seeds, to metamorphose into a mini-Silicon Valley. The starting points for building an entrepreneurial academic community would be a dynamic research culture and an environment that respects intellectual property. Without these two elements, there may be no technologies to transfer in the first place!

Below, I discuss the technology transfer systems in other countries. I want to emphasize that despite their varied forms, all of them provide the same set of core services in the form of IP and business advice and financial support for patent applications.

USA: University technology transfer in the US has its historical roots in the successful agricultural extension activities of the early American land grant universities, after which our University of the Philippines was patterned. Its administration is primarily through in-house technology licensing offices, which have responsibility for technology disclosures, patent applications, technology marketing and valuation, licensing negotiations and revenue collection. In recent years, these offices are either being complemented by university-industry relations units or are evolving to include some of their functions. This is evidence of increasing recognition that technology transfer is part of a much broader process of establishing university-industry linkages.

Other means of technology transfer are through external patent management firms and establishment of university-related foundations. However, these external entities, while shielding the university from business risks and legal liabilities, are not as widely used by US universities as licensing offices. For one, they choose only a certain percentage of invention disclosures to commercialize, leaving the university the responsibility for the rest. Obviously, they tend to choose those inventions with the greatest potential commercial value. To avoid this "cherry-picking," some universities retain the authority to decide which technologies to give to the external entity. But this strategy also requires establishing some form of in-house licensing or patent office to prescreen invention disclosures. Moreover, there were major publicized disagreements between certain universities and their partner research corporations in the past, which may have contributed to their negative image. The use of external firms or foundations remains a viable option, despite these disadvantages, as shown by UK, Australian, and German universities.

Right now, the US has one of the most vibrant systems of university technology transfer. There is a professional organization called the Association of University Technology Managers or AUTM which is composed of hundreds of licensing professionals from different universities. The success of the system has been credited to the famous Bayh-Dole Act of 1980, an act that gave American universities the right to obtain titles on inventions arising from federal-funded academic researches. It provided the academe incentives to set up their own technology transfer units and to maximize the transfer of their intellectual properties to industry. Indeed, according to the law, it is their mandate to do so. If they fail, then the federal government with its march-in rights can take these patents and commercialize them on its own.

The success of the Bayh-Dole in the US, with other countries like Japan and Germany adopting some of its provisions, has generated interest and debate in some of our academic institutions and within DOST. Their question is: "Will a similar legislation improve our local technology transfer system?" I have met some people who believe that it will and some who believe otherwise. On the other hand, most are between these extremes. Personally, I believe that there are lessons that we can learn from the history of the Bayh-Dole Act but these lessons must be placed in the right context. The Bayh-Dole's primary contribution was to remove the greatest obstacle to the process: bureaucracy. By giving the responsibility for technology transfer to universities, the Bayh-Dole removed the need for private firms to navigate the maze of oftentimes inconsistent rules on licensing by the different federal agencies that fund academic research. For example, licensing an invention that was funded by NASA, the NIH and US-DOE would have been a legal nightmare! Today, a firm would just have to go to an appropriate office in a university.

Further, the law allowed exclusive licensing of inventions from research funded by public funds. As a result, more firms were encouraged to license these technologies. However, not all inventions are licensed with exclusivity, as in the case of recombinant DNA (and other basic technologies that will impact several industries). But exclusivity has its inherent business advantages that make technologies more attractive to potential licensees.

Similar to the US, a large part of academic research in the Philippines is funded by government funds, and a Bayh-Dole-type law will greatly impact technology transfer. But the government, particularly the DOST, and not universities, has already developed specific mechanisms for technology transfer (IP Guidelines and TAPI or Technology Application and Promotion Institute), although UP also has an emerging system. Such legislation will facilitate the transfer process, especially now that other government agencies are designing their own IP rules and policies. However, it is highly unlikely that the DOST will give up its primary role in the transfer process. While the DOST system needs improvement, it has shown itself capable of successful technology transfer from university research, as in the case of the herbal medicines lagundi and sambong from UP Manila.

There is no evidence, local or foreign, to suggest that internal units within universities will necessarily be much better than a government agency or any other external entity in transferring technology. It may be that a DOST-led approach may add another layer of bureaucracy; however, based on my own observations, the "distance" between a UP professor and a DOST research council may be "shorter" than the distance between UP faculty and the UP administration. Nevertheless, there are good reasons for maintaining the autonomy of academic institutions, particularly in shaping their research agenda. A reduced role in managing their own intellectual properties may defeat the very purpose of academic freedom, with grave consequences. (A compromise then has to be made!) Regarding exclusivity for public-funded researches, I personally believe this is a non-issue in the broader context and is best decided for each case based on acceptable criteria.

Japan: Prior to the 1998 Technology Transfer Act, tacit and undocumented transfers of technologies and knowledge occurred primarily between an individual researcher and a single, large firm with which he/she had had informal relations. Japanese university professors then had ownership of intellectual properties arising from their own research, with the government having a share only if the research was supported by a specific allocation of public funds. With the aim of improving the transfer system, the 1998 Act encouraged the creation of technology licensing offices (TLOs) by, among others, providing seed funding for their operations.

These licensing offices are usually separate legal entities servicing either a single national university or a private university. There are also area-based TLOs that serve several universities and research organizations in a particular locality. They provide a range of services—patent applications, marketing, even business incubation—to individual researchers who choose to assign their inventions to them. On the other hand, companies have to pay membership fees to most TLOs in order to gain preferential access to undisclosed patent applications. More details are given by TLOs to an interested investor only after the two parties have signed a confidentiality agreement. A policy change in the future may require researchers to disclose inventions and assign patents to their respective universities.

United Kingdom: For a long time, commercialization of technologies from publicly funded research was the monopoly of the government through the British Technology Group or BTG. The BTG was subsequently privatized, and is now owned partly by several universities. These universities have, on their own, also commercialized their technologies without using BTG's services. In particular, King's College London and Oxford have their own research corporations. Before its privatization, the BTG also provided equity funding for academic researchers and institutions with marketable technologies. But with the growth of the venture capital industry, this function seemed redundant. The BTG is now focused entirely on technology licensing and sales.

Germany: Before 2002, university professors and their assistants had the "university teacher" privilege that gave them exclusive rights to exploit their patents. As a consequence, universities did not establish their own transfer units until a reform of the law required professor-inventors to transfer their rights to their respective universities. Six years before, the Ministry of Education and Research waived the government's right to collect shares of the license income from government-funded researches and allowed exclusive licensing of technologies arising from them. To support these reforms, the federal and state governments then vigorously established in-house licensing offices, and technology licensing bureaus and companies serving several universities within a city or a state. With the recent formation of sector-specific IP management firms that specialize on certain technology areas, a new technology transfer setup is evolving.

Australia: According to an Australian patent attorney whom I met in an international workshop, Australian universities traditionally own all the intellectual properties that they generate, unlike in most other universities. Thus, the problems of IP ownership as seen in the US that were supposed to be addressed by the Bayh-Dole were non-issues in Australia. Also, even the public universities are not subsidized by government funds. As a result, they are required to generate their own income from their activities and assets, primarily through university-owned corporations, like the UniSearch of the University of New South Wales. In recent years, exploiting intellectual properties has become the dominant contributor to total revenues generated by these corporations. As a consequence, these corporations have become more like patent management-type firms rather than full-blown business units.

ASEAN: Several universities, like those in Thailand and Malaysia, have established TLO-type offices and even university-owned firms/foundations (as in Australia) for commercializing technologies. A success story would be that of Thai Chulalongkorn University's University Industry Relations Unit or UIR, which has responsibility not only for technology transfer but also for managing collaborative projects with industry and providing business incubation services to university staff and students. In its first year in 2003, it already earned 7.7 million Thai baht from its six patents. On the other hand, in Malaysia, many of these business units have not been effective because of lack of resources and appropriate skills.

Our own University of the Philippines has also been trying to develop an effective means of technology transfer. Over the span of two decades since it first formulated an IP policy, the various constituent campuses and the system administration have experimented with in-house IP offices, a not-for-profit foundation and, more recently, a technology licensing office. The UPLB-BIOTECH has its own patent agent. Based on my own research, UP has at present a total of twenty-six pending patent applications, and five pending utility model applications from its large campuses in Los Baños and Diliman.

Many of these patent applications had been filed even before the creation of the system-wide licensing office and are presently managed by different administrative units. It will take time before a more unified system evolves. UP may consider adopting the system used at the multi-campus University of California which is a decentralized system of independent licensing units from the different campuses and major laboratories, coordinated by a lean system-wide licensing office.

For other universities thinking of formalizing their IP policies and creating units for technology transfer, I hope I have been able to provide a wide variety of options and lessons. Some of the issues I discussed are unique to a multicampus public university like UP and may not be relevant to others. Universities that are privately owned and financially independent may be more flexible in designing their own systems, unconstrained by government regulations. It may be that university administrators feel that their faculty is not yet producing enough research output to merit establishing TLOs. This is an important concern even in UP, where some people feel that not enough research is being done. To achieve economies of scale, an external TLO owned by a consortium of several universities may be an option.

Printed in two parts on June 2 and June 9, 2005

Fe Esperanza Espino

RITM: Working for the Public Good for Twenty-five Years

The Research Institute for Tropical Medicine (RITM), the research arm of the Department of Health, was established in 1981 as a special project of the then Ministry of Health. It was a development project of the Japanese government to the Philippines through the Japanese International Cooperation Agency (JICA). It was authorized to undertake activities in the diagnosis, control, and prevention of tropical diseases that were the major causes of mortality and morbidity in the country. The significant contributions of RITM to the study of infectious and tropical diseases of public health importance are discussed below.

Acute Respiratory Infections (ARI)

Intensive studies on important bacterial causes of childhood pneumonia (*S. pneumoniae* and *H. influenzae*), manifestations, risk factors, and treatment began in the 1980s in Quezon City, Muntinlupa, Pasig, and Bohol. The effectiveness of the World Health Organization (WHO) ARI case management protocol in reducing deaths due to pneumonia was studied in Bohol from 1984 to 1990. The result was the impetus for the DOH to implement the Control of Acute Respiratory Infections (CARI) program. From the treatment-based program, preventive methods were studied.

Acute respiratory infection vaccine studies in Cabuyao, Laguna, and Bohol for the past ten years addressed issues on vaccine-preventable diseases like pneumonia and meningitis, training of community health workers, improved vaccination coverage, surveillance of side effects of vaccination, and cost-effectiveness of vaccination.

Dengue

Community and hospital-based surveillances provided epidemiologic, clinical, laboratory, virologic, immunologic, and population susceptibility information essential for strategies in disease control and prevention (including vaccination programs). This information underscores the necessity of a multisectoral approach to dengue control, and the strengthening of laboratory capacities for diagnosis and continuous epidemiologic and clinical surveillance of the disease.

Research Coordinating Unit, Research Institute for Tropical Medicine, Alabang, Philippines.
fespino_2000@yahoo.com

Diarrheal Diseases

RITM studies contributed to basic knowledge on local epidemiology and etiology (bacteria, viruses, parasites) of diarrheal disease in the community and hospital settings, and practices of mothers of infants with diarrhea. Internationally accepted laboratory procedures for detection of enteric pathogens such as human rotavirus, antibiotic resistant *Shigella flexneri*, and *Vibrio cholerae* were pioneered in RITM. A nationwide, laboratory-based food-borne disease surveillance program, including prevention, control, and risk communication was established.

Human Immunodeficiency Virus (HIV)/AIDS

The first case of Acquired Immunodeficiency Syndrome (AIDS) in the Philippines was treated in RITM in the mid-1980s. The institution took the lead in raising local awareness of HIV and AIDS through research, training of health care providers, and care of HIV/AIDS patients. The first surveillance activities among commercial sex workers were carried out from 1989 to 1990. The natural course of HIV disease among Filipinos was described. An algorithm for HIV testing and national guidelines for HIV/AIDS treatment were also developed.

In 1994, monotherapy for HIV became available, and affordable antiretroviral drugs were made accessible to many HIV/AIDS patients. The RITM was the National Reference Center for HIV Testing until 2000. A national assessment program for HIV-testing was developed to provide quality control of HIV tests performed by private laboratories. The National Control Program adopted the use of pooled sera for HIV testing for its surveillance program.

Leprosy

Prevalence surveys in the National Capital Region in the early 1990s disclosed active transmission in the areas surveyed, delayed diagnosis of leprosy, and under-reporting of the disease. Seroepidemiological studies from 1988 to 1996 in Culion Island identified individuals at high risk of developing leprosy—household contacts of patients with leprosy. This focused the attention of disease control to this group of individuals. Chemoprophylaxis of these individuals was initiated and a reduction in incidence of leprosy was observed over the years. In 2005, pilot field evaluations of a rapid diagnostic test showed promise for the leprosy control strategy–detecting cases in communities.

Malaria

The biological (parasite, human host, and vector), epidemiological, social, temporal, and geographical dynamics of malaria transmission in the Philippines were better understood through field-based studies in Morong, Bataan from 1992 to 1998. These led to the simultaneous development and evaluation of new strategies for malaria control from 1993 to 2002 in Morong, in Tayabas, Quezon, and in Agusan del Sur. These were, among others, community mobilization and organization for malaria control, malaria microscopy diagnosis by nonmedical technologists, computerized malaria health information system, and health program management by local governments. In 1997, the interruption of malaria transmission was documented in Morong, Bataan. The revision of malaria treatment guidelines in 1992 was based on therapeutic efficacy studies from 1999 to 2001 in various provinces.

Rabies

In 1990, it was demonstrated in Sorsogon that rabies can be cost-effectively controlled by a massive vaccination of dogs. The nationwide implementation of rabies vaccination in 1998 was made possible through RITM's pioneering use of intra-dermal administration of rabies vaccine. This reduced costs of rabies post-exposure vaccination. Consensus meetings were initiated to update the national guidelines on animal bite management and human rabies.

Schistosomiasis

Community-based studies in Leyte in 1981 looked at the impact of praziquantel treatment on the prevalence, incidence, and morbidity of schistosomiasis. Cost-effective measures for schistosomiasis control were also developed. These were the basis for recommendations to the national control program for more comprehensive approaches to eliminate schistosomiasis in the Philippines. Effects of treatment on childhood growth and development, resistance to re-infection, mechanisms of liver disease, anemia, and cognitive impairment were more understood.

Fundamental components for the ongoing global efforts to develop vaccines against the parasite, and immunologic and developmental predictors of resistance were studied. These were hormonal changes during infection in puberty and processes of immune reaction, immunity, and nutritional status. A rapid diagnostic test for schistosomiasis (SJ-URIDIP) was also developed.

Tuberculosis

In the early 1990s, the Camalig Anti-TB Organization in Albay was among the first of RITM's community-based projects that attempted to establish local health program management. In 1993, a rapid diagnostic test using nested PCR was developed; it showed great potential in the early detection of TB meningitis. Clinical, radiological, and pathological features of TB infection in people with HIV infection in the Philippines were described in 1994. These helped in early detection of tuberculosis and other mycobacterial infection in people with HIV.

A study in Sta. Cruz, Laguna from 1995 to 1996 established a network of transporting specimens to a central laboratory. From this study, a new TB strain, called Manila family, was discovered. This strain was also identified among Filipino immigrants living in the United States.

Viral Hepatitis

A diagnostic kit for hepatitis B from locally purified surface antigens and antibodies was developed in 1983. The kit is composed of the following tests: RPHA (reverse passive hemagglutination) assay for detecting viral surface antigen (HbsAg); PHA (passive hemagglutination assay) for detecting antibodies to viral surface antigen (anti-HBs); and HI (hemagglutination inhibition) for detecting antibodies to viral core antigen (anti-HBc). Recently developed is a monoclonal antibody-based enzyme linked immunosorbent assay (ELISA) for detection of HBsAg.

The Other Significant Contributions of the RITM: Clinical Trials

Clinical trials have contributed to the use of vaccines and other biologicals in the prevention and treatment of rabies, childhood pneumonia, dengue, and influenza, and the management of diarrhea with oral-rehydrating salt solution. Other studies have contributed to the development of treatment protocols and alternative drugs for adult pneumonia, malaria, leprosy, and intestinal parasitism.

Training and Postgraduate Courses

Medical doctors in pediatric, internal medicine, and family medicine residency programs, and infectious disease fellowship programs spend from one to six months of their training years in the institute's hospital ward and laboratories. During this period, they gain further experience in the management and control of infectious and tropical diseases. Regular postgraduate courses and infectious disease treatment guidelines serve as means through which the institute disseminates information gleaned from experiences, research, and projects.

All of these undertakings would not have been accomplished without the generosity of individuals, groups and organizations, national agencies, local and foreign colleges, universities and other similar institutions, colleagues in the DOH, and provincial and municipal local government officials. We share their commitment to improve the health and well-being of our fellow citizens.

April 6, 2006

Part 4

On Being A Scientist

Mikaela Irene Fudolig

"*Pinagpapala ang pinagpapalà:*" The Role of the Scientist in Nation Building*

Let me start my speech with this quote: "*Pinagpapalà ang pinagpapala, at pinagpapala ang pinagpapalà.*"

For those who didn't get that, let me literally translate it in English: "Blessed are those who shovel, and those who are blessed are made to shovel." Of course, it loses its lyricism, so I repeat: "Pinagpapalà ang pinagpapala, at pinagpapala ang pinagpapalà."

From the various talks that I have attended, I noticed a general theme: that aside from being good students, which is the primary reason why all of us are here today to be recognized by the College of Science, we should have leadership capabilities and social responsibility. Of course, the fact that we're from UP sets great expectations.

Do I agree with that? I will make this clear: yes, I do agree. Academic excellence alone is a joy that you will most likely keep to yourself, and if not put to good use, then the University has invested on something for nothing.

The question now is: how do we, as scientists, help in nation building?

I have noticed, again, from the many talks that I have attended, that the common idea of "community service" is Sangguniang Kabataan, Red Cross, Gawad Kalinga. Opinion leaders view community service, which they correlate with nation building, as using physical energy to help the poor. You want to do community service? Solicit money from your congressman and donate a school building. You want to do community service? Help during calamities. You want to do community service? Build houses for the poor.

Again, let me make this clear: These ways are indeed community service. But are these the only ways to do community service? Should community service simply be giving something for nothing? Should community service necessarily involve a lot of legwork? Should the effects of community service be immediate?

Let me ask you now: When Michael Faraday discovered magnetic induction, was that community service?

When James Hutton developed and Charles Lyell promoted the theory of uniformitarianism, was that community service?

National Institute of Physics, University of the Philippines, Diliman, Quezon City, 1101, Philippines.
mikkifudolig@yahoo.com

* 2007 UP Diliman College of Science, Bachelor of Science Valedictory Address

When our very own Alexander Edward Dy made it possible for amoebiasis to be tested based on salivary IgA instead of stool, was that community service?

Current conventional wisdom would answer: NO. Faraday had been criticized before for discovering something without practical use. And so what if slow geological processes occurred eons before and continue until now? And how can Alexander Dy's amoebiasis test serve the poor in the squatters' area? His method will definitely not give jobs to them. It wouldn't give them shelter. And it's not FREE.

But Faraday's discovery of magnetic induction is what led to using AC power in our homes. It is the reason we can power this microphone. It is why you have lights at home and in the classroom. It is why you can power your refrigerator.

The concept of uniformitarianism did nothing to help the poor, but it helped gain more understanding of the Earth. And it is uniformitarianism that influenced Charles Darwin in formulating his theory of evolution. And I think you know how influential Darwin's theory of natural selection has been.

Mr. Dy's amoebiasis test would probably not be given for free. It would not give them shelter, and, most probably, wouldn't give jobs, at least not to the usual recipients of charity. But if amoebiasis can be diagnosed faster simply by getting the saliva of a patient, something which can be readily obtained, then more amoebiasis patients would be cured. More lives would be saved.

The community service of scientists is often underestimated. Our discoveries are often tagged as having no practical applications, of no use in calamities, and of no immediate help to the poor. If Faraday had concentrated on donating blood, if Hutton and Lyell focused on building houses, if Alexander Dy, now magna cum laude, insisted on tutoring every single kid in his barangay FOR FREE, then they would be considered by the majority as excellent servers of the community. But they would not have done what they have done. Where would we be now?

Fellow scientists, do not be disheartened. Our efforts may be devalued by those who seek immediate, visible, and tangible results. But the fact remains: science drives the technology that makes lives better all over the world. Our devotion to our craft, our unceasing search for our holy grail, that piece of knowledge that will change the way things are, is as much community service as the more popular and immediately recognized forms of giving. Let not the pressure to be recognized make us stray from our efforts to improve the life of humanity in the best way we can.

Thank you.

May 10, 2007

Dranreb Earl Juanico

Cease Not to Envision a Bright Future That Barely Resembles Today*

I am extremely honored to be standing in front of an audience consisting of brilliant and potentially productive talents, each of whom is presently facing a crossroad in his or her journey through life. Allow me to take this opportunity to congratulate the graduating batch of 2007, as well as our proud parents, relatives, and loved ones who may even have traveled a long way to get here and witness this momentous event of our lives. I also would like to salute the professors, the instructors, and the research advisers who, by serving.as role models, have been instrumental to our success.

A week ago I met a former neighbor I have not seen in a long time. He asked me, "How have you been?" I replied, "A week from now I will be a doctor." Then he innocently remarked, "May I take the pleasure of being your first patient?" I quickly responded by saying I will not be the type of doctor he had thought; but his remark struck me with a sad reality I have almost forgotten—that in this country, medicine is the only profession that people know would earn you the right to prefix "Dr." to your name. I feel honored to dare stand against that norm along with nineteen fellow outstanding PhD graduates of the College of Science this year.

Earning a PhD degree is a daunting task that entails a considerable amount of discipline and sacrifice. PhD study eats up at least five years more of your youth and your productive time after obtaining a college diploma. It usually conflicts with your employment and financial obligations to your family. Sometimes it even meddles with your personal life. In the College of Science, the bar in obtaining a PhD has even been raised higher by the requirement of at least one international scientific publication—a requirement that had not been in place just a few years back. With my experience at the National Institute of Physics, the publication requirement has caused some students to postpone graduation year after year.

I pondered the question, "What is the rationale for enforcing a requirement as stringent as writing a scientific manuscript that would be visible to researchers all over the world?" Each of us here knows the answer, which I believe is this. We Filipinos, as a people, are seeking to affirm and regain that inherent confidence in ourselves—a

---

National Institute of Physics, University of the Philippines, Diliman, Quezon City 1101 Philippines.
djuanico@gmail.com or earl@nip.upd.edu.ph

* Response in behalf of the 2007 MS and PhD Graduates of UP Diliman College of Science

confidence subdued by centuries of colonial rule. Unsurprisingly, we all are moved by the winnings of Manny Pacquiao, not because he is an outstanding boxer, but because he is a Filipino. With him we associate a national pride in our creative potential for contributing something important to the world that we would be recognized for. Our contribution as scientists, I suppose, is in communicating novel scientific discovery, theory, and ideas through writing and publishing.

We have been acquainted with the scientific method even as far back as when we were just beginning to memorize the multiplication table. Standard textbooks and teachers say it starts with an observation; then, an inference; next, experimental validation; and lastly, conclusion. Nevertheless, the scientific method taught in school had a missing piece we had not been explicitly instructed about—that is, publication of our results. Scientific research is incomplete unless technically-reviewed results are published. But even though our country boasts of more than 90 percent literacy rate, writing does not come naturally to most of us, especially if English is the required medium of communication.

Filipinos are generally not literature-oriented people, so I was told. Few people write because few people care to read. When a pedestrian sees that conspicuous MMDA signboard that says, "*Huwag tumawid, nakamamatay*," the pedestrian usually still crosses the road as though the signboard—and even the rampaging vehicles—did not exist. I suppose this is not due to a lack of time for reading the sign, and neither to a lack of ability to read, but rather because of the ingrained aversion to reading (and writing) in our culture. For anyone, such as you and I, who is raised in such a culture, the barrier to publish is considerably steep.

Another common excuse against writing is inaccessibility to sources of information. In this Age of Information, however, that ceases to be an excuse. The Internet has opened the gateway to a vast ocean of scientific knowledge by providing unprecedented information access. Through inexpensive Internet access, geographical boundaries are dissolved and economic disparities are leveled. With all things equal, I believe that scientific advancement is not exclusive to the Americans, the Europeans, or the Japanese; Filipinos are as capable of partaking in scientific inquiry as "genius" and "prodigy" can emerge anywhere regardless of race. Now that knowledge has become easier to acquire than ten or twenty years ago, it is about time that the world should know what the Filipino is truly capable of. It is time to tell the world that Filipinos are known not just for Manny Pacquiao, or Imelda Marcos, or domestic helpers, but also for ingenuity in crafting original scientific ideas that could potentially revolutionize our understanding of the world and perhaps our way of life.

In the past, most graduates who wanted to pursue scientific careers would seek their lot abroad—particularly in prosperous countries that have progressive scientific traditions. It was somehow a reasonable choice because our country did not have the institutions and facilities for advanced scientific research, in addition to minor funding and support from the government. But that situation has gradually changed. More importantly, we have a growing pool of scientist-educators equipped with world-class expertise to mentor students.

The Philippines has been steadily gaining recognition from the international scientific community. A decade ago, when I entered the University, we were but a speck on the map of international science. Now, we are still a speck but a more persistent one that does not easily wash away anymore. That is progress and we owe much of it to our professors in the College of Science, most notably the present dean, who voluntarily chose to remain in the country and teach science to our generation despite unfavorable circumstances, and even when opportunities in private industry and overseas were beckoning temptingly. Their sacrifices were essential in putting up a tangible reason for bright young minds to stay. I am truly honored for having been a student of these outstanding mentors. We should all be fortunate to have been trained by the greatest Filipino scientists of today.

Our nation calls for the establishment of a solid scientific tradition; several examples attest to a significant correlation between an established scientific tradition and the economic prosperity of a nation. In a country of 88.7 million people, we are among the rarity who chose to travel the path less traveled. Let us help demonstrate that scientific progress can be done nowhere else but here in our homeland, where the pasture is yet to be grazed upon.

In the subsequent days or weeks, we would be making decisions that would chart the course for the rest of our lives. Some may decide to work in industry where salaries are substantial; while others may decide to sustain the brain drain and migrate abroad. I also hope that some of you may consider pursuing scientific careers in the country as a viable option. But whatever option you decide to take, never cease to envision a bright future that barely resembles today; that is the only reason that keeps us, scientists of this nation, advancing forward.

Mabuhay tayong lahat! Maraming salamat po.

May 3, 2007

Joyce Sarah Ibana

Finding My Niche in Science

THE EXODUS of Filipinos abroad for greener pastures is true not only for domestic helpers, nurses, teachers and information technologists, but also for science researchers. We have "lost" a lot of our young brilliant scientists to more progressive and developed countries, for which I think, they cannot be condemned. Sometimes we really have to go in order to learn and contribute more to science. I myself am about to leave our country on a Fulbright scholarship for a PhD in the USA. Inasmuch as I am grateful for what I have learned here, I find it necessary to learn more about my field from a more advanced country. Hence, I am leaving, but, I am leaving with a grateful heart that I have stayed long enough to appreciate science by working in different institutions in our country that offer growth to young minds and provide service to our fellow Filipinos. And just before I go, I would like to share with you my story of how I found my niche in science in the country that I have chosen to serve—the country I intend to come home to.

My story starts at the Research Institute for Tropical Medicine (RITM). RITM is the cluster of white buildings on top of the only hill in Filinvest Corporate City in Alabang. You will not miss it if you only look to your right when passing the Alabang overpass across Zapote road. Before 2003, its existence was not well known to many, except for those who were bitten by dogs. RITM's most illustrious moment came when the incidence of SARS came to the Philippines. This is after the family of Adela Catalon, who died of SARS, and other suspected SARS victims were brought and quarantined in the hospital wing of the institute. During this time, RITM and its staff were featured on television and in newspapers with an image that made the institute and staff the most fearsome place and people in the country.

As for me, I had known about the existence of RITM since 1995 when I used to go there as a product specialist of a company that offered products for molecular biology research. I arrived there after passing a bumpy dirt road that had not yet been developed by Filinvest Corp., to troubleshoot PCR machines, discuss the advantages of a top-of-the-line DNA sequencer, and promote different laboratory products. It was just the beginning.

In 1998, I came to know more about RITM and got involved in their research activities when I joined the Marine Science Institute in UP Diliman (UPMSI) as a science research specialist. During that time, I started working on collaborative projects between the UPMSI and RITM. The first project I worked on involved searching for potential

Louisiana State University Health Science Center, New Orleans, Lousiana, USA.
joycesaibana@yahoo.com

antimalarial compounds from marine sponges. At UPMSI, we collect marine sponges (by scuba diving in different sites in the country) from which we extract, purify, and characterize various compounds with biological activities that have therapeutic potential. The project ended with a candidate that we still hope to pursue for further studies.

The funding for the first project ended but my involvement in RITM continued when I worked on my masteral thesis. The immunology department head, Dr. Luz P. Acosta, the director of RITM, Dr. Remegio M. Olveda, and my mentor from UPMSI, Dr. Gisela P. Concepcion, arranged for my "adoption" in the RITM. Through this arrangement, I was able to work in the institute on the development of an immunoassay for detecting a tumor antigen in serum of cancer patients. In this project, we came up with designs for a local alternative to imported cancer monitoring kit that is cheaper and less hazardous than its imported, radioactive counterpart. There are still a lot of things to be done in the project, but as it was, I feel so blessed to have completed my masteral thesis and discovered my interest in the field of immunology.

After my thesis, UPMSI and RITM forged another collaboration on a project involving the production of a monoclonal antibody against a tumor antigen as part of an ongoing multidisciplinary, multi-institutional research program known as AMOR (Antibody and Molecular Oncology Researchers; the acronym coined by Dr. Concepcion). The AMOR program aims to develop therapies for breast cancer and is supported by the Philippine Council for Health Research and Development (PCHRD) of DOST. Aside from UPMSI and RITM, AMOR involves researchers from the UP Diliman National Institute of Molecular Biology and Biotechnology (NIMBB), Institute of Chemistry, Institute of Biology, Far Eastern University and the Ateneo de Manila University.

In the six years that I have been working as a researcher, I have witnessed the collaboration between UPMSI and RITM grow stronger. Now, we have more projects involving immunology researches and have extended our partnership with the National Kidney and Transplant Institute (NKTI). The tripartite collaboration of UPMSI, RITM and NKTI aims to develop locally an immunosuppressant for transplant patients that has a targeting antibody component and an immunosuppressive compound isolated from a marine organism.

From all these projects I got involved in, I came to realize that right now, there are a lot of things that have to be done and can be done in our country as far as science research is concerned. I also realized from my experiences that we do have institutions and mentors here who are dedicated to making our country a nurturing place for the growth of inquisitive minds. However, it is still not financially rewarding to work as a scientist in our country as in other developed nations. And so, I hope that our government will exert more effort in providing a better scientific environment to encourage Filipino scientists abroad to come home. I do not think it is too much to ask for more incentive for science researchers. At present, we are earning a lot less than college graduates who opted to work in the industries or even in call centers. This I find a bit discouraging.

I am leaving the country now because there is a lot in science that I think I need to know and will learn from more advanced nations. Certainly, I find the prospect of making a significant contribution to improving the health condition of our fellow Filipinos particularly motivating and worthwhile. At least for now, and from where I stand, I still think and feel I have found my niche in science in the country I would like to come home to.

July 1, 2004

Mark Windell Vergara

A Research Assistant's Notebook

EVEN BEFORE the May 2004 elections, our country has been undergoing an "economic depression" and since then, more and more people have been losing their jobs or have been taking on jobs for which they are overqualified or underpaid. Since then also, I have been asked a lot of times if I have a job, followed by what my job is.

I am a science researcher. To describe what a science researcher does, I often tell stories of my adventures, or should I say, my misadventures in the field. It could be the time when I was confronted by a drunken policeman in Palanan, Isabela, holding a rifle and asking for my *cedula* (residence certificate) because he wasn't sure I was a Filipino. I sure had a hard time explaining why I didn't have my *cedula* with me in my wetsuit. It could also be when I had to wait for sea snakes to leave the transect line before I could proceed with my seaweed surveys in Polillo, Quezon, and I had to stop and wait from time to time whenever they would decide to return. Another one was when we were told by our boat tender that he had just asked a fisherman to postpone throwing dynamite in the water because we were still underwater conducting our survey. Illegal fishermen had to stop their activities and wait for us to finish because we accidentally chose to survey on the tiny island where they were storing and mixing the dynamite they used for catching fish. Still another one was when we were told by a fisherman in Hinatuan, Surigao del Sur just before a survey that sharks no longer existed in their waters—because the crocodiles had eaten them all!

I usually crack jokes about my job but when the discussions become more serious, I start by saying, "But seriously, I love my job!" and that's when I start getting those little smiles from people that I know really mean "Yeah, right!" It is a fact that compensation for science researchers, especially in government institutions, is what most would call "very disappointing." But I do enjoy this job. I love how I learn a lot in the laboratory and in the field, and how the janitor, secretary, administrator, cash officer, market person, cook, dishwasher, muscle man, utility man, technician, liaison officer, writer, public speaker, scientist, and supervisor all fit the job description of a research associate like me. I love learning new things every day and knowing that what I'm doing is contributing to scientific knowledge and to an improved quality of life for my fellow Filipinos.

Conducting marine research definitely has its difficulties but the rewards absolutely far outweigh the difficulties. I was a fresh graduate from UPLB when I joined Project 1 of

---

The Marine Science Institute, University of the Philippines, Diliman, Quezon City, Philippines.
vergaramark@yahoo.com

the four-year Pacific Seaboard R&D program funded by the DOST back in 1999. Project 1, entitled "Assessment of Reef Resources of Eastern Philippines," was implemented by the Marine Science Institute of the University of the Philippines (UPMSI). One of the highlights of Project 1 was going on research cruises on board the UP Visayas' TR/V *Sardinella* during the summer. I can't explain how excited I was on that journey. We were scheduled to be at sea for twenty-one days to survey the upper half of the eastern side of the country. Imagine doing two to three dives a day in areas which most people only dream of, or only see in postcards or magazines, or, more often than not, which have not been explored by divers.

For the four-year duration of the project, the macrophytes study alone (seaweeds and seagrasses) was able to establish 141 seaweed survey/collection sites along the eastern Philippine coast and we discovered a new species of seaweed in Sorsogon. We also found a seagrass species previously thought of as rarely occurring, to occur in fifteen more sites around the country. We also found that an economically important seaweed, a good source of agar, is abundant along the eastern coast of the country. We were always working very hard, but we were also diving in pristine sites, and had great encounters with *dugong*, dolphins, sea snakes, seahorses, large fishes and the like.

One of my most unforgettable experiences in the field, however, was when we conducted a reconnaissance survey on Calagua Island in Camarines Norte in preparation for a research cruise in the summer of 2001. What was supposed to be a three-hour boat ride took six hours due to bad weather. The swells were so high that during our lunch stop, the boatman had to tie his boat together because it was breaking in two. From there, the almost unbearable journey lasted for another four hours. This time we (another research associate and I) were even hungrier than before we had our lunch because we threw up more than what we had eaten. We didn't even bother rinsing our mouths and faces after throwing up because the huge waves that were getting into the boat would wash it off for us anyway.

Upon reaching the island, the locals immediately gathered around us and asked if we were the ones sent by the government to help them find a cure for the new disease that was damaging the crops. It turned out that a new disease, other than the common "ice-ice disease," was infecting their cultivated seaweeds. The collapse of the seaweed farming industry had driven its residents into poverty. The people on the island could no longer pay their electricity bills (a big generator provided electricity to the whole island) because their source of income was either reduced or totally lost due to the new disease. The whole island had been without electricity since the outbreak of the new disease. We regretfully had to inform them that although we did work for the government, the most we could do for them was bring samples of the infected seaweeds to someone who might be able to help them. That night we slept with a heavy heart, not because we slept on the floor, but because we pitied the residents of that poor island. This experience just made me love my job even more, knowing my work could somehow help improve the lives of the people who depend on the marine environment—say, 50 percent of the Philippine population at the least.

Three years after that experience, I'm still with the UP Marine Science Institute as a graduate student and part-time research associate. I keep telling my friends that there is hope in what I do—that I will not become part of the statistics of Filipinos who have chosen to forego their dreams and passion in exchange for "more comfortable lives" by migrating to Australia or Canada or by joining well-paying call centers. I'd like to prove to people that there is hope for science researchers and do just like what Confucius said, "*Choose a job you love, and you will never have to work a day in your life.*" Since I love my job, I don't consider it work!

December 23, 2004

Eduardo Padlan

It Is Harder to Earn a PhD in the Philippines than Abroad!

A few years ago, I overheard a colleague say "... *sa* (name of an Asian country) *lang ang PhD niyan...*" I should have said something because I have learned long ago that it doesn't matter where one earns his/her degree; what really matters is what one does with it. I know people who earned their PhDs from very well-known universities, but who did not accomplish much afterwards, and vice versa.

In the Philippines, we have progressed far enough so that we are now routinely granting PhDs in many disciplines. Still, many of our students, especially the young and the unattached, opt to go abroad than stay in the country to get their PhDs. Is it because we believe that a PhD from the Johns Hopkins University (just for an example) is more prestigious than a PhD from the University of the Philippines (just for an example)? Is this another example of our "colonial mentality"? Are we afraid that somebody might say of us, "*... sa Pilipinas lang ang PhD niyan?*"

Well, I have news for you: *It is harder to earn a PhD in the Philippines than in the United States or Europe.*

That may be too sweeping a statement to make. But let me compare graduate programs here and abroad. The basic requirements are more or less the same: formal coursework, an oral candidacy exam, completion of a thesis, and successful defense of it.

The number of formal courses required varies. In the University of the Philippines, forty-five units of formal graduate courses (fifteen three-unit courses, or equivalent) beyond the BS degree are required (see, for example, *http://www.dilnet.upd.edu.ph/~mbb/academic.htm* for the Molecular Biology and Biotechnology Program; I believe this is true for all programs in UP). The Biophysics Department at Johns Hopkins (*http://www.jhu.edu/~pmb/program-curriculum.html*) requires five core courses, as does the Biophysics Program at Harvard (*http://www.fas.harvard.edu/~biophys/ ProgramOfStudy.html*); the Department of Molecular Biophysics and Biochemistry at Yale (*http://www.mbb.yale.edu/gp/gp_04a_fyc.htm*) requires at least seven courses, of which four are specifically required, while the Department of Biochemistry and Molecular Biophysics at Columbia (*http://cpmcnet.columbia.edu/dept/gsas/biochem/ gradwork.html*) requires six (four core courses and two electives).

At what point the oral candidacy exam is taken also varies. At UP, the qualifying exam is taken after all the coursework is done. At Johns Hopkins and Columbia, the

Formerly with the US National Institutes of Health and currently with The Marine Science Institute, University of the Philippines, Diliman, Quezon City, 1101, Philippines.
edpadlan@yahoo.com

student is evaluated at the end of the first year; at Yale, the oral candidacy exam is taken at the beginning of the second year; at Harvard, the qualifying exam must be completed by the end of the second year.

(Here's a surprise from Johns Hopkins (*http://www.jhu.edu/~pmb/program-curriculum.html*): "Proficiency in biochemistry, cell and molecular biology is evaluated formally with an oral examination at the end of the first year. Students who have never taken courses in these areas are welcome, but not required, to take such courses during their first year. Tutorials and self-directed study provide alternative avenues for preparing for this evaluation." In other words, one does not need to take formal courses to take the qualifying exam! This is not surprising because Johns Hopkins is patterned after the universities in Europe and there (I know this is true in the UK), there are no (repeat: NO) course requirements for the PhD.

But the biggest difference between getting a PhD here and abroad is the ease with which thesis research can be accomplished. The labs abroad are relatively well-funded, all the equipment that a student needs to perform experiments are usually in the lab or just next door, and supplies and consumables are literally within easy reach. (In contrast, we are all well-aware of how it is in the Philippines.)

Graduate school in the US (and in Europe, too?) is (usually) free! Not only are one's tuition and other fees paid for, but one also gets a living allowance that could be big enough to support a family (not luxuriously, of course). In the Philippines, we (usually) pay.

So, why then do many stay and get their PhDs in the Philippines? I would suspect it is mainly for personal reasons. It takes an average of five to seven years to get a PhD in the US (although some have done it in less time, for example, Dr Enrique T. Virata, the very first university professor in UP, who got his PhD in Mathematics from Johns Hopkins in two years!). It is not easy to be away from one's family for five to seven years and be in a strange land at that. It cannot be because we are afraid that we cannot compete with foreigners. Let us not forget that the Filipinos who have been accepted in universities in the US had competed for slots with students from the US and from the rest of the world—and had won. Indeed, many of our countrymen have done exceptionally well in their studies abroad—and even afterwards. *Walang sinasanto ang Pinoy!*

Could we, maybe, make it as easy to get a PhD here as it is abroad? Or, do we want to be able to say, "*... sa US lang ang PhD niyan...?*"

I realize that most of the PhD holders in the Philippines at the moment, including myself, got our degrees abroad. I do not mean to belittle our degrees. If for some reason I have offended anyone by this article, I can only say, *"Batu-bato sa langit, ang tamaan huwag magagalit."*

October 27, 2005

Rafael Guerrero III

What It Takes to Be a Scientist in a Developing Country

Scientists are people who make science their lifetime work or career. To be a scientist, one must adhere to the rules of the vocation and its code of conduct, as in any profession.

Science, defined as "systematic knowledge based on study and observation," is the way by which mankind has gained its stock of knowledge about itself, its surroundings, and the laws of Nature. It has helped man provide for his basic needs and sustain his further development and survival.

The basic tool used by all scientists in the pursuit of science is the scientific method. It consists mainly of five steps: (1) identifying the problem or question, (2) formulating the hypothesis or possible explanation, (3) testing the hypothesis by gathering facts and making observations to validate or disprove it, (4) analyzing the facts or information gathered, and (5) arriving at a conclusion or finding that accepts or rejects the answer to (1).

There are various branches of basic and applied science. I chose the field of biology (the study of life), and specialized in the study of animals (zoology), more specifically fisheries (science of aquatic life).

I started my career in science with a BS in Zoology from the State University. Teaching was my first assignment. It was after I had worked for my MS in Applied Zoology that I learned the discipline and rudiments of scientific research through my thesis.

My skills and expertise in science were further honed with the dissertation I did to earn the degree of PhD in Fisheries Management in the US.

Today, with aquatic and marine resources R&D administration as my main concern, I am still very much into scientific work and related activities such as field and table research, research monitoring and evaluation, and research output utilization.

So what does it take for one to be a scientist?

Advanced degrees are important and very helpful for a professional but are not necessary for one to be a scientist. All it takes is for one to apply the scientific method and get the approval of peers on his/her methods and findings. In addition, I believe that a good scientist should have the virtues of commitment, relevance, and excellence.

Philippine Council for Aquatic and Marine Research and Development, Department of Science and Technology, Philippines.
pcamrd@laguna.net or aqubios@laguna.net

What is commitment? It is putting oneself fully to the task or chosen field of endeavor and making contributions for the advancement of such a field. Such a commitment can be for a certain part of one's lifetime or for its whole.

When I chose to solve the problem of uncontrolled population growth of tilapia in ponds for my dissertation, I committed myself to applying the scientific method and carrying out the work even after the publication of my paper and until the final results benefited fish farmers.

In close relation to commitment are the sterling qualities of integrity, humility, and creativity.

Yes, "honesty" is also "the best policy" in science. Intellectual honesty and staying true to one's scientific works and reports are the hallmarks of a good scientist.

Humility is the quality that a scientist should have in admitting mistakes and in willing to be corrected and/or criticized constructively by peers.

Creativity is what makes some scientists stand out from the rest. It is that special gift or effort that makes the ordinary extraordinary and makes the real difference in one's field.

Let's now talk about relevance. No one wants to be irrelevant or disconnected from reality. While science can be neutral or benign for "science's sake," most scientists in a developing country like the Philippines have to be not only scientifically focused but economically well-oriented as well. This is because with limited resources, the overlying concerns of development in our society are paramount and critical. Thus, relevance to the needs of our people (40 percent of whom live below the poverty level) for food security, productive and sustainable livelihood, and income generation may get higher priority for science and technology support than pure or basic science.

Finally, there is the value of excellence or the quality of being the best you can be in your field of science. A passion for excellence is one in which you strive to improve your scientific skills and know-how through continuing education, publications (read or published) and interaction with peers in meetings or through the Internet.

From my view as a scientist, it takes commitment, relevance, and excellence to be one and remain one. Do you have what it takes to be one? Only you can answer that.

April 3, 2006

Gil Jacinto

Edgardo D. Gomez: On "Ease Back Mode"

WHO IN the world is Edgardo D. Gomez? Indeed, who in the world of marine science in this country, Southeast Asia, and the international marine scientific community does not know the *Kapampangan* Edgardo D. Gomez?

Married to Ambassador Phoebe Gomez and transferring their genes to their only son, Jed, Dr. Gomez is not new to the world of excellence. After all, he graduated with a double degree, a Bachelor of Arts in Social Science and Bachelor of Science in Education with a major in English from De La Salle University (*summa cum laude*) and received the Gold Medal for General Excellence. He went on to obtain his MSc Marine Biology at the St. Mary's University in Minnesota, and got his PhD in Marine Biology from the Scripps Institution of Oceanography, University of California, San Diego—one of the best marine science institutions in the world, if not the best.

Ed Gomez is used as benchmark to measure the best, and this is what prevailed through all the years of his professional life. Under his academic leadership and with a clear and unwavering vision of marine science in the country, Ed Gomez was the principal force in initiating, pioneering, and developing the then Marine Sciences Center in UP Diliman into the Marine Science Institute today. As founding director of the Marine Sciences Center with only a piece of paper as "asset" and with a status of "informal settler" at the former Natural Science Research Center in the mid-1970s, he went on to recruit and entice staff and build an institution—an institution where the values of honesty, openness, and collegiality prevail because Ed Gomez lived these values and expected no less of his staff. Yes, Ed Gomez was director for almost all of his academic life—"easing back" only around six years ago to let the younger generation carry the baton. Those of us who have been administrators for even a fraction of that time will appreciate that this is no mean feat. For Ed, it meant foregoing the life of the typical academic and lucrative job offers at the early stages of MSI's development, and keeping the home fires burning, sometimes alone, while many of us were pursuing our PhDs.

It was not always smooth sailing for Dr. Gomez and the institute, as there were also detractors and seemingly insurmountable odds. Yet, Ed pursued his vision of putting MSI and the university on the world map, and succeeded.

His optimism, focus, determination, composure, and self-confidence, and his ability to handle the unexpected with grace and style were all crucial ingredients in institution-building.

The Marine Science Institute, University of the Philippines, Diliman, Quezon City, 1101, Philippines.
gilj@upmsi.ph

Last November 30, 2005, we officially marked Ed Gomez's retirement, two years after the "mandatory" age of sixty-five, and twenty-three days past the date of his sixty-seventh birthday. We reminisced with Ed the good old days and also the more difficult stages of his professional life, and what he has been to many of us and those who could not join us in person. We rejoiced and look forward with anticipation to Ed Gomez in "ease back mode," aware that he is already pursuing the next phase of his professional life as university professor emeritus.

Colleagues, students, and friends came together for an afternoon of celebration and tribute to a practicing botanist and de facto "gardener" of the College of Science (UP Diliman), outstanding administrator, a recipient of the United Nations Global 500 Roll of Honor, Presidential Lingkod Bayan Awardee, Pew Marine Conservation Fellow, author of more than 130 technical publications, academician of the National Academy of Science and Technology, and recipient of UP's Concepcion D. Dadulfalza Award for Distinguished Achievement—university professor Edgardo D. Gomez.

December 8, 2005

Edgardo Gomez

An Odyssey to Excellence*

When first approached to be your Commencement Speaker, my first impulse was outright rejection because of my commitments and because I felt that I do not hold a high enough position to merit the honor. I declared that commencement speakers were usually chosen from among high government officials, heads of large corporations, or outstanding UP alumni, none of which I am. I asked for some time to think about it, but I soon received a call from our Chancellor convincing me to accept. Among the considerations he mentioned was for me to share with you my experience with the Marine Science Institute, which, to paraphrase him, "from a small beginning developed into something great, a center of excellence which can serve as an inspiration for young people. Being world class is a reality in MSI, not just a notion." Hence, I will take you on "An Odyssey to Excellence," a journey of some thirty years in about fifteen minutes.

The baccalaureate graduates were not yet born in 1974 when the journey started. Resulting from a turf controversy between two campuses, a committee was formed under then President S. P. Lopez to resolve the issue of where and under whose jurisdiction should a marine science unit be established? The committee had just completed a two-page charter for a "Marine Sciences Center" (MSC) for approval by the Board of Regents when, fortuitously, I came looking for a job. As a naive PhD fresh from the Scripps Institution of Oceanography in California, I got the job to make those two sheets of paper into a reality.

The Early Years (1974–1980)

No room, much less a building, came with those two sheets; not even a stick of furniture. Initially working from my desk as a new assistant professor in Marine Biology at the then College of Arts and Sciences, I set about the task of looking for available space. Because there was none at Palma Hall or in any of its pavilions, I was directed to the then Natural Sciences Research Center, which also turned out to be full, but where I was told that I could wait for a social science project to be completed so that the vacated room could house the administrative office of the new MSC, as the Marine Science Institute was then known. Securing that room was no piece of cake, as it involved

The Marine Science Institute, University of the Philippines, Diliman, Quezon City, 1101, Philippines.
edgomez@upmsi.ph

* Commencement speech at the 95th General Commencement Exercises, University of the Philippines Diliman, last April 23, 2006

a skirmish with the incumbent social scientist who reluctantly vacated the natural science facility. It then took months to obtain furniture for an empty room, because all that the UP could provide, after an appeal to the president, was one desk and one chair.

But what good is an office for a research center if there is no laboratory space? I had to yield my laboratory room as a faculty member of the CAS to the first recruits, two young ladies who started research on seaweed chemistry. But that gesture was well rewarded because within a few years, the two researchers managed to obtain a grant from the Ministry of Natural Resources that included the construction of what is now the Seaweed Chemistry Building. The next mode was to convince regular faculty members of university units to do research with us, by providing them with small research grants but requiring them to use their own laboratories since I had none to offer. This was a slow process but it allowed for their recruitment into the future Marine Science Institute. Desperate for more laboratory space, particularly for my own research, I combed the campus and consulted many people. My desperation must have been so obvious that the chairperson of the Zoology Department pointed me to an abandoned greenhouse. That out-of-the-way structure soon became the nerve center for coral reef research. From activities initiated there, the first national inventory of the coral reef resources of any country in the world emerged, a project funded by the Ministry of Natural Resources that included the construction of a small "Coral Building," (which now houses the Office of the Campus Architect). We tried to get support from the Office of the President, but the focus at that time was the creation of a new UP campus in the Visayas anchored on a flagship unit, the College of Fisheries.

From Benign Neglect to Development Focus (1981–1993)

The struggling research center with a minuscule budget saw a window of opportunity when a proposal for a $20 million World Bank loan was being prepared for the development of fisheries and the new campus. With encouragement from the World Bank consultants, I approached the dean of the College of Fisheries if we could be part of the proposed development and share some of the resources. To my shock and surprise, I was told in no uncertain terms that the whole $20 million (today worth some $48 million or ₱2.5 billion was only for the College of Fisheries. Hearing that message, I took my hat off and waved goodbye.

With our small budget we plodded on, never losing sight of our goal and mission. Within a few years, to our great surprise, we were being wooed with some force, to move with the rich but hapless unit. My few researchers and I sent the clear message to the president that under no circumstances would we move to the Visayas for a number of reasons that I do not have time to elaborate here. As a consequence, our reward was to be treated with benign neglect for the rest of the incumbency of the president at that time.

But the stars and the planets were in alignment for us. Come 1981, the fledgling MSC had the daunting task of hosting the Fourth International Coral Reef Symposium, a challenge that had fallen on our lap when I attended the third symposium in Miami in 1977. For within a few years of our existence, the earnest research that had started in the abandoned greenhouse had begun to be noticed by the international scientific community. By a fortunate occurrence, we planned for the conferment of a Doctor of Science degree, *honoris causa*, on the late Professor Francisco Nemenzo, father of our former UP president, an unsung Filipino hero but well-known for his pioneering studies on corals. And as chance would have it, that academic exercise happened to be the first official function of the newly appointed university president, Senator Edgardo Angara. After the conferment, he graciously hosted a reception at the Executive House for the international coral reef scientists who had come. What was both fortuitous and fortunate for the occasion was the lavish praise from the foreign guests for the extraordinarily well-organized symposium, something that they did not expect in a developing country.

Subsequent to that event, I had the privilege to call on the new president. When he asked me how he might help us, I told him that we needed a marine laboratory by the sea. We already had a five hectare piece of land in Bolinao, Pangasinan, but no budget to construct a laboratory. He apologized that he could not give us the money immediately but promised to put us in the infrastructure development program for the following year.

The year 1983 saw a ₱2 million appropriation for a marine laboratory, but with the Aquino assassination that year and the consequent political upheaval, the proposed building shrank into a two-story box with a dozen unfurnished rooms. But again, the gods were with us. By the mid-80's, after having waited a full decade, we qualified on our second attempt for a UNDP institution development grant worth $1.3 million. This grant was more than matched by the Philippine government counterpart that allowed us to complete not just the unfinished laboratory building, but an additional eight, yes, eight other buildings, from administration to dormitories to diving locker. Even as these were being put up with extra-budgetary resources, President Angara instructed then College of Science Dean Roger Posadas to make the construction of our headquarters in Diliman a priority. This was a welcome surprise because in the first master plan of the College of Science Complex, there was no Marine Science Building.

Why was this happening to us? Was it because we were the first and only academic unit in the university and the country that had an all PhD faculty? In 1986 when the former research unit, the MSC, was transformed into the degree granting unit that is now the MSI, we decided that a doctoral degree was a *conditio sine qua non* to be a faculty member of the new institute. This was possible because virtually all those faculty members from the former Departments of Botany and Zoology who wanted to transfer to the new institute held doctoral degrees, and the young research assistants of the MSC who had gone abroad for their graduate studies began to return with their doctorates.

This combination was ideal because the experienced faculty were complemented by bright young PhDs eager to do research. Indeed, from day one of the institute, virtually all the faculty, old and new, had publication credits. This was and is the distinction of the MSI. All the faculty without exception are published and publishing, giving us an international recognition unmatched by any other unit until very recently.

The time had come to inquire about being designated a National Center of Excellence under a scheme established under President Marcos a decade earlier to recognize and promote scientific achievement in the university. Hands down, the MSI qualified to be the first new scientific center.

The Trials (Betrayal?) of 1994–1996

Sailing the ocean blue is never smooth, for stormy weather sometimes lurks in the horizon. So it happened that in the mid–90's, as we cruised along, a threat to our scientific activities arose from most unexpected quarters. Silently, nay surreptitiously, forces were mobilizing to construct a heavy industry, a mega-cement factory, 500 meters from our seaside marine laboratory in Pangasinan, an excellent facility barely completed with foreign and national funding. And who was behind this development? None other than a member of our own Board of Regents. And who would not help us fend off this looming threat? None other than the president of the university!

Being in this university is oftentimes a paradox. We pride ourselves of many achievements, and for good cause. But we also sometimes witness inanities unheard of in a civilized country. Thus it was that I first got wind of the proposed cement factory not from UP officials but from a public announcement at the municipal beach next to our marine station during a routine field trip.

The details of ensuing events are too numerous to recount here. Suffice it to say that neither the Regent nor the President, officials who are mandated to look after the welfare of their constituent units, was inclined to assist. Rather than make a big scene on campus and waste time, my decision was to ignore them and bring in other people to help us. Thus, all the constituent units of the university, except for the one from which they came, made public pronouncements condemning the proposed development. Many national and international scientific bodies came to our support. Three major international funding institutions committed to environmentally sustainable development lent their assistance. But most of all, the decent people of Bolinao opposed a project that was perceived as detrimental to their health and well-being. Hence, after more than a two-year controversy, the Department of Environment and Natural Resources "denied with finality" the request of the cement factory proponents for an Environmental Clearance Certificate. Common sense and decency had prevailed.

The Past Decade with the Changing of the Guard (1997–2006)

The year 1998 brought about the recognition of the MSI as the first and only Center of Excellence in Marine Science under the Commission on Higher Education (CHED) COE/COD scheme of promoting and rewarding scientific excellence. The MSI continued its march forward as a vibrant, productive, graduate department of the College of Science. For this period, we produced about twenty ISI-indexed publications per year, an average of 1.2 per senior staff, which is world-class. Its ability to draw research funding from national and international sources remains its strength, each year drawing in twice the amount of the total budget that it receives from Quezon Hall.

I stepped down as director at the beginning of the new millennium after leading it for a quarter of a century. The institute's present vigor is manifested in its being designated in 2004 as one of four international Centers of Excellence by the Global Environment Facility/World Bank Project (GEF/WB) on Coral Reef Targeted Research and Capacity Building for Management. Under this program, the once threatened Bolinao Marine Laboratory has become a greater nexus for international coral reef research. The benefits include the upgrading of facilities, including broadband complemented with wireless internet connection, a new PABX telephone system that allows researchers to reach international contacts, plus a number of other vital equipment. Equally important, more than half a dozen graduate students are being supported under this program. Three senior staff of the institute serve as members of international working groups in coral reef research with additional research funding.

Conclusion

How have we been able to achieve this triple designation as "Center of Excellence?" We, in the Marine Science Institute, are ordinary Filipinos coming from all over the Philippines, born in this beautiful and blessed land, raised and educated in this society with its particular history, collective memory, political institutions, and economic status. Many have written volumes on how these aspects of our life have held us back from realizing our true heritage: an excellent nation at par with the best in the world.

I have recounted to you this afternoon the "Odyssey to Excellence" of the Marine Science Institute. What made the difference? In desperate times as we find ourselves today, we continually seek models which may serve to inspire us to improve our personal condition and the environment around us. I believe that it was in consideration of this that I stand before you this afternoon to share some experiences and some trials that may be of value to you as you venture forth in a very competitive and challenging world, armed with a new degree. To conclude my speech, I shall provide you some of the guideposts that we followed on our odyssey, which may be useful for your continuing journey in life. The first four are concerned with institution-building, which you may not be doing till later in your lives, but which you should know about. The rest are more personally relevant to the younger graduates.

First, there must be a vision, as lofty as possible, that must be pursued relentlessly. Ours was a three-stage goal: first to be the best marine science institution in the country, then in Southeast Asia, and finally in the tropical developing world.

Second, we created a nurturing, supportive environment where all could do their research. Early on, we decided in our recruitment policy, that the only crabs that would be allowed in the center would be crustaceans. Respect, collaboration, and unity are essential.

Third, dedication to work is essential to productivity. Our work ethic was the forty hour week at one's place of work, which did not normally include one's home. Deliver what you promise to establish your credibility, especially if you are going back to the sponsor for continued support.

Fourth, recruit only the qualified to be your collaborators, preferably honor graduates. Thus, at one point, the MSI included one summa cum laude, one magna cum laude, and half a dozen cum laude graduates in its senior staff. Multiply this brain power with the extra time dedicated to work, and the resultant "IQ hours" of the institute as I call them, are unmatched by any department.

Fifth, basic virtues like honesty, fortitude, and humility are not to be compromised. One should make a personal commitment to cultivate these in one's youth, and to interact with people who espouse these virtues. It was painful to let go of several PhDs that scored low in this regard, but it was necessary for the health of the unit in the long run. We must have the moral strength and great faith to withstand political and corporate pressures if we are to be truly successful in life.

Sixth, a positive attitude towards life is an important ingredient to success. Too much time is spent bellyaching in a typical Filipino work setting, when one's attention and energies should be focused on opportunities, opportunities to remove obstacles and move ahead, or really genuine opportunities to race forward in pursuing one's goals.

Seventh, and finally, especially for young people in an age when we are bombarded by contradictory ideas, trust in Divine Providence and follow the Golden Rule. This exhortation may seem a bit strange in this secular campus, but my past religious training has left its influence. Examples are too numerous to list but let me just mention two. I have spent the better part of my life serving others, principally my students, my colleagues in the institute, and our nation at large. Going into retirement last year after thirty-one years of service, I was awarded two international research grants that have brought me more funding than I ever had in the years past. And, in recognition of my key role as the Coordinator of the Philippines/Southeast Asia Center of Excellence in Coral Reef Targeted Research and realizing that the UP would cease to pay my salary upon my retirement, the sponsors of that GEF/WB program saw to it that I was fully compensated for my time rendered.

But the greatest compensation that I have is not the material benefits that many aspire for. Rather, it is the recognition that I enjoy among peers in this country and all over the world, something that money cannot buy, a respect that buoys up the spirit even in these trying times, a satisfaction of fulfillment that I have served my people well, often at great cost. This, my dear graduates, is my parting message to you: Serve others well before you serve yourself, and the Man Upstairs (God, Allah, Yahweh) will look after you! Congratulations and Godspeed!

Printed in two parts on June 8 and June 15, 2006

Eduardo Padlan

My Odyssey in the World of Molecules

Embarking on My Journey

"Man is simply a collection of molecules" and "can be understood in terms of molecules."

—Linus Pauling

We have been fascinated by the various parts of our anatomy since time immemorial. As we developed more and more sophisticated tools, we learned more and more about ourselves and we began to make the connection between our actions and the components of our bodies. We are now probing deeper and deeper into our anatomy and we have begun to describe ourselves in terms of molecules and even of atoms. Nowadays, we ascribe not only our physical attributes, but sometimes also our mental capabilities and idiosyncrasies, to the molecules that make up our system. To be able to do this, we have found that we need to understand how those molecules work.

The key to the understanding of molecules and their function is structure. The visualization of the detailed structure of molecules did not become feasible until the early part of the last century. Since then, three-dimensional structures for all sorts of molecules, from simple ones consisting of only a few atoms to complex ones with hundreds of thousands of atoms, have become available. This became possible when physicists brought their tools, in particular, spectroscopy and crystallography, to bear on the problems of biology. It is the continuing cooperation of biologists, chemists, and physicists—and, more and more, mathematicians—that might eventually lead us to a full molecular understanding of our selves.

This is a narrative in two parts of some of my adventures and misadventures in the world of molecules. In the first part, I recount my early work. In Part 2, I describe how I am trying to contribute to the utilization of the structural knowledge that we now possess of some interesting biological molecules. Naturally (and I hope I will be forgiven for it), I will focus on my (few) successes and not dwell on my (many) failures.

Formerly with the US National Institutes of Health and currently with The Marine Science Institute, University of the Philippines, Diliman, Quezon City, 1101, Philippines.
edpadlan@yahoo.com

My Introduction to the World of Molecules

Armed with a bachelor's degree with a major in physics, earned from the University of the Philippines in 1960, I enrolled at the Johns Hopkins University in Baltimore, Maryland in 1964 to learn biophysics. I chose for my thesis adviser, Professor Warner E. Love, the crystallographer in the faculty. Warner, as we called him (we were not particularly formal at Hopkins and we called our professors by their first names, sometimes by their nicknames), was interested in the structure of proteins, their function, and their evolution. At that time, the three-dimensional structure of only two proteins—myoglobin and tetrameric hemoglobin—had been determined (both obtained by X-ray crystallography). The elucidation of the three-dimensional structure of a protein was quite an accomplishment then and John Kendrew and Max Perutz were awarded the Nobel Prize in 1962 for their work on those two proteins. Myoglobin and hemoglobin, whose primary function is to bind oxygen reversibly, turned out to have the same tertiary structure—soon termed the "myoglobin fold." But, those two are mammalian proteins. Many other organisms, including many invertebrates, some prokaryotes, and even some plants, have hemoglobin. Do those nonmammalian hemoglobins also have the "myoglobin fold"? Warner Love wanted to know.

For my PhD thesis, I decided I would determine the three-dimensional structure of an invertebrate hemoglobin. Because I was young and foolish (and ambitious, as Dean Rhodora Azanza put it two years ago when she heard me describe my first venture into research), I chose to work on a giant hemoglobin—one with a molecular weight of four million. [Hemoglobins range in size from about twelve thousand (truncated myoglobins), to seventeen thousand (myoglobins and monomeric hemoglobins), to sixty-seven thousand (tetrameric hemoglobins like the one we have in our blood), to several million (the giants with 180 subunits).] I was going to use a combination of X-ray crystallography and electron microscopy. So, I picked a second thesis adviser, Professor Michael Beer, the resident electron microscopist. I proceeded to isolate the giant hemoglobin from a marine annelid [I bought a couple of thousand worms from a bait company, cut them up with scissors, squeezed the blood out from the carcasses, strained and centrifuged the debris out, and "purified" the hemoglobin by ammonium sulfate fractionation—a very crude protein purification procedure (we didn't have chromatography then)] and set up the hemoglobin to crystallize. It didn't. Even after weeks of waiting, all I got was "mud" (my term for the amorphous precipitate that formed in my crystallization vials). I isolated some more hemoglobin and tried again. Again, all I got was "mud."

But while waiting for the giant hemoglobin to crystallize, I had started working on another hemoglobin—from another marine annelid, *Glycera dibranchiata*—one that had the more manageable seventeen thousand molecular weight. This hemoglobin quickly yielded beautiful crystals. And just as quickly, I dropped the giant hemoglobin project (and Michael Beer) and focused on Glycera hemoglobin. Soon enough, I had a three-dimensional structure at low resolution. This is an understatement. At that time, data collection took months and we had to write many of the computer programs we needed, and our computer was very slow (millions of times slower than present-day personal computers) and had scant memory (many millions of times less than what the computers of today have). Computer graphics was still many years in the future and we had to draw electron-density contours by hand and build our models by hand. Nevertheless, we managed. It was obvious, even at low resolution, that Glycera hemoglobin has the "myoglobin fold." (And so does every other hemoglobin whose structure has been determined to date.) Warner had his answer. We published the Glycera structure in *Nature* and it caught the attention of a science writer who featured our results in an article in the *London Times*. Needless to say, I enjoyed the (brief) publicity. All this happened in 1968. Some years later, we pursued the structure analysis of Glycera hemoglobin to "atomic resolution," which allowed us to assign a position to every atom in the molecule. The paper that we published on the atomic structure of Glycera hemoglobin did not go unnoticed. Shortly after that paper came out, I received a letter from Linus Pauling asking permission to include a figure in our paper in a book that he and his son, Peter, were writing. Of course, I consented. Receipt of that letter from the scientist I admire the most thrilled me no end—and it still does.

Early Adventures in the World of Antibodies

In 1971, I joined the group of David R. Davies at the National Institutes of Health (NIH). David (we were also not very formal at the NIH) had become interested in antibodies and his group had determined the structure of a human cryoglobulin (an antibody-like molecule produced in large amounts by a patient who was suffering from multiple myeloma and which crystallized in his serum when the weather turned cold, causing the clogging of blood vessels—the poor patient had to have extremities amputated before he mercifully died of the disease). When I arrived in the lab, David had just started a collaboration with Michael Potter of the National Cancer Institute (NCI). Mike Potter had a collection of many hundreds of mouse myeloma proteins, many of which had known ligand-binding specificity. Although Mike Potter's proteins were produced by plasma cell tumors, the fact that many were found to bind common antigens lent support to the belief that they were actually true antibodies. Mike Potter provided David's group with Fab (the antigen-binding fragment) from several of those mouse myeloma proteins and David Segal, a protein chemist in David Davies' group and who was also a Hopkins PhD, proceeded to crystallize them. The first one to crystallize was McPC603 Fab, which has phosphocholine-binding specificity, and I was tasked to determine its crystal structure.

The crystallographic analysis of McPC603 Fab was not easy. Although large crystals were readily obtained, they were not very good diffractors, i.e., X-ray data from the crystals were not of high quality and were difficult to collect. I spent many hours taking X-ray photographs of crystals (on many occasions, I would expose the film for a week) and preparing heavy-atom derivatives (needed to determine the structure). And after suitable heavy-atom derivatives had been obtained, it took me several months to collect the necessary data for a structure analysis. All in all, it took me a year to determine the structure of McPC603 Fab at low resolution. We published the low-resolution structure of McPC603 Fab in *Nature* New Biology and it caught the attention of another science writer who featured our results in a short article in *Nature*.

We then pursued the crystallographic analysis of McPC603 Fab to higher resolution, with Dave Segal doing most of the data collection and analysis. We soon had an electron-density map that was of high-enough resolution to allow us to build an atomic model of the molecule. It took me six months to build the model of McPC603 Fab and another six months to measure the coordinates of the atoms. (The model now sits in the NIH museum.) We published our findings and were soon invited to several meetings to present our results.

The early '70s saw the flowering of the structural analysis of immunoglobulins. Structures of various myeloma proteins and fragments had become available from the laboratory of Roberto Poljak at Johns Hopkins, from Alan Edmundson's at Argonne National Laboratories, and from Robert Huber's at the Max-Planck Institute in Martinsried, Germany. The amino-acid sequence data on immunoglobulins, which were starting to accumulate, could now be correlated with three-dimensional structure. We were now able to visualize the hypervariable regions—the Complementarity- Determining Regions (CDRs) which Elvin Kabat and T.T. Wu had proposed to be the segments that form the antigen-binding site of antibodies. Since the structure of McPC603 Fab that we had determined had phosphocholine bound to it, we were able to confirm the Kabat-Wu proposal. The structure of antibodies was the subject of several science reports published in *Nature, Science, The Journal of the American Medical Association, National Geographic*, and various foreign newspapers. Those were heady days.

More structural information on antibodies and their interaction with antigens soon became available and it did not take long for the structural knowledge to be put to good use.

Heading for Home

"Science without Humanity" (One of the Seven Deadly Sins)

—Mahatma Gandhi

Ventures into the Realm of Molecular Medicine

We study science to learn (more) about ourselves, our world. What we learn, we must put to good use. Of particular usefulness is the knowledge obtained from the structural studies of molecules of the immune system, especially antibodies, to which I had made some modest contributions.

By the mid-70s, structural knowledge had started to accumulate on antibodies and their interaction with antigens (foreign substances). The structures of whole antibodies were determined, as well as of various isolated parts of the molecule. The most studied part was the Fab (the antigen-binding fragment), by itself or, in many instances, in complex with antigen. It soon became clear that different antibodies had very similar structures, except in the region used for binding antigens. (The antigen-binding site is formed by the Complementarity-Determining Regions [CDRs], as originally predicted by Elvin Kabat and T.T. Wu.)

The binding of antibody to antigen is very specific. In fact, antibodies are known to be able to distinguish between two molecules that differ by only a few atoms. A notable example is the case of the blood group substances A and B, both large molecules, but which differ only in that the group B substance has a hydroxyl group at one position in its terminal sugar, while the group A substance has an acetylated amino group at the same position. This small structural difference is enough to preclude blood transfusion between individuals of these blood types (individuals with blood type A have antibodies to group B substance and vice versa, and those antibodies can cause the death of the recipient if transfused with as little as 30 ml of the wrong blood type).

The exquisite specificity of antibody-antigen interactions makes antibodies very useful in medicine. For example, antibodies could be used to target tumor cells—and only tumor cells—so that cancer treatment with antibodies would have fewer side effects than, say, chemotherapy which kills cancer cells as well as many other (normal) cells. So, we only need to produce antibodies against the various cancer types and we have a cure for cancer. Alas, it is not that simple.

While large amounts of specific antibodies could be easily produced, they are often obtained from nonhuman sources, e.g., mice. There are disadvantages to using murine and other nonhuman antibodies in human therapy. First, our immune system will recognize those antibodies as foreign molecules and will try to get rid of them. Second, the cells of our immune system, which cooperate with our antibodies in mounting an effective response to an invasion by foreign cells or substances, do not interact with nonhuman antibodies as well as they do with our own antibodies. There is then a need to make the nonhuman antibodies look like human molecules—we need to "humanize" them.

Various procedures have been developed to humanize a nonhuman antibody while preserving its specificity. One obvious method, called "CDR-grafting," developed by Greg Winter's group at Cambridge, England, retains the CDRs while replacing everything else with human parts (some non-CDR residues usually need to be kept in order to maintain the structure of the CDR segments). But we have shown from our structural studies that not all of the residues in the CDRs are involved in antigen binding.

Therefore, my group proposed to retain only those amino acids which are involved in the interaction with the antigen (the Specificity-Determining Residues, or SDRs, as I have named them). Another method that I proposed is to replace the residues on the surface of the nonhuman antibody with amino acids that are usually found in human antibodies, so that the nonhuman molecule is given a human "veneer" and is thus camouflaged. Several antibodies have been humanized by me and by others using the SDR-transfer and veneering techniques.

The first molecules I helped humanize were antitumor antibodies. By the early '90s, I had forged a collaboration with Jeffrey Schlom (NCI) whose group had developed several murine antibodies against human tumor markers (molecules on the surface of cancer cells). A particular human tumor marker, TAG 72 (Tumor Associated Glycoprotein No. 72), that is targeted by some of Jeff Schlom's antibodies, is present in many different cancer types, and in a large percentage of the cancer cells but only rarely in normal tissues, so that the antibodies specific for that marker have great therapeutic (and diagnostic) potential. I was asked to design humanization protocols for some of Jeff Schlom's murine antitumor antibodies.

The first antitumor antibody that we worked on was CC49. The target of CC49 is TAG 72. First, we humanized CC49 by CDR-grafting. Then, we guessed the possible SDRs of CC49 (we do not have a structure for the complex of CC49 with TAG 72 so that we could not identify with certainty which residues of CC49 are actually involved in the binding to the tumor marker) and transferred only those to the human template molecule. Those procedures worked, so that now we have humanized versions of CC49 that have very much reduced immunogenicity and with their specificity for TAG 72 intact.

However, humanization of antibodies often results in reduced affinity for antigen. Such was the case for both the CDR-grafted and the SDR-transferred versions of humanized CC49. We tried to offset this by designing a molecule that had four ligand-binding sites for TAG 72. That tetravalent molecule has an effective affinity for TAG 72 that is many times better than the humanized versions of CC49 and even the original antibody.

The second antitumor antibody that we humanized was COL-1, which is specific for carcinoembryonic antigen (CEA). COL-1 also was initially humanized by CDR-grafting and later by retaining only those parts of the CDRs which we felt contained the residues responsible for its specificity for CEA. Since CEA is found in normal tissues also, although over-expressed in tumor cells, COL-1 is not as tumor-specific as CC49. Nonetheless, the work on COL-1 is worth pursuing and is being pursued.

The work on CC49 and COL-1 involved two Filipinas who were postdoctoral fellows in my laboratory: Dr. Ameurfina D. Santos of the National Institute of Molecular Biology and Biotechnology (NIMBB), UP Diliman, and Dr. Noreen R. Gonzales, formerly of Ateneo.

I was involved in various other structural ventures, including work on a pathological hemoglobin, on other molecules of the immune system, on some aspects of allergy and autoimmune disease, and some enzymes, as well as in computer analyses of various biological phenomena. Some ventures resulted in publications; many were failures.

Going Full Circle and into the Future

I retired from the NIH in 2000. But even before then, I had started to spend time in the Philippines, teaching, giving talks, and collaborating with local scientists. I was a Balik Scientist in 1998 and I spent a month in UP Diliman lecturing on Molecular Immunology and Protein Crystallography. I have been coming back ever since, often spending three months a year here, so that I have started to call myself a Balik-nang-Balik Scientist. I must confess that the sight of the bright faces of budding Filipino scientists—eager to learn, eager to contribute—inspires me to continue my journey in the world of molecules. While I can no longer do "wet" science, I can still contribute with studies done *in silico* (in the computer).

There are diseases that could be analyzed through sequence analysis, e.g., prion and other "conformational" disorders. A paper on the analysis of "mad-cow" disease has resulted from one such analysis done in collaboration with Dr. Gisela P. Padilla-Concepcion of the Marine Science Institute (MSI), UP Diliman; that paper attracted the attention of a science reporter who wrote about it in *The National Post*, a Canadian newspaper. Giselle and I are continuing our studies of protein conformational disorders. With Ma. Pamela C. David, a member of Giselle's group, we recently found a possible structural explanation for the lack of direct transmission of scrapie, the prion disease of sheep and goats, to humans despite the fact that we have been eating sheep and goats since biblical times (the prevailing notion is that scrapie has been transmitted to other animals, e.g., "mad-cow" disease, and we then get the disease from eating infected parts of those other animals). A paper describing that analysis has been published. Giselle and I have also published a paper proposing an explanation for the apparent susceptibility to prion disease of certain individuals who have a particular version of the prion gene.

We have also analyzed antibody sequences in an attempt to explain the high affinity of antibodies for their specific antigens. (Yes, not only are antibody-antigen interactions noted for their exquisite specificity, they are also characterized by high affinity.) This work was done principally by Pam David, with the help of several other research assistants in Giselle's lab, and a paper describing the analysis is in press. Other sequence analyses are in the works. In fact, we feel that we are doing enough *in silico* experiments to justify establishing a "Virtual Laboratory of Biomolecular Structures"—something we formally did a few months ago.

And then there's our major effort (which got its start when I visited Giselle's lab at MSI in 1998) to generate "molecular guided missiles" against breast cancer. This is the

DOST-funded AMOR (Antibody and Molecular Oncology Researchers) project which involves a number of scientists from different disciplines from several institutions. The objective of AMOR is to produce antibody constructs that will specifically target breast cancer cells and kill those cells, either by direct action of the antibodies or through the delivery of cytotoxic drugs.

In recent months, Giselle and I have started looking at the other side of the antibody-antigen coin—we are now into vaccines. I think I have developed a possible strategy for designing vaccines against constantly mutating pathogens—the influenza virus, the cold virus, the AIDS virus, and the malarial parasite are examples of such pathogens. At the moment, we are gearing up to see if the strategy can succeed against influenza. If the strategy works, then we have a means to combat not only naturally mutating pathogens, but also intentionally mutated ones.

There are lots of molecular studies that I can do in the Philippines. Very importantly, I am doing them with Filipino colleagues. I'm home!

Printed in two parts on June 22 and June 29, 2006

Eduardo Mendoza

An Odyssey of Learning

When people ask me why I moved from a management position at Microsoft to scientific research three and a half years ago, I like to reply by joking that my strategy to stay young is simply "radically changing my profession every twenty years or so." Indeed, from the vantage point of becoming senior citizen card-eligible this year, I can describe my doings till now in four phases. Childhood in Malabon was followed by a second phase of fascination with "pure" mathematics, which started in high school at Ateneo (1959) and led to graduate studies and teaching in Germany (at the universities in Bonn and Wuppertal). Growing interest in computer networks led to Phase 3 in the Information and Communication Technology (ICT) industry (1980-2002), where my professional activities evolved from software development to technology management, culminating in the leadership at Microsoft of a consulting services organization with over a thousand employees in thirty-six countries. I began the fourth phase in October 2002 with a move back to academe, following my interest in studying complex systems, particularly in the life sciences, and a long-standing desire to contribute to science and technology education in the Philippines. Nowadays, I do research and teach at the Physics Department of the Ludwig-Maximilians-University (LMU) in Munich, and the Department of Mathematics and the Department of Computer Science at UP Diliman.

Learning in Malabon and Loyola Heights

If asked about things I like to remember in my childhood, I would be tempted to wax nostalgic about the lovely fishing town in the '50s and early '60s: its quiet fishponds sparkling in the sunlight, the open fields over which giant kites ("female" *gurions* and "male" *saranggolas*) engaged in spectacular "dogfights" on weekend afternoons and the lush mango trees my father (an engineer with a passion for nature and a "green thumb") grew in our lot in Maysilo. But there are still many lessons to be learned from the social and economic development since—in particular, the growing poverty and environmental degradation—the odyssey of my hometown is one too complex to discuss here.

Even now, it is hard for me to comprehensively describe all the things I learned from my parents, Anastacia (1906-2000, a math teacher and businesswoman) and Domingo (1904-1998, a practitioner and professor of mechanical engineering). They were "born educators" because they lived what they preached. It was very easy for me to understand

Physics Department and the Center for NanoScience, Ludwig-Maximilians-University in Munich, Germany and Department of Mathematics, University of the Philippines, Diliman, Philippines.
Eduardo.Mendoza@physik.lmu.de or mendoza@lmu.de

the importance of discipline and hard work for achievement because I saw examples of it daily. The professional careers of my six elder siblings and the enduring closeness of our family ties testify to the supportive environment they created. I was very happy that before I left for Germany, they were chosen UP Parents of the Year in 1964. My father always impressed me with his broad interests and his "lifelong learning" attitude. My mother was tremendously resourceful and had strong business acumen. Despite their caring attitude, they also knew when "to let go" and let their children try out new things on their own.

My first "big step out" was the transfer in Grade 4 to Ateneo de Manila in 1956: I initially attended the local school, St. James Academy (run by Maryknoll Sisters) like my elder siblings. But I perceived a strong favoritism toward female pupils among the teachers and when I complained to my mother about this, she challenged me to try to get admitted to Ateneo. There I learned—as a less "well-to-do" pupil—to earn some recognition from my classmates through academic excellence. But beyond that, I really embraced the Jesuit ideal of liberal education, enjoying the substantial library resources and availing myself of the diverse extracurricular activities offered. This learning environment led to my acceleration from Grade 6 straight to high school and four years later my becoming the Ateneo High School Class 1963 valedictorian.

Learning Multidisciplinary Mathematics in Germany

My specific interest in mathematics emerged late in high school: it was my math teacher in senior year, Fr. William Campbell, who awakened my interest in a number of open problems in number theory, including (of course) Fermat's Conjecture. In my first two years of college, driven by my interest to learn more about these fascinating questions, I took fifteen units of advanced math courses. One subject I really liked was Topology, which is coined from the Greek words *topos*, place and *logos*, study and is a branch of mathematics concerned with spatial properties preserved under continuous deformation (stretching without tearing or gluing). Years later, but really without any conscious design or strategy on my part, my PhD research would deal with a problem related both to Fermat Conjecture and a famous problem in three-dimensional topology, the Poincare Conjecture.

After sophomore year, I decided to go to Germany. There were two main reasons for this move: at the time, the few good Filipino mathematicians were abroad, so I knew I had to continue my studies elsewhere sooner or later. Secondly, I wanted to try out a new educational system, and the German *Freiheit der akademischen Forschung und Lehre* (which allowed students lots of course choices, self-pacing of studies, and even voluntary class attendance), seemed to promise more opportunities for individual development. But back in 1965, maybe it was mainly an eighteen-year-old looking for independence and adventure.

The move to Germany was a unique and, in many ways, defining learning experience for me. Learning to stand on one's own in a new social environment—especially coming from the usual caring Filipino family context—was a big challenge. But it enriched my own development in such a way, that when my older son Felix told us eight years ago that he wanted to do something similar—go to college in the US—my wife and I immediately consented, with only one condition on my part—it had to be a good college. (In the meantime, he has graduated *magna cum laude* with a double major in Physics and Mathematics from the University of Pennsylvania and is currently pursuing PhD studies in Physics as a Research Fellow of the California NanoSystems Institute at University of California Santa Barbara). Independence for me almost meant financial independence: during my first two years in Germany, I had to depend on my parents' support. But I was determined to get a scholarship—as I had at Ateneo. Due to my outstanding *Vordiplom* (BS Math equivalent) at Heidelberg University in 1968, I got one from the *DeutscherAkademischer Austauschdienst (DAAD)*. Since then, I have been able to maintain my financial independence.

The transfer to Bonn University after the *Vordiplom* was the highlight in my learning mathematics. The *Sonderforschungsbereich Theoretische Mathematik* (a Center of Excellence of the German Science Foundation DFG) had just been established: this institution enabled a steady stream of world-class visitors of different disciplines, and this opportunity of interaction began to attract the best students to Bonn. The Thursday afternoon *Oberseminar* (Research Seminar) conducted jointly by G. Harder (number theory), F. Hirzebruch (topology), W. Klingenberg (differential geometry) and J. Tits (algebra), all recognized leaders in their fields, uniquely enabled a creative, challenging atmosphere for the different mathematical disciplines to interact. The big problems could be tackled from different viewpoints, leading to new ideas often deepened in discussions during the ensuing tea (and coffee) hour. Another unique component of Bonn's mathematical research was the annual *Mathematische Arbeitstagung* (Mathematical Workshop), an informal week-long gathering of the best mathematicians in the world, where only the opening talk was fixed beforehand, and the remaining talks volunteered in several plenary assemblies during the week. A number of important results like Faltings' proof of the Mordell conjecture were first "announced" at such sessions. The Max Planck Institute of Mathematics in Bonn was established in 1981 as the successor of the *Sonderforschungsbereich Theoretische Mathematik*. It now ranks together with the Schools of Mathematics at Princeton's Institute of Advanced Study and at the *Institut des Hautes Etudes Scientifiques* (in the Paris suburb of Bures-sur-Yvette) as one of the leading centers of mathematical research worldwide.

In this community, I came to understand that, more often than not, really good mathematics emerges from problems whose solution demands contributions from diverse mathematical disciplines ("multidisciplinary mathematics"). I was lucky in my PhD research under G. Harder to experience this firsthand: I began with an algebraic problem, motivated by questions in number theory. An important step to the solution was viewing the problem from a three-dimensional topological and geometrical perspective. Finally, I identified an appropriate combinatorial structure (a "cell complex") in that three-dimensional space. Providing a solution involved five different mathematical fields! This structure is now often called the "Mendoza complex" (a term introduced by K. Vogtmann of Cornell University in 1985) in the mathematical literature and, according to a recent review by J. Schwermer of the University of Vienna, "opened the way for quite far-reaching investigations regarding the cohomology of Bianchi groups and related questions in the theory of automorphic forms for these groups." It is surprising (but also gratifying) that even after over twenty-five years, my thesis is still being cited (in nine new publications in the period 2004-2006).

The ability to perform multidisciplinary mathematics is often an indication of the maturity level of a country's mathematical endeavor. But establishing a community capable of doing this takes time, as the Bonn experience shows. In the Philippines, after over thirty years of focus on combinatorial mathematics, I am convinced it is time to try to move on toward more "multidisciplinary mathematics" research and I look forward to mentoring interested Filipinos, particularly because even this kind of "pure" mathematics is becoming relevant to applications in biology and medicine. However, long before the insight of this research's relevance to "real-world" problems, a growing interest in applying mathematics in general propelled me in 1980 into the global ICT industry. Part 2 will describe this twenty-two-year sojourn from mathematics to software development, technology management, knowledge management, and back to science.

Learning to Manage Technology and Knowledge in Large Organizations

My move to the ICT industry resulted, to a large extent, from more than a decade of struggling with questions of the social relevance of my scientific research. While a part of the lively mathematical community in Bonn, I was also engaged by basic social issues raised by the sociopolitical movements in Europe (the student revolts of the late '60s) and in Asia (the Vietnam War, martial law in the Philippines). Through lots of discussions, reading, and even attending lectures in the social sciences and economics, I learned about the importance of innovative technology for economic and social progress and the often-neglected process of turning scientific knowledge into such innovations. While an assistant professor at Wuppertal University, I became interested in the theory of error-correcting codes and started collaborating with communications engineers. Through that, I began to learn about computer networks and the important underlying software and recognized the field as a promising bridge for me from science to technology. I soon figured out that the best opportunities in this field were outside the universities, so around March 1980, I started applying to different ICT companies and planning my "big step out" of academe.

I did not take the job that paid best (though the 50 percent higher salary would have alleviated family budget worries: my wife Rosemarie had stopped teaching when we moved with our two kids from Bonn to Munich in 1981), but took that where I felt I would learn most—which was as a systems software developer at Siemens in Munich. The first one-and-a-half years in product development were quite interesting for me, as I learned professional software development. I was with Siemens from October 1980 to December 1983. Then I decided to move on because after mastering the initial "learning curve," I found the business processes, especially the decision-making, not sufficiently customer-driven, inflexible, slow, and cumbersome. (Siemens has learned from these mistakes and since then transformed itself into a more agile organization). I also had the chance from October 1982 onwards to be part of the thirty-person team tasked to develop the new Bildschirmtext Externer-Rechner Software, slated to debut at the Internationale Funkaustellung in September 1983. This effort was organized more like a project—tight schedule due to the fixed delivery date, cooperation across functional groups and departments, close contact with pilot customers, integration and management of external resources. We succeeded in delivering the software and I ended up being the project leader for a third of the team—and getting a special bonus—but the biggest benefit for me was that I knew that that kind of work would suit me best. That was why I moved to SCS (Scientific Control Systems), a subsidiary of British Petroleum, and at that time the leading information technology consulting group in Europe in January 1984.

My roles in consulting evolved quickly: software developer, project team lead, project manager (of several projects), department head, division director, director of a multinational consulting organization. I had to master these tasks without any formal business education, except for short workshops and a weeklong Executive Education course at Harvard Business School in 1997. I attribute my successes to a willingness to take some risks, learn a lot (particularly from mistakes) and the insight that consulting, management, and education are essentially all about ENABLING PEOPLE to achieve their goals—be they clients, colleagues or students. The task of the consultant, manager, and educator really is to help create the environment—the right mix of learning and doing—for their "customer's" success. In my ICT career, I always tried to choose challenges in INNOVATIVE parts of organizations LEADING in their fields—a balance meant to satisfy my interest in learning new things as well as my goal of maximizing impact. At SCS, we pioneered in the Office Automation Group the use of personal computers in large enterprises in the mid-80s. I joined the leading German systems house Softlab (now part of the BMW Group) in 1986 to help develop heterogeneous, enterprise-wide solutions, integrating PC, UNIX, and mainframe systems for large companies. We innovated by using portable software and emerging standards as much as possible. At

Microsoft, by building an enterprise-oriented consulting services "from scratch" (there were less than eighty consultants worldwide when I joined in July 1991 compared to several thousands nowadays), we contributed to its transformation from a "desktop software" company to a provider of enterprise solutions.

About ten years ago, when *Kamalayan*, a Filipino community magazine in Munich, asked me about work at Microsoft, I wrote: "I like working for the leading software company in the world, hence being able to influence technology and market developments. Microsoft has long-term vision and a unique dynamic culture combining "personal ownership," teamwork, and "zealousness" to learn and continuously improve. The drawbacks? Commitment and hard work are part of the Microsoft culture and occasionally I have less time for my family and friends than I would have liked. Today, I would add "competitive focus" as a further key characteristic, to emphasize the market-driven and customer-oriented characteristics of this organization. I remember the following statement of Bill Gates at a managers' meeting: "Our current products will be obsolete in three years. We must ensure that we are the ones to make them obsolete." Comprehensive analyses of key competitors in the different markets kick off every December the planning for the following fiscal year (which begins in July). This top-down analysis is followed by intensive weeks of subsidiary visits by the Executive Team (the legendary mid-year reviews), a key element of which is a region-by-region bottom-up analysis of competition. The results form one of the key inputs for the strategy work and business planning in the following months. Microsoft exemplifies in many ways what management scientists call a "learning organization": this synchrony of personal and organizational "philosophies" certainly contributed to my successes there. Personally, the period June to September 1997 is particularly memorable: it began with my older son Felix's secondary school graduation as one of the Top 3 of his class, continued with my receiving the Microsoft President's Award from Steve Ballmer at the Microsoft Global Summit in Orlando, Florida in July, and ended in September with the Bavarian team of my younger son Philipp winning the German Youth Team Tennis Championships.

Learning to Cultivate Interdisciplinary Multinational Science

People are often surprised when I tell them that from the very beginning, my plans were to stay only ten years at Microsoft. (Mathematically, it's trivial: two-and-a-half years at SCS, five years at Softlab, ?? at Microsoft). Kidding aside, my hope was that by 2001, my sons would be finishing their studies and the family would have a solid financial basis, so that I would have time and energy to engage in activities contributing to science and technology education in the Philippines. Throughout my career in the ICT industry, I was able to come at least annually to Manila, not only to visit my family but also to give talks at various universities and conferences. I missed my 1991 plan by about

10 percent (not bad for a ten-year plan), but the extra year was valuable in identifying the focus of my future work.

My interest in the theory of complex systems dates back to the mid-'90s and was nurtured by attempts to apply some of its techniques to the complex, dynamic consulting organization I was managing. These attempts had mixed success, but books like *The Complexity Advantage* by Kelly and Allison, which documented Citicorp's forays into similar activities, later vindicated the notion that "the Science of Complexity can help business achieve peak performance". (quote from the book cover). To deepen my understanding of complexity theory, I attended in July 2001 a conference "Dynamical Networks in Complex Systems" in Kiel, Germany. This conference opened entirely new perspectives for me: the talks demonstrated the power of the network paradigm in understanding complexity and pointed me to important applications in biology and medicine. Participation in a further conference in November 2001 at the IHES near Paris, where I also met outstanding mathematicians like Mikhael Gromov and Christoph Soulé who were working on biological problems, convinced me that modeling biological networks was a worthwhile scientific endeavor. By year's end, I started planning the transition back to science and technology education, both in Manila and Munich (where my family is based). It really seems that "luck favors those who are prepared": in April 2002, I got to know Joachim Rädler, an Experimental Biophysics professor at LMU interested in collaborating with modelers and willing to host me as a Visiting Scientist in his laboratory, and a month later, Ric del Rosario (then chairman of the UP Diliman Mathematics Department), who encouraged me to become an Adjunct Professor there. Even Microsoft contributed—by sponsoring my participation in several conferences (in Manila, Boston, and Paris)—to facilitating the transition to the emerging, broadly interdisciplinary field of Systems Biology. My essay "Mathematics of Cells" (*Star Science*, April 7, 2005) described the opportunities for Filipinos in this field as well as initial joint projects between computational scientists here and experimental groups in Germany, which exemplify new forms of multinational Internet-based collaboration.

Lessons Learned and To Be Learned

The results of the first three years of research and research management activities have more than exceeded my expectations—thanks to numerous talented Filipino students and researchers who have recognized and started to pursue the opportunities in computational systems biology. I am confident that we can report on many success stories and a few lessons learned in Filipino forays into this area of world-class science in the near future.

What I have learned so far is to value scientific knowledge of all kinds—I inadvertently experience the "unreasonable effectiveness" (to paraphrase the physicist

Eugene Wigner) in combining different pieces of science and applying the result to diverse problems. Not only the "pure" mathematics of my doctoral research is becoming useful for modeling biological membranes, even the consulting approach I learned in my ICT career is applicable to forging effective experimenter-modeler collaboration. But the will to turn science into innovations must be there. For this, the best motto I've found so far stems from SCS, the first consulting company I joined in 1984: *Das Machbare denken, das Denkbare machen. (Explore all possibilities, execute the best possible.)*

I look forward to many more lessons to be learned.

Printed in two parts on July 6 and July 13, 2006

Gisela Padilla-Concepcion

An Interview with Dr. Baldomero Olivera, Harvard Foundation's 2007 Scientist of the Year

Dr. Baldomero "Toto" M. Olivera, Distinguished Professor of Biology at the University of Utah, was recently selected as the 2007 Scientist of the Year by Harvard Foundation of Harvard University for his "outstanding achievements and contributions to American science." Dr. Olivera has distinguished himself as one of the world's leaders in marine drug discovery and neuropharmacology. His work in the last thirty years on peptides produced by venomous *Conus* snails has led to the elucidation of key molecular mechanisms that underlie nervous system function. Studies on conotoxins led to the development of the first commercially available marine drug in the world known as Prialt or Ziconitide, a natural peptide from the venomous fish-hunting snail *Conus magus*. Prialt has been found to be more effective than morphine against chronic pain. Several *Conus* venom peptides are in preclinical and clinical trials for pain, epilepsy, and myocardial infarction, and many conotoxins are widely used in neuropharmacology research.

In my interview with Dr. Olivera, he talks about his research on marine biodiversity and shares his views about science in the Philippines.

How can your studies on marine snails contribute to human health and welfare?

The studies on marine snails have revealed a unique source for drugs. A future problem for human health is that the so-called "drug pipeline" has dried up, meaning that drug companies are finding it harder and harder to come up with novel, effective medicines. As the number of older people in the world increases and the discomforts and illnesses of old age become more common (conditions like arthritis, Parkinson's disease, Alzheimer's), there becomes an ever greater need for new drugs. The studies on marine snails have not only directly provided a source for drugs, but perhaps more importantly, have also revealed how to more efficiently discover new leads for drugs not just from the snails, but from all animals.

The Marine Science Institute, University of the Philippines, Diliman, Quezon City, 1101, Philippines.
gpconcepcion@yahoo.com

Why should we explore the marine biodiversity in the Philippines?

The source of these new drugs is animal biodiversity. The Philippines is in the "golden triangle" of marine biodiversity, where biodiversity in the ocean is greater than anywhere else on the planet. I believe that as Filipinos, we need to understand and appreciate the unique marine biodiversity found in our own homeland; this is of value to society, quite apart from any potential practical applications such as being a source of drug discovery.

How can your studies help the Philippines, including our students, researchers, and rural communities?

I hope that the science we have done can be used to help students become really interested in developing a "scientific culture." I don't think that just because a community is rural, it should preclude young students from being exposed to science as a philosophy, and to learn how to deal with local problems and to explore their own economic improvement by using science. All too often, the lack of a scientific culture leads to superstition, with people feeling powerless to effectively address problems that arise. A scientific culture leads to self-empowerment, a sense that problems can be solved if one is rational and thinks things through.

How should we educate our youth and Philippine society in general about the benefits of science to society?

I think that the key to education about the benefits of science is to do it as early as possible. In the very early grades, students are curious and open; this is the best time to introduce science to them.

What does it take to develop a high-value drug like Prialt from marine sources?

Unfortunately, these days in the developed world, it takes an enormous effort and a huge amount of money to develop a new drug (over $500 million). However, it seems to me that it should be possible to do it much more cheaply in the developing world, and yet meet high scientific and ethical standards. I believe that if creative people think about this problem, we should be able to get drugs for Asia that don't necessarily have to go through the very expensive process required in the United States and Europe.

Is there a way for the Philippines to participate in and benefit materially from marine drug discovery and development?

Again, I think that creativity is required so that the Philippines can participate and benefit materially from marine drug discovery and development. This should be done by trying to get the best minds talking together: businessmen, entrepreneurs, economists, finance people, and experts from the pharmaceutical and biotech industries, and on intellectual property and regulatory matters.

What can the Philippine government do to promote and support research and product development in the area of marine drug discovery?

I believe that there are multiple initiatives that the Philippine government can do to promote and support marine drug discovery. A focused effort to improve the research infrastructure is a clear requirement. The existence of academic programs to encourage research at the leading universities that have research potential is a second obvious initiative. A third is flexible funding to allow researchers and academics who are involved in research programs to connect with scientific laboratories in their field; this does not necessarily mean long, extended fellowships, but sufficient flexibility to allow the coordination of research efforts in the Philippines with complementary scientific programs all over the world. I do not believe that it would take that much to make our research groups part of the international mainstream; one key is to figure out how to reward creativity and productivity, and to decrease rigid bureaucracy.

How can we harness the expertise of foreign-based Filipino scientists to contribute to our scientific and economic development?

I think that there are more and more foreign-based Filipino scientists in the autumn of their careers who want to help out whenever they can. Indeed, many of these scientists are willing to spend their own money for this effort. At the minimum, such scientists should be a resource for connecting local labs to the international scientific community—thus, if a local scientist happens to have a specific technical problem, an e-mail to a group of appropriate foreign-based Filipino scientists could help quickly find an international expert who could address the problem.

What are the major problems and obstacles that you see for the development of S&T in the Philippines? How can they be overcome?

One major problem is that funding for science is still inadequate, and the research infrastructure has to be improved. A second problem is the bureaucracy at all levels. In addition, the seniority system makes it difficult for young people with the best ideas to explore their ideas effectively. I think one way to address these problems is to have the most creative scientists get together, to have a pot of research funds for providing flexible funding for the most deserving groups (and if possible, preferentially to scientists who are young).

What are the priority R&D areas that the Philippine government should support in the medium and long term?

I believe that a nascent biotech industry that provides products initially aimed at ASEAN countries (and potentially the rest of Asia) would be a great initial goal.

What are some prospects and areas of competitive advantage that you see for Philippine science?

My hope is that our generation of scientists will be able to initiate a biotechnology initiative based on discoveries from the biodiversity of the Philippines. In my own research, the obvious application has been biomedical. However, I believe that with a little imagination, we should be able to explore all kinds of other applications. Some of the most promising applications could even be based on pests that bedevil us. For example, the traditional practice of kaingin is really no longer defensible given the density of the population of our islands, and its threatened ecosystems. In theory, however, we can convert all of the dry plant material during the dry season (that is traditionally burnt down) into a source of energy—one key to this may be the termites that presently eat up our homes. They have the ability to convert dried wood into sugar, and it is relatively simple to convert sugar to ethanol. Working on the enzymes from termite guts that accomplish this is the type of biotech research that takes advantage of the biological material which surrounds us. Such termite enzymes could be the basis for converting useless dried plant material into biofuel.

Based on your life experience, what qualities and values would you like to instill in our youth that would lead them to follow in your footsteps in science?

I really think that a sense of curiosity and a love of the natural world are important values. In an age when young kids are bombarded with media of all types and have instant access to the Internet, this has become increasingly difficult. It is harder to appreciate nature if you have been staring at videos for hours.

Can you recount briefly highlights and breakthroughs in your life that you think contributed to your success as a scientist?

I can think of three incidents in my life that, in retrospect, were critical. The first was that my parents moved out of the ancestral home in San Juan to a new house where we were isolated, with no neighbors at all. Since I had no brothers and sisters, I really had an enormous amount of free time. I think having a lot of solitude early in one's life is an enormous advantage. It led to my fascination with all the creatures that were wandering around in the garden, and also gave me time to read a lot of books.

Another event was that I spent second grade in the San Francisco, California public school system. There was a teacher named Miss Uhler who had the second grade kids do experiments. I remember that it was a very simple solubility experiment, but for me it was a discovery. I went around testing everything I could get hold of to see if it were soluble in water or not. This sense that this was something you could determine yourself, something that you didn't know before, is one reason why I'm a scientist today.

The third and most obvious scientific influence in my life was a dedicated high school teacher: I was uniquely lucky to have for my first teacher in chemistry and zoology, the late Dolly Hernandez [later the head of ISMED (Institute for Science and Mathematics Education Department) in UP Diliman]. There's nothing more important than an inspiring teacher. The Philippines is filled with dedicated teachers, who really labor under conditions where they get little appreciation and almost no monetary reward. Representing all of the students who ultimately benefited greatly from these underappreciated and dedicated efforts, I'd like to thank them. I tried to thank Dolly Hernandez in particular by naming a shell after her, Turris dollyae—we are analyzing this snail to see if we can get promising pharmacologically active compounds from it.

You have been contributing to Philippine science significantly in the last thirty years in various ways. How can we accelerate scientific progress further in the Philippines?

I hope that we can work in a much more interdisciplinary mode, which is necessary these days for rapid progress, and in particular, for rapid translation into the applications directly useful to society.

How can the Philippines catch up in the advancement of science with its Southeast Asian neighbors and other countries such as India and China? Should we follow their route or should we explore other ways?

I believe that we should examine and analyze what worked for other Southeast Asian neighbors, and countries such as Korea, India, and China. There is no point in re-inventing the wheel; the more we know about how others succeeded, the more efficient it will be to bring progress in science to our own country. However, I think we should also recognize that as Filipinos, we have our own culture, our own particular idiosyncrasies. Therefore it will be necessary to take advantage of the Filipino character and to understand our culture and our character thoroughly enough so we can adjust our plan to develop science in a manner consistent with our strengths.

March 29, 2007

Amador Muriel

Waking Up to a New Law

I once woke up to discover that the Philippines was under a new law—it was called Martial Law—in 1972. Luckily, I was in New York and insulated from any imagined repercussion. With that law, I simply protested.

Many years later, in 2004, in Geneva, I woke up to a new law, which for lack of any name I will simply call "A Scaling Law." It happened this way. I was looking at the data observed by my colleague Luke Hinkle. The data simply described the flow of argon gas through a capillary tube, of the type used by chemists to transfer liquids from one beaker to another. I think it is called a pipette.

The gas was impelled to flow through the tube by pressure from a tiny tank. The gas exited out to the atmosphere. However, the pressure was accurately controlled to be at the critical pressure when the gas flowed from laminar or regular or smooth to turbulent. With complete control of the critical pressure, the gas could not make up its mind whether it was laminar or turbulent. When the flow was laminar, it flowed more. When it was turbulent, it flowed less. Not being able to make up its mind, the gas periodically became laminar or turbulent, a flow which I called oscillatory, but which Luke called vacillation. His description was actually more appropriate if the gas had a mind. This intermittent behavior was known to exist in some random fashion, but we were first to show it could be periodic, not intermittent. Furthermore, the transition to turbulence was always accompanied by sound, that was also known, except we were again first to discover that each gas had its characteristic sound. For argon, it was a frequency of 210 Hertz, or 210 cycles per second. Each gas had its own characteristic sound.

With the data, I decided to build a theory, using elementary quantum mechanics. The result was very simple, and could be explained using a simple example.

Instead of our carefully controlled apparatus, let us use a balloon at high pressure. If we open the mouth of the balloon, argon gas, say, will rush out, surely in turbulent fashion. If the flow rate *fo* (say gallons per minute) at pressure *Po* is measured, what would be the new flow rate *f* at a new higher pressure *P?* My first theoretical calculation, using quantum mechanics—never mind what that is—was simple: $f / fo = (P/ Po)^{1/3}$. Then I went back from CERN (*Centre Europeen de Recherche Nucleare*) to my studio where I met my guest Sergei to eat dinner.

Next day, Saturday, at 5 a.m., I realized that I had the data to test the equation. I took a cold shower—our boiler gave up two weeks earlier—took the thirty-five-minute

Formerly with the European Organization for Nuclear Research (CERN),
Amador.Muriel@cern.ch or amador_muriel@msn.com

bus trip to CERN, and worked all day to see if the data fitted. It did, and at 5 p.m., I met again with Sergei, who was fast asleep when I left twelve hours earlier. By then he had already jumped into the lake, as he missed a cold lake in Siberia.

He asked me, "Where have you been? I looked for you in your living room, and could not find you." The living room, I may explain, is Place Bourg de Four, the square next to my tiny studio, which I call that because that's where I meet everybody. My studio has a simple bed that falls out of the wall; there is no place to entertain.

I went to the kitchen whiteboard (Rule: every dwelling I have has a whiteboard) and wrote the law, and I exclaimed that we have discovered a new scaling law. He said, if true, it was significant, and I said that it had to be true because it was so simple and cute. Furthermore, the discovery of any scaling law for fully developed turbulence is a minor cause of celebration in research circles. To celebrate, I offered to jump into the lake with Sergei.

When he went back to Russia, he tested the exponent, which was not 1/3, but varied from one gas to another. My first version of the scaling law was wrong, but we did discover more accurate scaling laws where instead of the exponent 1/3, or index 1/3, we discovered mass-dependent indices. And the law could be written as $f / fo = (P/ Po)^n$ where n is a function of mass. Theoretically, I must now challenge students and myself to derive that index. But already, engineering tables could be prepared for indices for various pure gases and mixtures, my gift to the engineering world that I entered as a freshman in the UP College of Engineering before I was seduced by physics. To vex your physics teacher whom you never liked anyway, ask the teacher what the index should be for smooth laminar flow—it is ___.

And that's how we discovered an empirical law for fully developed turbulence. Except that I had to be wrong first and thoroughly enamored with simplicity and beauty.

June 23, 2005

Amador Muriel

Tracking Einstein's Ghost

Every theoretician's fantasy is to approximate the life of Albert Einstein. Some fancy becoming a patent clerk somewhere; he was that when he published his *annus mirabilis* papers in 1905. Others simply marvel at his originality. I am one of the latter. But 2005 is the Einstein Centennial and I indulged myself.

The first Einstein celebration I attended was at the New England Section Meeting of the American Physical Society where I read a paper on the role of molecules in turbulence. I thought people would find it curious that after Einstein's paper on Brownian Motion, which settled the debate between Ernst Mach and Ludwig Boltzmann in favor of the existence of molecules, my paper would offer some historic significance, as I now advocate that turbulence could not be explained without the use of molecules. It is a very unpopular idea—nearly one hundred percent of engineers and turbulence researchers do not think they need quantum mechanics to explain turbulence. I applied Einstein's ideas two ways: first, with the molecules of Brownian Motion, and second, with photons of the Photoelectric Effect. My attendance of the MIT meeting simply yielded dinner that is best forgotten.

The second Einstein celebration was sponsored by the European Physical Society in Bern, held from July 11 to 15. Bern went all out for Einstein, on placards, on souvenirs (I bought Einstein T-shirts so my grandchildren can sleep in them) and an Einstein exhibit at the Bern Historical Museum. At the museum, the Swiss had no hesitation displaying pictures of Einstein's women. Einstein's day bed was in display; yes, I know that it is really good to lie down and fall asleep if ideas do not come so quickly. On the second day of the meeting, my poster was scheduled. Why my talk became a poster escaped me, then I read that to allow young physicists to present new ideas, poster sessions were scheduled. How a sixty-five-year-old Filipino physicist could become a youngster escaped me; until I realized that I am not a member of the European Physical Society, and that I must be a young physicist who could not afford the membership fee. I have this thing about posters; they remind me of women on display in some streets of Amsterdam between the railway and the university—my mentor Max Dresden told me that sometimes students did not make it to class from the railway station. Anyway, I simply displayed copies of our latest papers in lieu of my prepared PowerPoint presentation.

So did my molecules have an impact? Maybe, since I was invited to the University of Oldenburg in Germany to give a talk and explore a visiting position. The turbulence group wanted to repeat all our experiments to date.

Formerly with the European Organization for Nuclear Research (CERN),
Amador.Muriel@cern.ch or amador_muriel@msn.com

Apart from the new scientific contacts, Gloria (my spouse) and I enjoyed a most interesting dinner with an Australian far-infrared astronomer, two Heidelberg nuclear physicists (obligatory Rizal conversation), and surprise, a connection with Mileva Einstein, the Serbian wife of Einstein who got all of Einstein's Nobel prize, by a divorce agreement that the two signed. Several years ago, I met a Serbian woman by the name of Milena, whose name triggered the following conversation:

"Hi, I am Milena from Serbia."

In jest, I asked, "Did you know Mileva?

"Mileva who?"

"Mileva Einstein."

"Of course, we grew up in the same street, at different times. We Serbians believe that Mileva is not fully credited with her assistance to Albert Einstein."

"We will have some time to discuss that. But do you know what happened to Lieserl, the love child?"

"No, we don't know about that."

That conversation inspired a plan to visit Novi Sad to speak with Milena's father, a student of Mileva's life, and debate the role of Mileva in Einstein's papers. As it turned out, my mentor was a member of an American Physical Society committee to investigate the role of Mileva; he looked at all the academic records of Mileva at the Zurich Polytechnique, to convince himself that Mileva was not involved in any of Einstein's works.

This time, at the dinner table were three Serbian professors of physics who were surprised about my readings on Mileva and Albert. They then invited me to lecture at the University of Belgrade and finally visit Novi Sad, which was mercifully spared from the American bombing. The Serbians explained to me that when a child is put up for adoption, as Lieserl might have been, all records of pedigree are sometimes erased, which could explain why there is no more record of this would-be centenarian woman, who would certainly share part of the Einstein estate. We have arranged to visit Novi Sad. Something tells me that my world line will intersect with Einstein's ghost once again.

October 6, 2005

Mafel Ysrael and Fortunato Sevilla III

Farewell to Dr. Gloria de Castro-Bernas: Exemplary Science Educator, Researcher, and Administrator

This month, the biochemistry and molecular biology community in the Philippines lost one of its most active and inspiring movers—Dr. Gloria de Castro-Bernas of the University of Santo Tomas.

Dr. Bernas was the immediate past president of the Philippine Society for Biochemistry and Molecular Biology (PSBMB) from 2003 to 2005. Her involvement in the society dated back to its early years, when the organization was known as the Philippine Biochemical Society. She participated enthusiastically in the activities of the organization which were geared toward updating and upgrading biochemical education and research in the country. She contributed significantly to the growth of the society into a prestigious and internationally recognized scientific organization.

Biochemistry occupied a high place of preference in the academic pursuits of Dr. Bernas. She obtained her undergraduate degree in Chemistry at the University of Santo Tomas, and postgraduate training in Clinical Biochemistry at the Massachusetts General Hospital in the United States. She pursued further studies in Biochemistry, completing her MSc at the University of Santo Tomas and PhD at the University of the Philippines in Manila. She carried out her doctoral dissertation at the Harvard Medical School and Boston University Medical Center.

As an educator, Dr. Bernas was based at the University of Santo Tomas where she pushed forward the teaching of biochemistry toward the modern molecular approach. Inspired and guided by her mentor Dr. Clara Lim-Sylianco, she made her students realize the relevance of the structure of biomolecules to their functions in a living system. Her students at the College of Science, the Faculty of Medicine and Surgery, and the Graduate School of the university remember the enthusiasm with which she taught them the mechanisms of biochemical reactions.

Mafel Ysrael: Department of Biochemistry, Faculty of Pharmacy, University of Santo Tomas, España, Philippines.
mcysrael@mnl.ust.edu.ph
Fortunato Sevilla III: College of Science, University of Santo Tomas, España, Philippines.
fbsevilla@mnl.ust.edu.ph

Research was also a passion of Dr. Bernas. She engaged her intellectual talent in tracing the molecular roots of cancer and in investigating the potential of Philippine plants in the management of cancer. She also conducted studies on the activity of some local medicinal plants in the growth and in the death of blood vessels, demonstrating the possibility of their application in controlling hypertension and cancer. Her research work was carried out mostly at the UST Research Center for the Natural Sciences. Her research outputs were published in a number of refereed journals and presented in national and international conferences.

As recognition of her academic achievements and leadership, Dr. Bernas was appointed chairwoman of the UST Department of Chemistry of the College of Science in 1994, and was later made the dean of the college in 1996. During her deanship, undergraduate degree programs in Computer Science, Information Technology, Information Management, and Applied Physics were introduced in the College. In 2002, she was named assistant to the rector for research and development. While holding this administrative position, Dr. Bernas formalized the establishment of research centers in a number of fields of studies, such as Movement Science, Cultural Heritage, Applied Ethics, and Contextualized Theology.

Family life was likewise a priority for Dr. Gloria de Castro-Bernas. At home, she was a dedicated and loving wife to Silverio. Together, the husband and wife team of Silverio and Gloria shaped their three children—Brice, Teresa Rowena, and Sheila Marie—to become professionals and achievers themselves. In the past few years, she assumed another role, that of being the doting grandmother to two energetic boys, Samuel Leonard and Zachary Abad.

To all her friends and colleagues, Dr. Gloria de Castro-Bernas will always be remembered as a woman of great courage and optimism. She had earned the respect and admiration of scientists, both here and abroad, for not letting her illness get in the way of her passion and drive in pushing science education and research in the Philippines. Her active, long-term participation in organizations such as the Organic Chemistry Teachers Association (OCTA), PSBMB, and Philippine-American Academy of Science and Engineering (PAASE) and her various research collaborations were a strong testimony to her commitment and support for the growth of science and technology in the country. Here are some of her friends' endearing thoughts about her:

> *"I have a great admiration for her leadership, dedication, and efforts in pushing scientific research in the Philippines. In my conversations and contacts with her, I found someone who was refreshing to talk to and who never lost optimism."*—Dr. Seville Detera-Wadleigh, PAASE

> *"I have respected Dr. Gloria Bernas since I first met her ten years ago. I wish to express my official and personal sympathy to you and UST for the loss of a great leader of science in your country."*—Prof. Takashi Okamoto, Nagoya University

"Those of us who participated in the UST-PAASE Research Symposium in June last year (before the PAASE meeting in Cebu) will remember how this courageous woman, with no regard for her illness, delivered a most interesting talk on the angiogenic activity of certain Philippine plants."—Dr. Judy Ribaya-Mercado, PAASE

June 1, 2006

Ben de Lumen

Rizal, the Scientist

Jose P. Rizal was a man of many talents and interests. For a man who lived only thirty-five years, his achievements are remarkable and numerous. The Rizal Centennial Commission listed 278 written works of Rizal including his two major novels *Noli Me Tangere* and *El Filibusterismo*.

Although Rizal is well known for his literary prowess, perhaps his accomplishments as a scientist are not well publicized. In this age where science and technology play a major role in economic development, his scientific achievements are relevant and inspiring. To talk about Rizal as a scientist, it is difficult to separate Rizal the natural scientist (one who practiced the natural sciences) from Rizal the social scientist and political reformer because he believed that knowledge should be used for enlightenment and liberation and not for oppression. In his choice of medicine as a career and during his education in Europe, he never lost sight of his goal: to serve his people and liberate them from years of oppression and injustice by the Spaniards.

Here are some highlights of Rizal's scientific accomplishments.

After five years in Europe, he went home to the Philippines in 1887. He operated on his mother's eyes to remove her cataracts—the surgery was successful and was the first of its kind ever done in the Philippines. His fame as an eye doctor spread quickly and people began coming to him for treatment from all over the Philippines and even from as far away as China. He opened a clinic, ordered equipment, charged moderate fees, and treated the poor for free.

After only six months, Rizal had to leave the country because his novel *Noli Me Tangere* had circulated and the friars were out to get him. He went back to Europe via Japan and the US. Here again, Rizal made some perceptive observations of the US then. After fifteen days crossing the Pacific, their ship was quarantined in San Francisco for a week although none of the passengers were sick and health clearance had been given. The authorities cited smallpox as a risk. He noted that there were a number of Chinese immigrants, the cargo silk had been unloaded without fumigation and the custom officers were not afraid to eat aboard. Rizal discovered the real reason for the quarantine. He wrote that: "America was opposed to Chinese immigration and since it was election time, the administration appeared strict toward the Chinese to obtain the people's votes."

He took the train across the US and made a number of stops along the way. He wrote his impressions of the US: "Undoubtedly America is a great country but it still has many

Department of Nutritional Sciences and Toxicology, University of California at Berkeley, CA, USA.
nitto@nature.berkeley.edu

defects. There is no real civil liberty. In some states the Negro cannot intermarry. Because of the hatred towards the Chinese, other Asians like the Japanese, being confused with them, are likewise disliked by the ignorant Americans."

In London, he undertook a project that he had wanted to do. As a boy, one of his uncles told him about a book written in the sixteenth century by a Spaniard that gave a truthful picture about early Philippine history. All accounts he had read thus far were written by prejudiced Spaniards seeking to justify Spain's colonial rule on the ground that the natives were "child-like savages." The book *Sucesos de las Islas Pilipinas* written by Antonio de Morga and published in Mexico in 1609 was available only in a few libraries, and a copy was in the British Museum in London! His plan was simple. He would study Morga and other writers who dealt with pre-Spanish Philippine history, compare them all and publish a new edition of Morga, with notes and comments by himself. Thus the truth about the Philippines would become available to his people and the Europeans who had learned about the early Filipinos through the prejudiced eyes of the Spanish colonizers.

Going through Morga's volumes, Rizal found that the Filipino people had been historically wronged. In the coastal regions where most of the islanders lived, their arts, industries, and energy had been at a high level when the Spaniards arrived. Morga described their skills in weaving, in metal work, in agriculture, in commerce, in navigation, in government, their fine ships (better than Spain's), their busy market places. It was a civilization that Rizal and the Filipino people could be proud of. More, it cut away the basis for Spain's claim to colonial rule. Rizal wanted to give the Filipino people back their past for he believed that a people without a proper understanding of their past was a people without a future.

The last major episode of his life was spent in exile in Dapitan, in northern Mindanao, where he was sent by Spanish authorities after he returned to the Philippines in 1892. As one author wrote, it was one of the most extraordinary exiles in human history. In Dapitan, there was no water system, no school, no street lighting, no hospital, the land was fertile but farming techniques were primitive. But Rizal, with his characteristic creativity and self-discipline, tackled these problems. In his four years in Dapitan:

He established a large and well-known medical practice where his patients came from all over the Philippines and from Hong Kong and other Asian cities.

He built a hospital.

He built a small house for himself and a large one for his family and visiting friends.

He bought lands and practiced scientific farming.

He set up a water supply system based on gravity.

He set up and taught in a school for local children.

He paid for the first street lighting system.

He beautified the town plaza and a made a giant relief map of the Philippines which is still preserved today.

He obtained from Kalamba an improved type of fishing net that helped the Dapitan fishermen improve their catch. He imported farm machinery from the US for himself and local farmers.

He subscribed to the magazine *Scientific American* and ordered medicines and pharmaceuticals from the US.

He collaborated with foremost scientists from Europe at that time. With his students, he collected specimens of plants, animals and ethnographic materials from Mindanao and sent them to his colleagues in Europe.

Some of the animal specimens were rare and named after him: A new species of frog named *Rhacophorus rizali*, a new species of beetle named *Apogonia rizali*, and a new species of lizard named *Draconi rizali.*

In Dapitan, as everywhere he stayed, Rizal followed a disciplined schedule. He had a brilliant mind, but the key to his productivity was planning and self-disciplined execution. He wrote to his Austrian friend Blumentritt how he spent a typical day in Dapitan: "I get up early at 5:00, visit my fields, feed the chickens, I wake up my people and start them moving. At 7:30 we take breakfast. Afterwards I treat my poor patients who come to my land. Then I dress up and go to town to treat the people there and return at 12 noon for lunch. Afterwards I teach the boys until 4:00 and I spend the afternoon farming. Evenings are used for studying and reading."

Finally, Rizal shared with us his philosophy and thinking about education and science. Within the limits of the circumstances in Dapitan, Rizal gave his students the key elements of his educational goals: academic knowledge, industrial training, ethical instruction, and physical development. He believed that moral values were as important as knowledge itself; indeed they were the only assurance that knowledge will be used to help and enlighten, rather than oppress men.

Einstein echoed similar ideas when he addressed the students at California Institute of Technology: "Concern for man himself and his fate must always form the chief interests of all technical endeavors—in order that the creation of our mind will be a blessing and not a curse to mankind."

PS - I have used a number of books and other publications at the University of California Berkeley library for this write-up. I would be happy to share these titles with anyone who wants to do further research on Rizal, the scientist. One of the few remaining original copies of *Sucesos de las Islas de las Filipinas* by A. Morga annotated by Rizal is at the rare book collection of the UC Berkeley Bancroft library.

July 20, 2006

Benjamin
Vallejo Jr.

Georg Joseph Kamel, SJ (1661-1706): First Biodiversity Scientist in the Philippines

Camellias are one of the most sought-after ornamental plants. While not very fragrant, the blooms are spectacular. These flowers are grown by plant fanciers worldwide, especially in its native lands, China and Japan. There are many cultivars of camellias, one of which is named after the Philippines' national hero, Jose Rizal.

While many people are familiar with camellias, very few know that the plant genus is named after a Jesuit lay brother missionary who worked in the Philippines from 1688 until his death in 1706. Georg Josef Kamel arrived in the Philippines in 1688 and spent the remaining part of his life describing the taxonomy and systematics of Philippine plants and animals as part of his scientific practice of pharmacy and medicine. To understand the world in which Kamel did his work, we have to recall that during this time in history, the medical sciences were just emerging from medieval superstition and alchemy. Kamel, who trained as a *botanicus et apothecarius* (botanist and pharmacist) in the Jesuit College in Brünn, Moravia (now Bruno, the Czech Republic), was one of the first medical practitioners to use science in the treatment and management of disease.

Kamel's scientific work in the Philippines cannot be understood without reference to pharmacy. European colonists in Asia and tropical America had to contend with unfamiliar diseases for which drugs from native plant sources were needed. The Jesuit historian Murillo Velarde describes how Kamel went about his task of dispensing medicines made from Philippine plants. He experimented on new formulations and noted the right dose for his patients through experiment. His success in treating diseases and concern for the poor made him beloved by the citizens of Manila.

From his arrival in 1688, Kamel began collecting and describing Philippine plants. In 1698, he shipped specimens and botanical reports to England but these were lost at sea which Kamel regretted. His reputation as a physician and pharmacist was well known even to the British in India. Kamel published his book entitled *Herbarum Aliarumrque stirpium in insula Luzon e Philippinarum* from 1697 to 1698. Copies of this book which we know now as the *Herbarum Philippinarum* in three volumes were sent to the eminent English naturalist John Ray in 1698 through an English medical doctor in Madras. This

Department of Environmental Science, University of the Philippines, Diliman, Quezon City, 1101, Philippines.
bmvallejo@up.edu.ph

was lost together with Kamel's botanical specimens that year. The following year, he sent another volume and this was the one published as an appendix to John Ray's *Historia Plantarum* of 1704. This is how Kamel became an internationally renowned scientist.

Herbarum is Kamel's major contribution to science. Parts of the *Herbarum* were published in the *Philosophical Transactions of the Royal Society* by another eminent English naturalist, John Petiver. Kamel's work on Philippine natural history went beyond botany. He wrote scientific papers on birds, corals, fish, crustaceans, and large animals entitled *de Monstris et Monstruosis*; quadrupeds, Turbinid mollusks, *Bivalvus et Univalvus;* insects and fossils. These works are evidence of the keen interest that Kamel had in biodiversity.

Among the more significant papers Petiver published in the *Transactions* is the one entitled *De Igasur seu Nox Vomica Legitima Serapionis* subtitled, "An Account of the Virtues of Faba St Ignatii." This paper relies on Kamel's description of the Saint Ignatius bean, a plant that is the source of strychnine. Kamel noted that this poison in low doses has medical properties. When he first collected this plant, he noted its distribution—its absence in Luzon but existence in abundance in Samar. The plant is named after the Jesuit Order's founder since it was first collected in Jesuit mission territories in the Philippines. Kamel was also interested in ethnobotany. He recorded the local names of Philippine plants from Luzon and even from the Visayas and Mindanao. Linnaeus noted this important fact in his works.

Kamel was foremost a scientist. His skepticism regarding numerous myths has been recorded. In the colonial Philippines of the late 1600s, praying mantises were believed to arise from leaves and twigs became worms. Kamel said, "I do not accept this because I have observed that the sexes are distinguished and the young are born from eggs."

The world now remembers Kamel for camellias. Linnaeus renamed the plant from the original genus *Thea* to *Camellia* in his honor. History does not record whether Kamel sent a specimen of the plant to any botanist in Europe at the time. This was unlikely since Kamel never went to China and Japan. There is no record that Jesuit missionaries in these countries ever sent Kamel a specimen. Linnaeus also never received a specimen from Kamel since Kamel died a year before he was born. Linnaeus named the plant simply in honor of Kamel's excellent work in botany.

Kamel died on May 2, 1706 in the San Ignacio College in Intramuros, apparently of dysentery. He was forty-five years old. He was buried in the first San Ignacio Church (a site now occupied by the Pamantasan ng Lungsod ng Maynila). This church no longer exists and fell into ruin after the expulsion of the Jesuits from the Philippines in 1768.

What does Kamel's life teach us about how science is done? While the historian Leo A. Cullum emphasizes Kamel's religious vocation, he also writes that Kamel placed his scientific work fully within his apostolate and the Christian virtues of charity and humility. Thus he writes: "Even the science he had loved without loving himself in it, playing an obscure role of unselfish charity with the poor at his door and the English savants of the Royal Society."

However, we can also learn from Kamel's life that science is essentially a communication and collaborative activity that transcends nationality and religion. This is why he corresponded with the leading scientists of his time. It is true that he did not have sole authorship of any of his works and he was content to collaborate with other scientists. Perhaps he viewed his scientific work in the context of his Catholic missionary work. Nonetheless, his scientific skills were clearly recognized and cited. Kamel's science teaches us that proper documentation and international publication, then as now, are the key to scientific credibility and progress. Kamel also shows us perseverance as a key scientific virtue. He was not discouraged by the loss of his scientific specimens and publications. A year later he rewrote his manuscript and sent them again. He challenged the medical and natural history myths of his time by experimentation and empirical observations. Aside from his botanical work, today he is remembered as one of the earliest exponents of pharmacy as a science. It is for this reason that the Czech Republic, UNESCO, and his hometown of Bruno commemorated his 300th death anniversary last May 2.

However, in the Philippines where he did much of his science, Kamel has all but been forgotten. I hope this essay makes some amends for our neglect of his memory. Even if I may be criticized for doing so, I call Kamel the first Filipino biodiversity scientist. Yes, he was Czech, but I cannot help but recall another "alien" Jesuit in the 1950s in the Philippines, Fr. John P. Delaney, who famously asked the regents of the University of the Philippines, "How much more should a man give before he becomes one of you?" I think in Kamel's case the answer is clear. He gave so much to the Philippines and its people. This first biodiversity scientist is certainly a Filipino.

January 4, 2007

Part 5

Immunology, Infectious Diseases, and Public Health Concerns

Eduardo Padlan

Antibodies: The Good, the Bad, and the Ungainly

When we catch the flu, our body makes antibodies against the flu virus. When we get a bacterial infection, our body makes antibodies to fight the bacteria. In fact, whenever any foreign substance—an antigen—manages to enter our body, we make antibodies to destroy it before it can do us permanent harm. (But we tolerate food! Thank goodness for that.)

Antibodies are a major component of our immune system. They are proteins that are designed to be good binders to all sorts of (big) molecules. When an antigen gains entry into our body, before long our immune system will have produced antibodies specific for that antigen and able to bind to it very tightly. Further, the binding of antibodies to an antigen brings into the fray other components of our immune system—cellular as well as other molecular components—and together they bring about the neutralization, destruction, and elimination of the foreign substance. Furthermore, the cells and molecules of our immune system are constantly on patrol, making sure that our body is free of germs and other harmful agents. They are also on the lookout for our own cells that have become abnormal or have become infected by viruses or other pathogens; our immune system gets rid of those cells, too.

All the cellular components of our immune system, including the cells which will eventually produce our antibodies—the so-called B cells—originate in the bone marrow. After completing their differentiation in the bone marrow, the B cells migrate to the periphery (the blood and lymphatic system) to become the antibody-producing plasma cells. While still in the bone marrow, the B cells are screened for "self-reactivity," i.e., they are tested to see if the antibodies they produce bind to our own molecules. If they do, those self-reactive B cells do not leave the bone marrow—they die there. (You can imagine what havoc they would produce if a host of self-reactive antibodies is let loose in our body!)

There is a further safeguard against self-reactivity. During their development, the B cells are stimulated by "helper T cells" (cells that originate in the bone marrow but complete their differentiation in the thymus). In the thymus, the developing helper T cells (and other types of T cells) are likewise screened for self-reactivity and the self-reactive ones do not leave the thymus.

Formerly with the US National Institutes of Health and currently with The Marine Science Institute, University of the Philippines, Diliman, Quezon City, 1101, Philippines.
edpadlan@yahoo.com

We then have an immune system whose components are ready to react to any pathogen, or harmful foreign substance, but not to our own molecules. Since we cannot know what sort of pathogens we will encounter on any given day, we produce millions of different B cells and T cells every day. Moreover, our immune system remembers. Memory B cells and memory T cells are produced so that we are ready to vigorously fight any pathogen that tries to invade our body a second time. We have a magnificent defense system.

But read on.

Sadly, our immune system is not perfect. Sometimes, antibodies and T cells (especially the killer type) are produced that do react to our own molecules. We then have what is called an autoimmune disease.

Some individuals, predominantly women, have antibodies to DNA or other components of the nucleus, which causes an autoimmune disease called Systemic Lupus Erythematosus (SLE). The antibody-antigen complexes resulting from SLE often clog up the kidneys and other organs of the body and could cause death. In individuals suffering from myasthenia gravis, there are antibodies that bind to the acetylcholine receptor and interfere with the transmission of signals from nerve to muscle, causing muscle weakness. Some unfortunate individuals produce antibodies to their own antibodies and this causes rheumatoid arthritis. One form of diabetes is caused by the presence of killer T cells that attack the cells in the pancreas which produce insulin. These are just a few examples of autoimmune disease. There are many others, most of them caused by wayward antibodies.

And then there is allergy. Allergy is caused by IgE—a type of antibody—which is usually found on the surface of mast cells and basophils. Mast cells and basophils have granules filled with histamine and other vasoactive compounds. The binding of allergen (an antigen that causes allergy) to the IgE causes the release of the histamine and the other compounds, resulting in runny nose, sneezing, and other symptoms of allergy. Why do we have IgE when it seems that all it causes is misery (sometimes, even death in severe allergic reactions)? Well, it turns out that IgE affords protection against parasites. The world is cleaner these days and we have learned to cook our food well, so that not many of us harbor parasites anymore. But we all still produce IgE—just in case—and more and more of us are suffering from allergies.

In addition to autoimmunity and allergy, there are other disorders due to antibodies.

When protein molecules are not very stable, they could (partially) denature and aggregate, and the aggregates could clog up vital organs. Antibodies are no exception. With the millions of B cells generated every day, it is highly likely that unstable antibodies would be produced. Fortunately, if an antibody does not find an antigen to bind to, or is otherwise nonfunctional (which it would be if it is not stable), the antibody (or, more correctly, the B cell that makes it) is eliminated from the pool. However, there are diseases in which the B cells (or plasma cells) become "transformed" and do not die resulting in cancer. If the antibody, or a part of it, that a transformed cell makes is unstable and has a tendency to aggregate, we have an additional disorder.

Antibodies are good. Without them, we will not last very long. (We carry ten times more bacteria, in us and on us, than we have cells in our body, and I shudder to think how many viruses are lurking inside us—all ready to pounce as soon as they sense that our immune system has weakened.) But, as you have seen, antibodies can be bad too. And the unstable ones cause more harm than good.

September 22, 2005

Eduardo Padlan

Survival of the Fittest? Not So Fast

The phrase "survival of the fittest" was coined by Herbert Spencer, a British social philosopher, to accentuate the differences in survivability of the different social classes. Inevitably, the notion fed into the belief of the existence of more advanced individuals and cultures, and served to justify colonization (of "inferior" societies) and imperialism.

To Spencer, the concept was equivalent to what Charles Darwin was then calling "natural selection, or the preservation of favored races in the struggle for life." The phrase was later adopted and widely used by Charles Darwin himself. The phrase can be applied, not only to individuals and groups, but to cells and molecules as well.

Survival and perpetuation of one's genes are every organism's most immediate goals. There is an unending struggle for survival among the many organisms in our world and the most able to compete—the "fittest"—have the highest chances of winning. The many different organisms that can be found in our world today are the result of eons of change, competition, and selection of the fittest (in other words, evolution), and the present-day organisms represent the forms that are best adapted to the prevailing conditions.

Who are the fittest? Are they the strongest? The healthiest? The most intelligent? All of the preceding? None of the preceding? In fact, the "prevailing conditions" are what define who are the fittest (and who will survive).

In a situation where there is shortage of food, those who do not require much food—the smaller individuals—will more likely survive. In situations where there is competition for territory, the better fighters—the bigger and stronger, or those who have better weapons—will more likely win.

If two individuals—one fat, one lean—are dumped in the ocean, one will probably float while hardly moving a muscle, whereas the other may sink straight to the bottom. If instead they are dumped in lion country, the cats will more likely end up feasting on meat generously marbled with fat, instead of lean meat. (I'm obviously trying to be funny here.)

But really it all depends on the "prevailing conditions."

That is true also in the cellular and molecular world of biology. The fittest among the molecules and cells, the ones chosen by Nature to prevail and to persist, are defined by the circumstances. Let me illustrate.

Formerly with the US National Institutes of Health and currently with The Marine Science Institute, University of the Philippines, Diliman, Quezon City, 1101, Philippines.
edpadlan@yahoo.com

Recognition is the hallmark of biology. Biological molecules and cells do not associate randomly—otherwise, our world will be chaotic. Rather, molecules (and cells) interact specifically with the entities (other molecules and cells) they are meant to interact with. There is a pattern to which we conform. There is order and structure. In situations where molecules are meant to bind one another, the molecules which best fit each other and whose interaction serves the intended purpose would be expected to be favored and kept during the course of evolution.

A part of us in which recognition is very important is our immune system. The system has to be able to distinguish between beneficial and harmful (food is good, germs are bad)—we tolerate the first, we rid ourselves of the second. The system also has to be able to distinguish between our own cells (and molecules) and foreign cells (and substances)—we leave our own cells alone, we kill the invaders. How does the immune system accomplish this?

Three types of molecules are primarily responsible for immune recognition: the antibodies, the T cell receptors (TCRs) for antigen, and the Major Histocompatibility Complex (MHC) molecules. A membrane form of antibodies is initially found on the surface of B lymphocytes (cells that complete their differentiation in the bone marrow) and then a soluble form of antibodies is secreted after the B cells have gone through several cycles of development (maturation). The TCRs are found on the surface of T lymphocytes (cells that complete their differentiation in the thymus and of which two types will be mentioned here: the helper T cell and the killer T cell). MHC molecules are present on the surface of all cells, except those which no longer synthesize protein (for example, red blood cells). Two types of MHC molecules are of interest here: MHC Class I, which is recognized by killer T cells, and MHC Class II, which is recognized by helper T cells.

When they are first formed (in B cells and found in membrane form on the surface of the cells), antibodies have low affinity (strength of binding) and are not specific for anything (binding is said to be specific if it is to only one ligand). When a foreign substance (antigen) gets inside us, there may be B cells that by chance are able to bind to it, even though only weakly. Those B cells will internalize the antigen, break it up into pieces (peptides) and then display the peptides on their surface, bound to MHC Class II molecules. Helper T cells will then come along and if there is a helper T cell whose TCR can bind to the MHC:peptide complex, the T cell will then secrete molecules directed at the B cell, causing the latter to divide. When a B cell divides, its antibody genes undergo hypermutation (mutation at a very rapid rate). Once again, by chance, antibodies on the daughter B cells may bind the antigen—more strongly this time. Those daughter B cells are then "selected," because of their higher affinity, to be activated by helper T cells. The cycle is repeated until, in the end, we have B cells producing antibodies that are specific for the antigen and are able to bind to it with high affinity. The "best" antibodies then are characterized by high affinity for antigen. (B cells that bind to our own molecules are "killed" before they leave the bone marrow.)

For TCRs, the criterion for which are the "best" is quite different. TCRs bind to both the MHC molecule *and* the bound peptide. If the binding of the TCR to the MHC molecule is too strong, a helper T cell could activate a B cell that is not displaying the right peptide, or kill the cell if the T cell is of the killer type and the MHC molecule is of Class I; either of these events would be disastrous. The "best" TCRs then are characterized by only medium-strength affinity. (As in the case of B cells, T cells that bind to our own peptides are "killed" before they leave the thymus.)

Many other examples abound that demonstrate that there is no easy answer to the question of what constitutes "fitness." So if someone asks which individuals, cultures, organisms or molecules are the "fittest" and best suited for survival, a good response would be: "It depends."

September 14, 2006

Cheryl Agdaca,
Cerrone Cabanos,
Mary Ann Torio,
and Babak Nakhoda

Antibody Humanization Technology: "Reshaping" the Future of Antibody Therapeutics

Our immune system has the ability to discriminate self from non-self substances or any material that is considered "foreign." These "foreign" substances are collectively called antigens, which trigger the immune system to produce antibodies against them. An antibody binds specifically to an antigen and then sends signals to the immune system's cells to sequester and destroy the foreign entity. This is one of the ways in which antibodies are involved in defending the body from infection by foreign organisms and in fighting intrinsic diseases, such as cancer, that arise from the body's own defects.

Antibodies can be isolated from the blood by a relatively simple purification process. These purified antibodies are sometimes used for crude therapy. However, this procedure produces a highly complex mixture of antibodies that bind to thousands upon thousands of different molecules. For treating specific therapeutic needs, such a mixture of antibodies does not suffice. Instead, scientists have found the need to develop individual monoclonal antibodies, antibodies sourced from a single cell, which bind specifically to the antigen of interest.

Today, monoclonal antibodies are produced routinely for both medical and research applications. In the medical world, they are used as diagnostics, to detect cancers or infection by certain bacteria or viruses; as therapeutics, to target foreign bacteria, viruses or cancerous cells; and possibly also as vaccines, to boost the body's immune response. (The specificity of antibody binding to antigen is due to the complementarity between the structures of the binding surface of the antibody and the part of the antigen to which the antibody binds. An antibody directed against an antibody to an antigen could conceivably mimic the structure of the antigen and thus the anti-antibody could potentially serve as a vaccine in lieu of the antigen.)

Monoclonal antibodies are usually produced in mice. However, when used in human therapeutic applications, mouse monoclonals present serious obstacles. First, mouse antibodies themselves are recognized as foreign substances by the human immune system and thus they provoke an immune response termed the HAMA (Human Anti-Mouse

University of the Philippines, Los Baños, Laguna, Philippines.
cerrone@lycos.com

Antibody) reaction. The HAMA reaction decreases the efficacy of the mouse monoclonal. HAMA could also induce fairly severe symptoms, such as allergic reactions, in the recipient. Furthermore, mouse antibodies are simply not as good as human antibodies in signaling the human immune system to destroy the malignant cells, bacteria or virus of interest.

There have been two major strategies that have emerged over the past few years in order to circumvent these limitations. These are: (i) producing fully human antibodies in mice or in the laboratory, and (ii) "humanizing" mouse or other nonhuman antibodies to make them compatible with the human immune system.

Humanization and other attempts to "engineer" protein molecules are now routinely done. This is because molecular biologists have mastered the trick of introducing foreign genes into practically any organism (there are even attempts to introduce genes into human patients to correct genetic defects). This allows the production of foreign proteins in more convenient hosts. Indeed, human proteins are now being produced in large amounts in bacteria, in transgenic (bearing foreign genes) goats, cattle, and other animals, and even plants (in tobacco, for example). Before a gene is put into a host organism, the gene could be modified so that the protein product would have altered properties, e.g., greater stability, improved function, etc. The protein product could be made shorter by making the gene smaller, or two genes could be spliced together to produce two proteins in tandem. A single amino acid could be changed, or a whole segment of the protein could be replaced. Parts of one protein could be grafted or transplanted onto another. In other words, the structure of the protein product could be manipulated at will. Sometimes, not just a single gene is introduced into a host organism, but an entire gene locus, e.g., the family of genes that produces the antibody repertoire of an animal.

In fact, fully human antibodies can now be produced in transgenic mice. These mice are engineered to synthesize human antibodies instead of murine antibodies when they are injected with a stimulatory molecule or antigen. Such fully human antibodies can also be produced through bacterium-infecting viruses, known as bacteriophages. Human antibodies produced either way are potentially attractive since they contain no mouse or otherwise foreign protein component. The drawback of these methods, however, lies in slow production time and greater expense. Moreover, there has been no convincing demonstration that they are in any way superior to humanized antibodies

On the other hand, the humanization of the murine antibodies is a more promising tool to overcome such limitations—by making the nonhuman antibody compatible with the human immune system while still retaining the same binding strength (i.e., affinity) and specificity of the original mouse monoclonal.

The first attempt toward humanizing murine monoclonals was to graft a portion of the murine antibody, known as the Fv fragment that is responsible for antigen binding and recognition, to a human antibody scaffold. This lowers the murine content of the monoclonal antibody and reduces the previously described HAMA complications.

Subsequently, a number of "chimeric" (part human, part mouse) antibodies (approximately 30 percent mouse, 70 percent human) had been constructed. The first chimeric antibody that has been approved for human therapy is Rituxan (for the treatment of non-Hodgkin's lymphoma, developed by BIOGEN IDEC).

An improved humanization technique (known as "CDR-grafting") was developed in which the mouse CDRs (Complementarity Determining Regions), the segments which form the antigen binding surface of the antibody, are transplanted onto a human antibody framework. This time, the resulting antibody is approximately 90 percent human. However, simple application of CDR grafting methods, although they provide a good yield of humanized antibody, often results in the loss of antigen-binding affinity. A few amino-acid residues outside the CDRs are usually also transplanted in order to maintain the antigen-binding properties.

Other methods have been developed for the humanization of nonhuman antibodies. For example, it is known that there are only a few amino acids in the entire antibody molecule that are critically involved in antigen binding. Consequently, scientists are zeroing in on the amino acids that are in direct contact with the antigen. This method, termed SDR-transfer, transplants not the entire CDRs but only those residues that are known or are thought to be involved in antigen binding. Since there are only a few of those residues, the mouse residues retained in the humanized antibody represent only a very small percentage of the total (often no more than three percent). Another humanization technique, termed "veneering" (literally, changing the outside appearance of the molecule), is accomplished by replacing the amino acids on the surface of the nonhuman molecule (the residues to which the antibodies of the human patient bind) with amino acids that are commonly found in human antibodies. A veneered antibody is thus camouflaged and may not be visible to the human immune system. This is akin to the technique used by schistosomes (worm parasites) to evade our immune system—these parasites coat themselves with our own molecules so that our antibodies do not "see" them as foreign.*

This technology that reshapes an antibody structure is sometimes called the rational or knowledge-based approach. In contrast, some scientists are also trying Nature's way in which amino acids in the CDRs are randomly replaced with every possible amino acid and the resulting molecules then screened for the desired antigen affinity and specificity.

Humanized antibodies produced by either method, or by combination of both, are promising. Successfully marketed examples of this class of antibodies include Zenapax (for the treatment of kidney-transplant rejection, developed by Roche), Raptiva (for the treatment of a severe skin disease known as psoriasis, developed by Genentech and Xoma), Myelotarg (for the treatment of acute myeloid leukemia, developed by Celltech/

* Editor's note: These two humanization techniques that were just described were developed by Filipino scientist Dr. Eduardo A. Padlan.

American Home Products), Herceptin (for the treatment of breast cancer, developed by Genentech), and Xolair (for the treatment of asthma, developed by Genentech), among others.

The technique of antibody humanization has now matured into a major antibody engineering technology. Experience over the last decades has resulted in the evolution of a number of different methods to convert the technology into reproducible practice. As the interest in the antibody field grows in both biotechnology and pharmaceutical sectors, so grows the demand to develop monoclonal antibodies that are surely effective and with minimal side effects. Humanization technology, indeed, plays a part in a renaissance in the use of antibodies as versatile therapeutic compounds.

April 28, 2005

Eduardo Padlan

Molecular Guided Missiles against Cancer

At the UP Marine Science Institute, we teach a graduate course called "Targeted Marine Natural Products for Improved Therapeutic Efficacy." I introduce the course by showing a cartoon from the comic strip, Beetle Bailey by Mort Walker. In that cartoon, two characters, husband and wife, are talking about medications. The husband asks: "Why so many pills?" and she responds: "This one's for your liver, this one's for the stomach, this is for your headache, and one for the pain in your arm." He then asks: "How do they all know where to go?"

How, indeed, do the medicines we take find their way to their intended targets? As they course through our body, the medicines encounter and probably affect other cells. If their effect on normal cells is deleterious, then we have untoward side effects. The ideal medication would be one that has an effect only on the target cells and no other.

There are medications that are very selective. For example, penicillin disrupts cell-wall formation in Gram-positive bacteria and since our cells do not have cell walls, our cells are spared the action of penicillin (although some of us develop an allergic reaction to it). But penicillin kills all Gram-positive bacteria—even the "good" ones. (Yes! Not all bacteria are bad. In fact, we harbor good bacteria in our gut, for example.)

So if we could identify a distinguishing feature in a cell that is diseased, we could focus on that feature and target that cell and other cells that have it. Some such "guided missiles" are currently being developed against cancer.

Cancer arises when a cell somehow escapes normal restraints—restraints that will cause the cell to die when it is time for it to die and keep it in the tissue or organ where it belongs. Instead, a cancerous cell proliferates (it keeps multiplying) uncontrollably and spreads to other parts of the body. Obviously, a cure for cancer would be one that kills such errant cells, or at least prevents their proliferation and spread. There are various procedures that are currently being used to accomplish that. One is by physical removal of the cancer cells, e.g., by surgery. Another is by blasting cancer cells with radiation. Yet another is by poisoning cancer cells with toxic drugs. A new technique that is being developed would deprive the cancer cells of nourishment so that they cannot grow. While these procedures have increased survival rates significantly, there are drawbacks. Surgery works very well when the cancerous tissue is localized and accessible, but not when the

Formerly with the US National Institutes of Health and currently with The Marine Science Institute, University of the Philippines, Diliman, Quezon City, 1101, Philippines.
edpadlan@yahoo.com

cancer has spread to too many parts of the body. Radiation treatment has to be very focused, otherwise too many of the normal cells in the vicinity will also be killed.

Similarly, treatment with anticancer drugs usually also kills any cell that shares the targeted characteristic (e.g., fast growth) of cancer cells. (The latter is the reason cancer patients undergoing chemotherapy usually lose their hair. A more serious side effect of chemotherapy is the concomitant killing of fast-growing cells in the bone marrow—cells that are essential components of our immune system—so that the patients become susceptible to infection.) In order to minimize collateral damage and other side effects, "molecular guided missiles" are being developed that target the cancer cells while sparing all others. This strategy works when a difference exists between cancer cells and normal cells. There are a number of known cases where that is true.

Chronic myeloid leukemia (CML) is a cancer of white blood cells. In most cases of CML, a chromosomal damage results in the production of an unusual molecule—an enzyme called *bcr-abl*—and it is this enzyme that causes the uncontrolled proliferation of the cells. A chemical compound has been designed that inserts itself, in a precise and tight way, into a groove in the enzyme and specifically inhibits the enzyme. The compound, marketed as Gleevec (Glivec in Europe), has been found to be effective in treating a very large number of CML patients.

Other molecules have been found that distinguish cancer cells from normal cells. These molecules, usually referred to as "tumor-associated antigens" or simply as "tumor markers," may be present only on the surface of cancer cells. Or more commonly, they are present in significantly larger amounts on cancer cells compared to normal cells—in the words of biologists, the molecules are "over-expressed" in cancer cells. Those surface tumor markers could be targeted by antibodies.

Antibodies are molecules of the immune system that bind tightly and specifically to their target molecules (antigens). The binding of antibodies to cancer cells can cause the lysis of those cells. Another way of killing cancer cells using a specific antibody is to attach a radionuclide to the antibody. The antibody would seek out the cancer cells and these would then be destroyed by radiation. Yet another way is to attach a highly toxic drug to the antibody and the internalization of the drug would then kill the targeted cancer cell. Alternatively, the antibody could be attached to a synthetic vesicle—a liposome, for example—that is filled with a toxic drug, or a combination of toxic drugs for greater effect.

Many antibody-based therapeutics are in the process of development and some have already been approved for human use. One is Rituxan, a chimeric (part murine/part human) antibody that is directed against CD20, a molecule that is over-expressed on the surface of B lymphocytes (a white blood cell) in patients with non-Hodgkin's lymphoma. Other therapeutic antibodies that have been approved for the treatment of non-Hodgkin's lymphoma are Zevalin and Bexxar, both radiolabeled murine antibodies (Zevalin with Yttrium-90, Bexxar with Iodine-131) and both also directed against CD20.

Another therapeutic antibody is Herceptin, a "humanized" murine antibody that is directed against HER-2 (or erb B2), a molecule that is over-expressed on the surface of breast cancer cells in 35-40 percent of breast cancer patients. (Antibodies are more easily obtained from nonhuman sources, usually the mouse, and have to be made to look like human molecules so as not to be rejected by the patient's immune system.)

Two antibodies have recently been approved for the treatment of colorectal cancer: Erbitux, a chimeric murine/human antibody directed against epidermal growth factor receptor (EGFR), and Avastin, a humanized murine antibody directed against vascular endothelial growth factor (VEGF). EGFR is important in transmitting signals from the outside to the inside of the cancer cell, instructing it to grow and multiply, while VEGF stimulates the growth of new blood vessels needed to support the fast-growing cancer cells. And so, Erbitux prevents the growth of the tumor and Avastin inhibits the growth of new blood vessels to the growing tumor, thereby "starving" the tumor. Other therapeutic antibodies in the pipeline are directed against other tumor markers.

An effective strategy of antibody-based therapy makes use of an antibody directly linked to a drug or linked to a drug delivery system that contains drugs. Myelotarg is an antibody conjugate used as a last resort in the treatment of Acute Myeloid Leukemia (AML). AML is the most serious of the leukemias and has the poorest prognosis. Myelotarg consists of a humanized antibody directed against the AML marker CD33 and that is linked to a highly toxic drug known as calicheamicin. For the treatment of breast cancer, researchers are developing an anti-HER2 antibody linked to a liposome that contains the anticancer drug doxorubicin.

A tumor marker that is a very promising target is TAG 72 (Tumor-Associated Glycoprotein No. 72), discovered by the group of Jeffrey Schlom at the National Cancer Institute of the US National Institutes of Health (NIH). TAG 72 is a "pancarcinomic" antigen, meaning it is present in many different cancer types and in a very high percentage of the cancer cells, but not in normal tissues (except in the endometrium and transitional mucosa of the gastrointestinal tract). A number of murine antibodies against TAG 72 have been generated; the one that is the most studied is "CC49." Various forms of CC49 have been constructed to improve its effectiveness. For example, I designed a humanized version of CC49. Further, the effective affinity of CC49 for TAG 72 has been dramatically increased by the construction of a tetravalent version of the molecule (done by Dr. Ameurfina D. Santos of NIMBB, UP Diliman, while on a fellowship in my laboratory at the US NIH). Furthermore, improved versions of humanized CC49 have been constructed (by Dr. Noreen R. Gonzales-McCurdy, formerly of the Ateneo de Manila University and also a former fellow in my laboratory).

There is currently a group of scientists in the Philippines called AMOR that is exploring the use of CC49 against breast cancer. The AMOR program was conceptualized and organized in 1998 when I first met Dr. Gisela P. Concepcion, who works on anticancer marine natural products at the UP Marine Science Institute. (She coined the term AMOR which stands for "Antibody and Molecular Oncology Researchers.") I had proposed that we combine antitumor antibodies with anticancer marine natural products to act like molecular guided missiles. She invited key researchers in UP Diliman and other universities to work with her and form the AMOR team. In 2000, the Department of Science and Technology approved major funding for the six-year AMOR program. AMOR is now in its fifth year. AMOR is exploring the use of CC49 by itself to target and kill breast cancer cells. AMOR is also using CC49, combined with cytotoxic Philippine marine and plant natural products, to target TAG72 in breast cancer cells and kill those cells. To date, Dr. Concepcion's project appears to show some encouraging results in the testing of AMOR prototype products in animals. The hope is that one day, AMOR will produce an effective targeted therapy using local natural products that can be used in combination with surgery for the benefit of Filipino women afflicted with breast cancer.

April 21, 2005

Eduardo Padlan

Antibody versus Antibody

Antibodies protect us from parasites, germs, toxins, and other foreign substances (antigens) that may enter our body. Binding of antibodies to an antigen causes the neutralization of the substance, its immobilization, and increased susceptibility to elimination by natural processes, or death in the case of an invading cell. Antibodies can be produced against virtually any foreign substance—actually, against any accessible part of the antigen—and biotechnology allows us to produce virtually unlimited amounts of any antibody whose properties we desire. Not surprisingly, antibodies have found many uses in medicine and industry.

Yet, despite the many benefits we derive from antibodies, and despite our body's stringent screening to prevent its occurrence, some wayward antibodies do sometimes get produced—and they do harm. For example, some of us produce antibodies to our own molecules—an autoimmune disorder—and we could die from it. A case in point is systemic lupus erythematosus in which the patient has antibodies that bind DNA, or other nuclear components, resulting in immune complexes (antibody:antigen complexes) that could clog up vital organs. Another apparent malfunction of the immune system is allergy.

The antibody type that is responsible for allergy is IgE. This antibody type seems to be our natural response to parasitic infestation. But even in the absence of parasites, we produce IgE—and we suffer from allergies. In fact, the incidence of allergy is increasing worldwide, possibly from increased pollution and exposure to unnatural food additives and other substances.

Mast cells (granulated cells in connective tissue associated with all blood vessels) and basophils (in the blood) have high-affinity receptors for IgE on their surfaces. Most of the IgE that we produce are soon bound to mast cells and basophils. There they wait for allergens (antigens that trigger an allergic response). When an allergen that the IgE recognizes comes along and the IgE binds to it (and is crosslinked), the mast cell (or basophil) "degranulates" and releases histamine and other vasoactive compounds (molecules that act on blood vessel permeability). The result is the sneezing, coughing, watery eyes, runny nose, and other symptoms that we usually associate, with allergy. Worse, local edema could occur that may cause our throats to close or even worse, a systemic reaction could result in a sudden drop in blood pressure (anaphylactic shock) that could mean death.

Formerly with the US National Institutes of Health and currently with The Marine Science Institute, University of the Philippines, Diliman, Quezon City, 1101, Philippines.
edpadlan@yahoo.com

We take antihistamines or steroids to relieve our allergies, and some of us even carry along ready-to-administer epinephrine in case of severe allergic reactions. Some of us have food allergies and we wisely avoid those foods. Many of us know what we are allergic to, although we do not know why.

Could we develop a universal treatment for allergies? Better yet, could we develop a vaccine against allergy?

Since we can generate antibodies against almost anything, can't we use antibodies to fight allergies? In fact, we can. There is now in the market, sold under the trade name Xolair, an antibody treatment for allergies. (Before I proceed, I must first assure you that I have no business interest in Xolair nor any connection with the companies that make it.) Xolair is an anti-IgE, an antibody against IgE. It is a "humanized" mouse antibody directed against human IgE. How does it work?

Xolair binds to soluble IgE and prevents it from binding to the high-affinity receptor on mast cells and basophils. Obviously, Xolair binds to that part of IgE that binds to the receptor, or at least near enough to it to prevent the binding. Xolair does not bind to IgE that is already bound to the receptor. (If it did, there will be mass degranulation of mast cells and basophils and shock will likely occur. The researcher who developed Xolair admitted killing hundreds of mice before finding an anti-IgE that worked.) But IgE, like all antibodies, has two-fold symmetry, so that it has a receptor-binding site on one side and another binding site on the other side. So, even if the IgE is already bound to the receptor, there is the other binding site that should still be available for binding by Xolair. But Xolair does not bind to IgE that is already bound. Why not?

The answer comes from the work of the husband-and-wife team of David Holowka and Barbara Baird at Cornell. David and Barbara showed several years ago that IgE, when bound to cells, is actually bent. Now, it is easily shown by modeling that bending the IgE will cause the second binding site to be occluded and inaccessible. That, then, is the reason Xolair can no longer bind to IgE that is already bound to cells. That's why Xolair can be used to treat allergy (without killing the patient).

Now, can we develop a vaccine against allergy? We actually tried several years ago—with some success. Birgit Helm and her group in Sheffield, England had determined the location of the binding site on IgE for the high-affinity receptor. At that time, I had built a model of IgE and Birgit asked me to look for a stretch of polypeptide in the receptor binding site that might be useful as a vaccine. I subsequently proposed a segment, engineered to be a cyclized peptide, and we had it synthesized and injected into a rabbit. It worked! Serum from the immunized rabbit was found to prevent the degranulation of sensitized basophils. It was clear that the rabbit had actually produced antibodies that presumably bound to the receptor binding site of the IgE. So, we have developed a vaccine to protect rabbits against human allergies. Big deal! What we need is an allergy vaccine for humans! No, we haven't developed one yet.

April 21, 2005

Eduardo Padlan

Viral Intelligence: A No-brainer

In the *Wizard of Oz*, the Scarecrow sings:

"...I would not be just a nuffin'
My head all full of stuffin'
My heart all full of pain
I would dance and be merry
Life would be a ding-a-derry
If I only had a brain."

Obviously, the Scarecrow believes that his life is meaningless because he does not have a brain and that life would simply be great if he had one.

And, of course, there is Man, who believes that he is the superior being on Earth because he has the biggest brain, relative to body mass. We believe that we are the dominant species because of our superior intelligence. We maintain that the bigger the brain, the higher the intelligence. (Einstein would dispute that—his brain was smaller than usual.)

But is having a brain really necessary to have a meaningful life? Do we even have to have a brain to be considered intelligent?

There are numerous creatures that do not have brains, for example, viruses, bacteria, protozoa, fungi, and others that do not have even the semblance of nervous tissue. Yet they exist and have existed for many millions, perhaps billions, of years and, in the battle for survival, they could defeat even us, the brainiest creatures on Earth.

Let's consider the viruses.

Viruses are very simple beings. They are total parasites—unable to survive on their own—and depend on host organisms for the molecules that they need to carry out the chemical reactions that they require, very often even the ones needed to reproduce themselves. We might even question whether they are alive or not. But they victimize all living things—plants and animals alike, even bacteria.

Viruses come in different shapes and sizes, and all are very small. The simplest viruses are nothing but a strand of nucleic acid (some viruses have DNA for their genetic material, others have RNA), enclosed in a protein coat. The genetic material of simple viruses codes for only a few proteins, sometimes just one (for only the coat protein, in the case of satellite viruses which require the presence of other "more complete" viruses). Some viruses encase

Formerly with the US National Institutes of Health and currently with The Marine Science Institute, University of the Philippines, Diliman, Quezon City, 1101, Philippines.
edpadlan@yahoo.com

themselves with lipid membrane, taken from the cell they had just victimized. Some viruses kill their hosts (we all know about the AIDS virus); some damage or disfigure their hosts for life (polio, smallpox, and others). Some lie "dormant" inside their hosts (sometimes for many years); an example is the virus that causes chicken pox, which could hide in our nerves and resurrect itself and cause shingles many years later. Some even insert themselves into the genetic material of their hosts. (Yes, we can be sure that we, too, have viruses inserted in our DNA.)

To perpetuate themselves, viruses move from host to host. They are very clever about this. The cold virus, for example, causes us to sneeze; the spray that results causes the virus to be dispersed in the air and when inhaled by an individual nearby provides the virus with a new host. The AIDS virus is even more clever. It is transmitted via body fluids and its main mode of transfer from victim to victim is probably the most pleasurable activity known to man—sex. But the hepatitis B virus surpasses the AIDS virus in this regard. The hepatitis B virus is also transmitted through body fluids, but it is one hundred times more infective than the AIDS virus.

And viruses have devised clever ways of evading our immune system. Some mutate and alter their "appearance" (their surface molecules) so that our immune system can no longer recognize the "new" virus. (Our immune system has memory and it remembers, by their surface molecules, germs that it had encountered before.) This is the reason why we can catch cold repeatedly, or get the flu every year. Some viruses hide inside our cells (like the hepatitis viruses that hide inside the liver and move from liver cell to liver cell, thus not exposing themselves to the immune system). In regard to evading the immune system, the AIDS virus is the most clever: it targets a very important component of the immune system, the helper T cell. Essentially, what the AIDS virus does is destroy the immune system, leaving its victim helpless. But the AIDS virus does not kill its victim until eleven years after infection (on average), so that the victim has ample time to transmit the virus to other unsuspecting hosts.

Can we say a virus is intelligent? Of course! It's a no-brainer. The Scarecrow has nothing to complain about.

September 29, 2005

Mishail Tupas

Viruses versus Humans: Molecular Evolution at Work

The common cold, the flu, autoimmune deficiency syndrome (AIDS), hepatitis, meningitis, measles, severe acute respiratory syndrome (SARS), hemorrhagic fever, certain cancers ... the list just goes on and on. These debilitating diseases, many of which are fatal, are caused by viruses, the smallest known parasitic microbe. The recent outbreak of SARS, caused by the SARS corona virus (SARS-CoV) originally infecting chickens; the scare in the increase of encephalitis, high fever, and other respiratory syndromes due to the West Nile virus; and anxieties over the flu because of the vaccine shortage, indicate how these tiniest of all microbes affect the quality of human life.

Viruses wreak so much havoc because they can easily sneak past the immune system, resulting in millions of deaths per year globally. Viruses are simply genetic information in the form of either deoxyribonucleic acid (DNA) or ribonucleic acid (RNA) surrounded by a protein coat. Most of the viruses that cause certain cancers are DNA viruses while those that cause common or chronic respiratory and immune diseases are RNA viruses, and include the human immunodeficiency virus (HIV), SARS-CoV, influenza virus, and the West Nile virus. Retroviruses like HIV are the most deadly of these, because once inside the host cell their RNA is reverse transcribed into DNA, where it subsequently integrates into the host genome and can remain undetected for years.

It is not surprising that new viruses with increasing virulence are discovered every year. Viruses undergo evolutionary changes in their genetic information, creating new viral particles that are unrecognizable to the host immune system. The process is part of what is now called molecular evolution. There are two main mechanisms for molecular evolution: genetic drift and genetic shift. Genetic drift is when genes are gradually changed over time and is the most commonly known form of evolution. Genetic shift is when a large amount of DNA or RNA is changed, which in viruses can be attained through re-assortment, or when two different viruses infect the same cell and their RNA get mixed up.

These molecular changes allow viruses to develop new strains in a short period of time, therefore displaying the fastest rate of molecular evolution. For example, the mutation rate among DNA viruses is around 0.1 per genome (viral genome is about 10,000 nucleotides) per replication, while the mutation rate of influenza and corona

Oakton High School in Fairfax, Virginia, USA.
mikotupas@cox.net

viruses, which are RNA viruses, is about ten times faster. The high mutation rate of RNA viruses compared to DNA viruses is responsible for their enormous adaptive capacity which often results in a new and sometimes more virulent strain.

A good example of rapid molecular evolution is displayed by the influenza viruses, RNA viruses in the family Orthomyxoviridae. Although there are three types of influenza viruses (A, B and C), it is the Influenza A virus that causes the flu pandemics known to infect humans, other mammals, and birds. Influenza viruses express surface antigens known as hemagglutinin (H) and neuraminidase (N) proteins which are glycoproteins found on the outside of the viral particle essential for viral reproduction and periodically undergo mutation. The hemagglutinin and neuraminidase proteins are targets for antiviral drugs and are also recognized by antibodies, i.e., they are antigens. The responses of antibodies to these proteins are used to classify the different serotypes of Influenza A viruses, hence the HxNx identification system.

Scientists believe that during the 1918, 1957, and 1968 flu pandemics, a mix of avian or swine and human flu viruses resulted in new H and N proteins, making it difficult for the human host immune cells to muster up a response because they could not recognize the avian or swine component. The 1918 pandemic alone resulted in the loss of forty million lives around the world. In January 2003, human diseases associated with Influenza A virus subtype H5N1, also known as A(H5N1) or H5N1, reemerged for the first time since an outbreak resulting in human deaths was reported in 1997. The bird adapted strain of H5N1is called HPAI A(H5N1) for "highly pathogenic avian influenza of type A, subtype H5N1" and is the causative agent of the H5N1 flu commonly known as avian influenza. Avian influenza technically denotes the disease, not the virus. It is the Influenza A virus that causes avian influenza, now recognized because of its global spread through multiple bird species as a significant pandemic threat.

The high mutation rates of the influenza and corona viruses enable them to infect new hosts across species and give them the edge in emergence. So far there are fifteen H proteins and nine N proteins that have been discovered, with a total of 135 different combinations of both proteins. This extensive pool allows the viruses to escape immune detection and infect new species, keeping vaccine manufacturers guessing as to the next prevalent strain for the coming year.

The human immune system is able to respond to these viral infections because they have cells that also undergo molecular evolution. In complex organisms such as humans, molecular evolution takes place at a rapid pace but is limited to particular immune cells such as B-lymphocytes. These cells are able to produce a large quantity and variety of virus-fighting antibodies through molecular changes in the immunoglobulin genes. In response to infection, these genes undergo rearrangement or recombination at very high frequency in a specific region known as the variable segment. This high frequency mutational event is known as somatic hypermutation, which results in the production of a large variety of antibodies. Thus, the B-lymphocytes are able to make a whole range of antibodies against a large number of infectious agents, thereby ensuring the survival of the organism. Mutation rates during somatic hypermutation can be as high as 0.001 per base synthesized per generation, which are still several orders of magnitude lower than the mutation rates of RNA viruses. One of the risks for an increased mutation rate and hyperactivation of the immune cells is the development of autoimmune diseases, where the host attacks its own cells, resulting in their death.

Molecular evolution in human B-lymphocytes is not fast enough to combat the rapidly mutating viruses. Humans have developed other means to overcome these pathogens through the use of vaccines that stimulate a faster immune response without resulting in immune hyperdrive. Although this is some form of adaptive response from the host, with the high mutation rates among viruses, it seems that availability of the appropriate vaccine may lag behind viral molecular evolution. With viruses replicating in a matter of minutes, they consequently have thousands more mutations than humans would ever have within a generation time of twenty years.

A generally accepted concept in the theory of natural selection and mutation is that organisms with short replication times will more often undergo mutations and pass it on to the next generation. The fact that the flu virus mutates at such an astonishing speed and the recent flu vaccine shortages prompted Klaus Stohr, head of flu surveillance at the World Health Organization, to declare in November 2004 that "a global influenza pandemic is closer than at any time in a generation." Molecular vaccines would be a way to equalize this situation because these vaccines can be changed more rapidly than traditional ones to counteract new viral strains. This is the future direction of vaccine therapy.

January 11, 2007

Eduardo Padlan

Viruses versus Humans: Can We Win the War?

VIRUSES, THOSE tiniest of parasites, manage to make us suffer again and again, and sometimes even kill us. Those pesky little critters do it by continually mutating, so that, although we have an immune system that defends us against their attack, we continue to be victims of viruses year after year. This is because our immune system only reacts to an attack—it only defends—it does not take the fight to the enemy. And as they say: In a fight, if we choose only to defend, the best that we can hope for is a draw.

But, we have big brains. We should be able to figure out how to win. Right?

Let us consider what we do to fight the flu virus.

The flu virus has two molecules on its surface which it uses to infect cells: hemagglutinin and neuraminidase. The hemagglutinin recognizes and binds to a carbohydrate moiety on the surface of a victim cell; the hemagglutinin is then activated (by enzymatic cleavage) and is transformed into a structure that allows it to be inserted into the cell membrane; the hemagglutinin, with the rest of the virus in tow, then enters the cell. Once inside the cell, the virus takes over the cell's biosynthetic machinery and dictates the production of many copies of itself. The newly formed virus particles emerge from the cell, but their hemagglutinins bind to the nearest carbohydrate moiety that they recognize—which is that of the cell they just emerged from. The neuraminidase comes into play at this point, cutting off the attached piece of the carbohydrate, thereby freeing the new virus to infect some other cell.

Knowing this much, scientists have devised ways to fight the virus. Antiviral drugs that act as neuraminidase inhibitors and M2 (another flu protein) inhibitors have been designed. These drugs can shorten the course of infection and make it less severe, but they cannot prevent infection. Other ways of interfering with viral replication have been developed. Molecules that prevent the initial binding of the hemagglutinin to the cell have been employed, for example, antibodies that we produce ourselves through vaccination, or ones that had been synthesized in the lab or obtained from other animals.

But, then, the virus mutates. We may find that our inhibitors no longer work, or that our antibodies are no longer effective. We are back to square one. The usual strategy for fighting the mutating flu virus is to produce a different vaccine every year. But, since we cannot really know what the virus will mutate into, we can only guess which viral strain(s) to use as vaccine.

Formerly with the US National Institutes of Health and currently with The Marine Science Institute, University of the Philippines, Diliman, Quezon City, 1101, Philippines.
Email: edpadlan@yahoo.com

Can we, for once, be a step ahead of the virus? Can we devise a means of fighting the virus even if it mutates?

Again it is said: A good way to know how to defeat an enemy in battle is to know beforehand how he fights.

Let's see what we can do about designing better vaccines against the flu virus. First, let us figure out how the virus manages to evade our immune system. The answer seems simple enough: when the virus mutates, it localizes its mutations in its "immunodominant epitopes."

Antibodies can be produced against any exposed part of the hemagglutinin and neuraminidase. But, like all molecules, some parts of the hemagglutinin and neuraminidase elicit a greater antibody response than other parts; in other words, some parts are more attractive to antibodies than others. Those are the so-called immunodominant epitopes. Now, antibodies are very specific. When the virus mutates and changes even just one amino acid in an immunodominant epitope, the antibodies that used to bind to that epitope may no longer recognize the altered molecule. Our immune system will produce new specific antibodies, but it takes four to seven days for this to happen; in the meantime, we suffer from the flu.

If we could locate the immunodominant epitopes in the hemagglutinin and neuraminidase and generate new molecules in which those epitopes are no longer immunodominant (we can do this by protein engineering, whereby we replace the residues in the immunodominant epitopes with amino acids whose presence results in lower attractiveness to antibodies), we could use those engineered molecules as vaccines and our immune system will produce antibodies that would be directed against other parts of the hemagglutinin and neuraminidase, the parts that the virus doesn't normally mutate. As long as the virus continues to localize its mutations in its immunodominant epitopes (of course, it cannot know that our vaccines no longer contain those immunodominant epitopes), we have a weapon against it. Nature is on our side in this regard. The virus tends to confine its mutations in its immunodominant epitopes because there are "hot spots" (highly mutable codons) in those epitopes. Further, the virus is under no selection pressure to change this tendency because the flu virus victimizes not just humans, but other animals as well, including chickens, ducks, turkeys, and wild birds, and cats, dogs, pigs, and horses, and even whales and seals. (Those poor animals don't know how to engineer proteins. Only we humans do.)

Can we win the war against viruses? Yes! (I think.)

January 18, 2007

Mary Anne Astilla

What You Don't Know Can Harm You

"Anything that can be eaten will be eaten" was the first statement Dr. Eduardo Padlan said during our seminar class at the UP Marine Science Institute. Our topic wasn't about predators and prey in the vast ocean but something more familiar to us—the human body. It wasn't a seminar course in Marine Ecology after all, but in Molecular Immunology.

It is quite amusing how we can go on with our daily tasks and not notice and feel the "battle" going on inside us. In connection with the first statement I quoted from my professor, it is unfortunate that we are, in a way, food—a source of nutrition and essential for the survival of a wide array of microorganisms. But fortunately enough, our body has defense mechanisms to overcome this attack by preventing the entry or containing the spread of these foreign materials collectively called *antigens*, once they are able to find their way inside our cells or in the body in general. The immune system, composed of humoral and cellular components, is responsible for our survival despite the ubiquitous presence of bacteria and other pathogens (disease-causing agents). This brings to mind an experiment in our Microbiology class back in college wherein we prepared sterilized agar culture media in petri dishes. We exposed one to the open air, others were made to come in contact with a doorknob, a money bill, and lastly, our own hands. It was a total revelation of the presence of bacteria everywhere when several different bacterial colonies were seen in all plates after incubation. It is unimaginable how many of those kinds we will have to come in contact with during our lifetime. It's a relief to learn what our immune system can do. The humoral arm of the immune system involves antibodies that circulate in the body fluids and specifically recognize antigens such as bacterial and viral components. There are approximately 10^8 different B cells in one's body each producing antibodies with different specificities at any given time. Antibodies cross-link with the antigens and form an immune complex that is engulfed by macrophages (a cellular component). Aside from the macrophages, the cellular arm is composed of cytotoxic T lymphocytes (CTLs), and T-helper cells. CTLs target and kill cells expressing fragments of the antigen on its surface. T helper cells direct antibody- and cell-mediated immune responses through the secretion of molecules called cytokines that have important effects on B cells, CTLs and other immune cells. However, despite the sophistication and complexity of our immune system, having been a product of evolution for quite a long

---

The Marine Science Institute, University of the Philippines, Diliman, Quezon City, 1101, Philippines.
m_astilla@yahoo.com

time, our body may still lose control and succumb to invaders. Most, if not all, pathogens are continuously evolving by the process of mutation to be able to crack the defenses of their host, develop a new mode of transmission, and expand their host choices.

Let's take, for example, the Human Immunodeficiency Virus or HIV, the retrovirus that causes AIDS. The retrovirus' genetic material is in the form of RNA instead of the usual DNA but it possesses an enzyme called reverse transcriptase that converts RNA to DNA. This allows the retrovirus to use the genetic machinery of the host cell for its own replication. Our immune system initially has the ability to contain HIV infection, as it is able to produce antibodies directed to the HIV. But after some time, the immune system is subverted and the person becomes susceptible to diseases caused by other pathogens, which normally would have been easily regulated by the immune system. The latter signals the onset of AIDS. The progression from HIV infection to AIDS tends to vary between individuals and the HIV strain. How is HIV able to get away from the offense launched against it? It is tempting to believe that there is such a thing as "viral intelligence" as Dr. Padlan would usually refer to it. It is interesting to note that this highly evolved virus seems to be using a very ancient tactic—a successful approach used in the Trojan war of ancient Greece—the Trojan Horse. Integrating the existing data on retroviral biology, researchers from the Johns Hopkins University composed of Dr. Stephen Gould, Amy Booth and James Hildreth, came up with a model of retroviral biogenesis and transmission and called it the Trojan Exosome Hypothesis. An article bearing the same title was published in the September 16, 2003 issue of the *Proceedings of the National Academy of Sciences* (PNAS).

Exosomes are small membrane bound molecules that are released into the external environment, which fuse with membranes of neighboring cells to deliver membrane and cytoplasmic proteins from one cell to another. Inter-cellular signaling by exosome exchange is important in many physiological processes, including lymphocyte activation and the development of immunological tolerance.

Previous models of retroviral transmission assumed that the binding and fusion of retroviruses, such as HIV, to host cells are completely dependent on the retroviral envelope proteins. This is not the case, however, as retroviruses can have many types of host cell molecules and are capable of receptor- and envelope-independent infections. The Trojan Hypothesis states that the retrovirus taps the preexisting, nonviral pathway of exosome biogenesis and exosome uptake for the formation of retroviral particles and transmission, respectively. As the cells involved in the process of immune surveillance and signaling circulate throughout the body, they are exposed to numerous exosomes, which may be loaded with retroviral particles. As it replicates predominantly in the immune cell population, it is able to evade attack launched by the immune system and at the same time disrupt the host's response to the infection, thereby enabling continuous replenishment of the virus population.

Despite the tremendous efforts in HIV and human immune system research, an ideal vaccine has not yet been made. This difficulty lies in the HIV's extraordinarily high mutation rate. This also allows the virus to evolve resistance to the drugs used for treatment. And this is further aggravated by the fact that HIV exploits the immune system designed to stop it and other infections. It is by increasing our knowledge of how the HIV and the components of our immune system interact at the molecular level that an effective approach to combat HIV infection and AIDS may be developed. As Sun Tzu stated in his book *The Art of War*, "If you know the enemy and you know yourself, you need not fear the results of a hundred battles."

December 29, 2005

Battling Tuberculosis on Many Levels

John Phillip Fadul

THE PROBLEM with tuberculosis is that you may already have it. Unknowingly, you may be harboring inside your body the agent that causes it, as you go about your daily routine. Imagine it quietly waiting inside your body, patiently biding its time until it finds an opportunity to catch you when you are unaware—and when you are weakest. Do we need any more reason to learn more about it?

Everyone has heard of TB. Quite possibly, you may know someone who has it. It is an ancient, highly infectious, and persistent disease that has plagued humans for centuries. Despite the availability of a vaccine and anti-TB drugs, it remains the leading infectious disease worldwide, especially in developing countries. It is responsible for killing more humans than AIDS, malaria, leprosy, and diarrhea. In 2000, the World Health Organization (WHO) estimated that there were 249,655 new cases of TB in the Philippines. An even scarier statistic is that more than a fourth of that number is projected to die from the disease. Also in the same year, it was reported that the Philippines had the seventh highest incidence in the world and the second highest in Asia. Because of these astounding figures, the Philippine government has vigorously renewed its commitment to fight the deadly disease.

TB is most often caused by the bacterium *Mycobacterium tuberculosis*. It is an immobile, rod-shaped bacterium which requires sufficient levels of oxygen to survive. This is why *M. tuberculosis* complexes are almost always found in the well-aerated upper lobes of the lungs. The bacterium is an intracellular parasite of macrophages, a kind of blood cell that functions in ingesting and destroying large amounts of foreign material. Unfortunately, humans are the primary host and the only real or natural host of *M. tuberculosis*. When talking, spitting, or coughing, a person with TB releases tiny, aerosolized droplets contaminated with the bacteria. An unsuspecting, healthy individual may inhale the expelled bacteria and contract the infection. (Luckily, one is unlikely to become infected by just one exposure.) The microbes are engulfed by macrophages present in the alveoli and they replicate in these cells for two to three days before spreading throughout the body.

TB does not immediately manifest itself upon infection. The disease has two stages, the first one being latent infection. It is characterized by the formation of hard, grayish nodules, called tubercles, in the lungs. But for most infected people, the macrophages

The Marine Science Institute, University of the Philippines, Diliman, Quezon City, 1101, Philippines.
johnfadul@yahoo.com

are able to contain the bacteria for life. (Inside the macrophages, the bacteria are not destroyed but continue to live in a dormant, spore-like state.) No apparent disease is noted; they suffer no effects and are not contagious. Since this stage is asymptomatic, no one suspects the presence of the microbe in the body. It is estimated that a whopping two billion people—one-third of the world's population—is currently in this state.

However, in ten percent of cases, the disease progresses into the second stage of active tuberculosis. Factors such as old age, malnutrition, or the presence of another disease—especially HIV/AIDS—enable latent infections to develop into active TB. In this stage, tubercles proceed to the air passages, blood vessels, and other parts of the body. Local tissue and cells of our immune system die, causing the formation of granulomas, sites where active microbes can survive. These granulomas expand with more destroyed lung tissue and may spread into the larger airways of the lungs. They may also liquefy as the disease worsens. allowing the bacteria to multiply and spread to create new lesions. Meanwhile, the patient experiences a plethora of symptoms, from chest pains and shortness of breathe to weight loss, from fatigue and paleness to painful joints and swollen glands. TB during this stage becomes very contagious.

To address the TB problem on a worldwide scale, the Global Tuberculosis Programme of the WHO developed DOTS (Directly Observed Treatment, Short-course) in 1993. More than just prescribing drugs and ensuring that their supply is uninterrupted, DOTS underscores the importance of self-reporting to health centers and of patient compliance with the treatment. Most importantly, it enjoins governments worldwide to go all out in the fight against TB. In this program, there is a standardized recording and reporting process that allows assessment of treatment results for each patient and of the overall TB control program. With DOTS, the patient, health care workers, public health officials, governments, and communities all share the responsibility of effectively treating and controlling the killer disease.

Treatment under DOTS consists of the daily administration of four drugs, namely, isoniazid, pyrazinamide, rifampin, and ethambutol, for two months; and a follow-up, thrice-a-week treatment of isoniazid and rifampin for another four months. The drugs affect different biochemical processes in the bacterium to inhibit its growth. Interestingly, some of these drugs enter the body in their inactive form. Proteins present in the intracellular environment alter the structure of these prodrugs and consequently activate them. Mutations in the genes coding for proteins involved in the target biochemical processes and in the activation of the drugs confer drug resistance, which is one of the foremost problems in treating TB. Drug resistance also arises when the full drug therapy is not completed. Therefore, compliance of the patient is highly important.

With this increased need for drugs to combat TB and its drug-resistant variants, there is an urgent need for investment in anti-TB drug discovery. A battery of rapid, low-cost, high-throughput screening methods for selecting new drug candidates against active TB is available to researchers. For this, initially only avirulent, i.e., non-disease-causing, strains of the bacterium, are used. This is to obtain quicker results and more importantly, to protect the worker from contracting the disease. Once compound leads have been identified, virulent and drug-resistant strains of the bacterium are then tested using high-level containment facilities. Among the methods used in anti-TB testing are the Agar Diffusion Method, the Microplate Alamar Blue Assay, BACTEC, and macrophage culture. The Agar Diffusion Method is a widely used method for drug testing. In this method, bacteria are grown in culture plates. Candidate drugs are then introduced; clear zones in the "lawn" of bacterial growth indicate effective killing of the bacteria. The Microplate Alamar Blue Assay is a high-throughput method that utilizes the dye Alamar blue for assessing cell viability. The nonfluorescent dye is originally blue but turns into a fluorescent pink in response to changes in the culture medium resulting from cellular growth. This color change can then be measured quantitatively. In the BACTEC system, the culture medium used contains a substance called palmitate, specially tagged with radioactive atoms. As *M. tuberculosis* multiplies, it breaks down the palmitate and releases radiolabeled carbon dioxide, which can easily be detected and measured.

Anti-TB testing may also be carried out in an in vitro (literally, "in glass") macrophage culture, to make testing conditions closer to the real situation in the body. In this technique, macrophages are grown in culture plates. They are then infected with *M. tuberculosis*, and treated with possible anti-TB drugs. After a prescribed incubation period, the macrophages are ruptured and the remaining bacteria inside them are counted. In another anti-TB test which is relatively new, compounds are tested to see if they inhibit the enzyme isocitrate lyase, since this enzyme is shown to be important in latent TB. Each method has its advantages and drawbacks; researchers have to consider factors such as the availability and cost of materials and equipment, time required until results are acquired, reproducibility of results, and biosafety concerns, before selecting which method is most suitable to their objectives. In all cases, compound leads are those that are shown to kill or inhibit the growth of the *M. tuberculosis* but not mammalian cells. After in vitro testing, a drug candidate is tested in vivo (in animals) to see whether the drug is effective as it circulates in the body.

The beginning of this piece was quite pessimistic; fortunately, we do not have to end on that note. The Department of Health, with the help of the WHO, has instituted structural reforms and has also provided a better system of technical support and drug procurement to control TB. Research interest in anti-TB drug discovery continues to grow, particularly in university-based research centers such as the UP Marine Science Institute where the research groups of Drs. Gisela P. Concepcion and Lourdes J. Cruz have published on anti-TB marine compounds. So you see, the war against TB is multi-armed. Equally as important as the availability of the anti-TB vaccine and drugs (which are imported) are government prevention and intervention through public health measures, local development and production of the BCG vaccine, rigorous research on anti-TB drug discovery, and a zealous drive to help stop TB—and stop it now. The outlook is actually quite good, especially for you who took the time to learn about this. As the saying goes, knowing is winning half the battle.

Printed in two parts on January 12 and January 19, 2006

Philip Ian Padilla

The Culprit in Gastric Ulcer: *Helicobacter pylori*

It has been two decades since Barry Marshall and Robin Warren first reported that the bacterium *Helicobacter pylori* is the culprit in gastritis and gastric ulcers. And for that reason, they were the recipients of the 2005 Nobel Prize for Medicine and Physiology. Before their seminal report, stress was said to be the main cause of gastric ulcers. The Nobel Committee in Stockholm, Sweden cited the two Australians for going against medical dogma at that time and turning peptic ulcers from a chronic debilitating disease to a treatable infection using a short course of antibiotics.

The pathologist, Warren, and the gastroenterologist, Marshall, teamed up to prove that this gram-negative, spiral-shaped bacterium was the cause of acute and chronic inflammation in the stomach. They found it hard to convince the larger medical community because conventional wisdom dictated that the stomach environment was sterile due to its extremely acidic environment. Interestingly, Marshall was able to culture the bacteria by serendipity. He accidentally left his culture specimens during one Easter holiday, giving the bacteria enough time to grow and reproduce. *H. pylori* is a tricky bacterium to grow and culture in stomach biopsy specimens because it is slow-growing in ordinary conditions where other bacteria can achieve optimum growth. And it thrives only in an environment with little or without oxygen. Despite being able to culture it, Marshall still found it hard to convince the scientific world. So he underwent endoscopy, inserting a fiberoptic tube to his stomach to ascertain he had no gastric lesions, ingested a test tube full of pure culture of the bacteria, and after two weeks, inserted the tube again to prove gastritis was present. Within five days of ingesting the bacteria, he had early morning nausea and vomiting and pictures from the fiberoptic scope showed acute inflammation. He did not wait for ulcers to develop. He medicated himself by taking a course of antibiotics, with the resulting eventual healing of the gastritis.

Although it took almost two decades for Warren and Marshall to be recognized for their work, their breakthrough led other scientists to find other chronic diseases that might be linked to the presence of microbes (rheumatoid arthritis, ulcerative colitis, etc.). The current field of *H. pylori* pathogenesis and its relation to gastroenterology has grown to include such novel experiments as the use of its toxin to study cell biology, rapid diagnostic techniques, and different vaccination strategies against the bacteria. There

University of the Philippines in the Visayas, Miag-ao, Iloilo, Philippines.
padillap@skyinet.net or padillap@nhlbi.nih.gov

are now annual international scientific congresses on *H. pylori* and even a peer-reviewed journal, *Helicobacter*.

Presently, *H. pylori* is known not only as the cause of stomach ulcers but also of gastric cancer and cancer associated with the lymph nodes that line the stomach. Researchers have documented that patients with previous infection with the bacteria were more likely to develop gastric cancer than the normal population. Treatment of the infection in patients with cancer of the gastric lymph nodes (MALT lymphoma) resulted in remission. It must be noted that *H. pylori* infection is a chronic one. It starts early in childhood and, if left untreated, can lead to a deadly consequence like cancer. This is why it becomes imperative to treat such an infection early.

Colonization of the bacteria is very high in developing countries like the Philippines. The bacteria, in contrast to the 50 percent incidence in industrialized countries, colonize almost 80 percent of the population in Third World countries. Transmission is thought to be by the oral-fecal route, which is why infection is clustered in members of one family. Poor hygiene and unsanitary conditions play a huge role in the spread and development of this infection.

Every time one feels an epigastric acid type of pain, early morning nausea, and vomiting, one needs to have a checkup with his/her doctor. It is understood, however, that other causes of such pain should be ruled out such as biliary stones or bleeding caused by NSAIDS (drugs that cause bleeding like indomethacin, ibuprofen, aspirin, and the like.) If peptic ulcer is suspected, one needs to do an endoscopy and document the presence of a lesion, and ideally, get samples for culturing of the bacteria. Most of the time, there is a negative culture of the bacteria so a urease test is done, too, in which one gets tested for the presence of an enzyme secreted by *H. pylori* called urease.

The latest recommended treatment modality for *H. pylori* infection involves two courses: a triple chemotherapy and quadruple therapy, depending on the condition of the patient. The former involves the use of three drugs (a proton pump inhibitor or PPI like omeprazole, and two antibiotics, clarithromycin and amoxicillin) given twice a day for one week. The latter is recommended if the triple therapy fails: four drugs (PPI, two antibiotics—metronidazole and tetracycline—and a bismuth compound) given for two weeks.

The bottomline on peptic ulcers is this: if you are suffering from it for a long time, ruling out other causes like NSAIDS, etc., you can be cured by eliminating the pathogen, *H. pylori,* with an antibiotic prescription from your doctor.

September 7, 2006

Philip Ian Padilla

Bacterial Protein Toxins

A toxin is defined as a poison that is of animal, plant or microbial origin causing disease when present even in low concentrations. Animal toxins are generally for predation (snake and spider venoms) and defense (bee and wasp stings). Examples of plant toxins are ricin and RCA (*Ricinus communis* agglutinin) from castor bean extracts (one seed can be fatal to a child). And then there are bacterial toxins that help these microorganisms colonize their hosts. This article will introduce the bacterial protein toxins and highlight their importance in biology and the medical sciences.

I have been studying the vacuolating toxin (VacA) since my graduate work in the Institute of Tropical Medicine in Nagasaki University. The causative organism for most gastric ulcers, *Helicobacter pylori*, produces it. VacA is a large molecule that induces large vacuole formation in the eukaryotic cells, eventually leading to cell death. Thus, VacA is very cytotoxic to mammalian cells, but it has its uses in biomedical science. In fact, with our discovery of a potential receptor candidate for the toxin in the laboratory, VacA is now a very valuable tool in studying cell vacuole formation, cell signaling from membrane receptors and pathways leading to programmed cell death (apoptosis). It is noteworthy that only the purified native form of the toxin has the ability to induce vacuole formation. Unlike other more common bacterial toxins, the recombinant or cloned VacA is inactive.

The most frequently used example of a bacterial toxin of medical importance is the potent neurotoxin produced by *Clostridium botulinum*: botulinum toxin or more commonly known by its drug brand name: botox. Ingestion of the toxin from contaminated food (food-borne botulism), infection of open wounds by the bacteria (wound botulism), and ingestion of bacterial spores by infants (infantile botulism) are considered medical emergencies because these can rapidly lead to paralysis and death. Nowadays, because of its ability to cause muscle paralysis, very tiny amounts of the toxin are being used by dermatologists and plastic surgeons to make skin wrinkles disappear. Expanding this "botox effect," it is now being used for a diverse number of neuromuscular disorders like crossed eyes (strabismus), involuntary muscular spasm of the neck (cervical dystonia), writer's cramp, facial spasm, tremors, lower back pain, epilepsy, tennis elbow, and even excessive underarm sweating. One area that is important to Filipino patients is its use to alleviate the debilitating involuntary muscular spasms

---

University of the Philippines in the Visayas, Miag-ao, Iloilo, Philippines.
padillap@skyinet.net or padillap@nhlbi.nih.gov

of *lubag* (X-linked dystonia-Parkinsonism or XDP), a hereditary lethal neuromuscular disorder endemic in Panay Island, specifically in Capiz province and its environs.

The tetanus toxin is another example of a neurotoxin. Infection of wounds by *Clostridium tetani* is the principal cause of tetanus (lockjaw). Its action is the direct opposite of botulinum toxin: instead of flaccid paralysis, it causes spasms of the muscle due to excessive firing of signals from the neurons (spastic paralysis). Both the tetanus and botulinum toxins have contributed to our understanding of how our nervous system functions. The tetanus toxin itself is used as an ingredient in vaccines to prevent tetanus.

Toxins are also defined and grouped according to their protein structure. The following are members of a family of toxins called A/B toxins which are composed of two distinct protein molecules (A and B) with the former serving as the vehicle for cell entry (by binding to its receptor) and the latter bringing about the biochemical event leading to the disease state: cholera toxin, Shiga toxin (Stx), *E. coli* heat labile (LT) and heat stable (ST) toxins, diptheria toxin, pertussis toxin, *Pseudomonas* exotoxin A, and anthrax toxin.

The cholera toxin is responsible for the excessive watery diarrhea ("rice water stools") in cholera, which is brought about by consumption of food or water contaminated by *Vibrio cholerae*. Other toxins that attack the gastrointestinal system, giving rise to some type of diarrhea, are the Shiga toxin (Stx) that is secreted by *Shigella dysenteriae* and the enterohemorrhagic *E. coli* (EHEC), and enterotoxigenic *E. coli* (ETEC) LT and ST toxins. Patients infected with *Shigella dysenteriae* (shigellosis) exhibit fever, abdominal cramps, and bloody diarrhea due to the effects of the Shiga toxin. Although symptoms could resolve after five days in adults, infants and toddlers can be so severely affected that they need hospitalization. EHEC with the specific serotype 0157: H7 is the main culprit in numerous fatalities involving food poisoning cases in the US and Japan because it can lead to severe renal failure (hemolytic-uremic syndrome). ETEC infection is the most common cause of diarrhea cases for tourists visiting Third World countries (travelers' diarrhea or TD).

The diptheria (*Corynebacterium diphtheriae*) and pertussis (*Bordetella pertussis*) toxins produce extensive respiratory mucosal cell death in diptheria and whooping cough. Another toxin that is produced by a respiratory pathogen is the *Pseudomonas* exotoxin A. *Pseudomonas aeruginosa* infection can lead to severe pneumonia. Biochemical studies on these toxins by virtue of their common structure have led to our understanding of how cellular signals are transmitted from the plasma membrane to the cytoplasm (the role of secondary messengers like adenyl cyclase in the activation of cAMP). And because of their ability to induce a strong immune response, members of the A/B family of toxins (whole or fragments) are incorporated in vaccine preparations to prevent the diseases that they cause, e.g., DPT vaccine (composed of detoxified diphtheria toxin, pertussis antigens and toxoid and tetanus toxin). Because of their ability to stimulate a strong immune response (adjuvant property), the cholera toxin and *E. coli* LT are being incorporated in experimental vaccines for flu, HIV, and even *H. pylori.*

The *Pseudomonas* toxin and Shiga toxins (Stx) have been used for cancer immunotherapy. They are used as "missiles" (or "magic bullets") that target cancer cells for destruction. By targeting specific receptors found in cancer cells, they bring about cytotoxicity. These toxins are usually attached to radioactive materials and antibodies that further enhance their lethality. Clinical trials are now ongoing to test how effective these are for leukemia and other related cancers.

The most well-known bacterial toxin in this post-9/11 world is the anthrax toxin because of its use for bio-terrorism. Anthrax can cause skin and pulmonary anthrax with the latter being almost fatal. A few weeks after hijackers crashed two commercial airplanes into the World Trade Center buildings in New York, letters with white powder were mailed to several TV and print media headquarters and the US Senate. The white powder turned out to be a weaponized form of the anthrax spores. Once inhaled, the bacteria multiply rapidly producing their lethal toxin. Five people died because of these letters, including two postal workers, with another seventeen people seriously injured. Interestingly, in contrast to the other members of the A/B toxin family, the anthrax toxin has three components: Factor I (edema factor), Factor II (protective antigen), Factor III (lethal factor). Each produces an adverse reaction to the host cell but is not deadly. All three factors combine to produce toxicity by attacking the lungs and white blood cells (lymphocytes). The anthrax toxin is currently being used as a tool to study the relationship between edema and cAMP, and its effect on the immune system, i.e., macrophage phagocytosis, contributing to our knowledge of how our immune system is regulated when a particular pathogen invades our bodies.

These microbes and their associated toxins have evolved to adapt in disrupting numerous biochemical pathways in eukaryotic cells. Despite the fact that these toxins are harmful to human beings, they have been used and exploited as probes in biomedical research to further our understanding of basic cell biology and microbial pathogenesis, and also to find and develop promising treatments for numerous medical maladies.

May 24, 2007

Cynthia Hedreyda

Microbiologists Deal with Very Small Organisms and Play Crucial Roles in our Lives

Several years ago, the much talked about fatal cases of bloody diarrhea in the US were caused by eating Jack-in-the Box hamburgers contaminated with a bacterium called *Escherichia coli* strain O157. I heard then US President Bill Clinton, in a television interview, blame a virus for the death of the infected children. Even the US president got confused or was not aware that a bacterium is different from a virus. About four years ago, when I invited Professor Winnie Monsod (UP School of Economics and GMA–7) to be the keynote speaker of a national convention of Filipino microbiologists, she mentioned that she had to search the Internet for a crash course in microbiology before the convention. Last year, I was invited by former Quezon City Vice Mayor Charito Planas to be a resource person in her radio program where she introduced me as a microbiologist. She emphasized that it was the first time she had invited a microbiologist in her program and was interested to find out what microbiologists do. This was when I realized that there is an urgent need to promote popular awareness of microbiology in the country. If outstanding and distinguished personalities like Prof. Monsod and Atty. Planas are not very familiar with basic microbiology and are not aware of what microbiologists do, then the same is true for so much more millions of Filipinos. How does one become a microbiologist? Where can we find microbiologists? What roles do they play in our community, the country, or the world in general?

Microbiology is the study of very small organisms (microorganisms) that can only be seen under the microscope. A microorganism may be made up of a tiny single cell (unicellular) like all bacteria, yeasts, and protozoa (amoeba, euglena, and paramecium). In the laboratory, several bacterial or yeast cells grouped together may be seen with the naked eye as a colony in an agar-based medium. Other species of microorganisms may be made up of more than one cell (multicellular) such as molds. Microbiology also includes the study of even smaller single entities (not cellular) called viruses, which contain nucleic acid (DNA or RNA) enclosed in a protein coat. Take note that there are useful bacteria, yeasts, and molds that are used to produce food, pharmaceutical, and industrial products. On the other hand, there are also species of bacteria, yeast, molds, protozoa, and several viruses that are agents of plant, animal, and human diseases. A microbiologist,

National Institute of Molecular Biology and Biotechnology, University of the Philippines, Diliman, Philippines.
hedreyda@laguna.net

therefore, is someone who underwent formal education in microbiology or related fields and training to work with any of the above microscopic organisms and/or viruses. A microbiologist is expected to be knowledgeable in basic microbiology techniques for all microorganisms, but may have focused work and obtained expertise on just one to a few species of bacteria, yeast, mold or a particular type of virus.

A microbiologist may be a scientist in an academic and/or research institution performing basic research on a specific microorganism or group of microorganisms. Microbiology covers a wide range of topics such as microbial physiology, microbial ecology, medical microbiology, food microbiology, virology, and microbial genetics. Basic microbiology research aims to gain knowledge and information about microorganisms, without emphasis on immediate use or application. Basic microbiology research generates information about a microorganism's structure, physiology, functions, and interactions (with members of the same species, with other species, or the environment). Basic microbiology research may be conducted to determine the ability of microorganisms to cause diseases, to arrive at species identification, to achieve characterization and to evaluate relatedness (phylogeny) and classification.

Microbiologists may also be involved in applied research using microorganisms toward the development of products and processes useful to man. Applied microbiology research may be conducted with any of the following objectives: to determine the ability of a microorganism to produce a useful product, to study the factors that favor microbial growth and efficient synthesis of the useful product, to generate a strain of microorganism with improved quality and/or quantity of product, to identify and eliminate the factors that will inhibit production, and to eradicate undesirable microorganisms that produce toxins or other unwanted products. Therefore, microbiologists may be interested in the growth and maintenance of live microbial cultures that produce desirable products or may be focused on preventing the growth of undesirable microorganisms such as those that cause undesirable flavors, food spoilage, and plant and animal diseases.

There are agricultural microbiologists with formal education in agriculture but work with microorganisms that enhance plant growth or cause plant diseases. Veterinary microbiologists (veterinarians by profession) work with microorganisms that are causative agents of animal diseases. Most medical microbiologists work in hospitals and handle microorganisms that cause human diseases and may be medical technologists, physicians, or nurses by profession. Medical microbiologists are responsible for identifying microorganisms that cause human diseases, obtaining crucial information necessary to suggest the best way to eliminate the disease-causing agent. They may also be tasked to determine microbial susceptibilities to drugs like antibiotics before doctors can recommend appropriate medication. Veterinary or medical microbiologists make use of killed and weakened (attenuated) disease-causing microorganisms or viruses to produce animal and human vaccines that result in protection against different infections. Food microbiologists are important in industries manufacturing different food products using microorganisms. Production of beer and wine, for instance, use the yeast Saccharomyces cerevisiae to produce the alcohol. Cheese production makes use of certain species of bacteria and fungi for cheese maturation and flavor formation. Food microbiologists may also be hired by companies involved in large-scale production of fermented products, including several traditional fermented foods like fermented vegetables ("burong mustasa" and "kimchi") and fermented fish and shrimp ("bagoong," "alamang" and fish sauce). Other food microbiologists are responsible for ensuring food safety or absence of unwanted microbial contaminants or microbial toxins from processed food.

Environmental microbiologists are concerned with using microorganisms to degrade household, agricultural or industrial wastes which will otherwise result in polluted land areas and bodies of water, if left untreated. There are also aquatic microbiologists or marine microbiologists who work on microorganisms that infect economically important fishes, shrimps, seaweeds, corals, and other marine organisms. Industrial microbiologists are needed in several industries that rely on microbial enzymes, including companies that produce digestive aids, meat tenderizers, detergents, stain removers, and leather products. In addition to these, microbiologists are needed in companies producing microbial organic acids, dyes, antibiotics, and amino acids. Microbiologists are even hired by pharmaceutical and cosmetic companies where they are tasked to handle microorganisms that produce pharmaceutical products. Moreover, microbiologists are also responsible for quality control to ensure that their products are free of microorganisms that may cause diseases or that their claims of antimicrobial active ingredients are valid. Even airline companies hire microbiologists to guarantee food safety of flight meals. In advanced countries, production of useful proteins by genetically improved microorganisms is conducted, including human insulin for diabetics, growth hormones, and antitumor factors.

Microbiologists may be trained to be teachers in high school or may decide to be mentors in college, responsible for educating young students on the principles of microbiology. These microbiology educators are instrumental in producing and training future microbiologists of the country. Some microbiology teachers are also involved in research and extension. There are microbiologists who have decided to be businessmen or biotechnologists, producing microbial-based products as cottage industries or on a commercial scale. Still other microbiologists work in service diagnostic laboratories or work as sales representatives promoting products used in microbiology laboratories. Therefore, microbiologists are needed in schools and universities as teachers, in several biotechnology industries as entrepreneurs, in farms and aquaculture, in hospitals and diagnostic medical laboratories, and for environmental management. The demand for microbiologists has not at all dwindled through the years.

How do you become a microbiologist? There are several possible routes to becoming a microbiologist. The direct route is to obtain a Bachelor of Science degree in Microbiology or a Bachelor of Science degree in Biology with Microbiology as major.

These programs, however, are only offered in about four institutions in the country. To be a more competent microbiologist, additional training and work experience with microorganisms after completing a BS degree are deemed necessary. Expertise in microbiology may also be further developed by seeking advanced degrees such as a MS and PhD in Microbiology here or abroad. There are also several opportunities for deserving young microbiologists to avail themselves of microbiology training abroad.

While it is true that only a few universities in the country offer BS, MS, and PhD in Microbiology, there are other routes to becoming a microbiologist. One may take BS Medical Technology and obtain additional training in microbiology to become a medical microbiologist. One can be a microbiologist by completing any BS degree in related fields like Agriculture, Food Science, Botany, Zoology, Chemistry, Environmental Science, etc., with microbiology as the minor and by working on microbiology after graduation. Additional microbiology training or degree will further enhance expertise in the field. Microbiologists need to undergo constant training and retooling for advanced techniques in order to get updated on new findings and procedures for the advancement of the profession. Therefore, one may have obtained education and expertise in a different field of science initially, but after training and relevant microbiology experience, one may find himself working on microorganisms and be considered a microbiologist. Constant retooling and updating are important for a microbiologist to be abreast with the fast advances in the field.

There are several microbiology-related professional organizations in the Philippines. The Philippine Society for Microbiology Inc. (PSM Inc.), for instance, boasts of over 3,000 members who are practicing microbiologists or are people in other fields but are very much interested in microbiology and biotechnology. To date PSM has three regional chapters: PSM Visayas, PSM Mindanao, and PSM Northern Luzon. PSM National and all three chapters hold annual scientific meetings for updating on new findings in the field. The societies also organize national and international symposia on microbiology topics. PSM hosts an international Asia-Pacific Biotechnology Congress every five years where microbiologists with different expertise from other countries are invited to participate. Other microbiology-related professional organizations with microbiologists among their members include the organization of medical technologists and an organization of health workers involved in infectious diseases. In addition to several microbiology-related societies, there is the Philippine Academy of Microbiology (PAM), an accreditation arm of PSM with Diplomates and Fellows who have qualified based on outstanding credentials in microbiology teaching, research, and extension as members. PAM has been responsible for evaluating and accreditation of Registered and Specialist Microbiologists in the country. To qualify for accreditation, one must exhibit enough knowledge on Basic Microbiology, Microbial Physiology, Microbial Ecology, Virology, Food Microbiology and Medical Microbiology by passing a comprehensive examination administered by PAM examiners.

I therefore encourage our young students to become microbiologists using any of the possible routes I mentioned. Microbiologists have the luxury of a wide range of job opportunities here and abroad. Most of all, being a microbiologist can be a very rewarding profession. Several microbial-based products have significantly contributed to agriculture and aquaculture. Advances in medicine and solutions to problems in animal and human health have relied heavily on the contributions of medical microbiology research. Microorganisms are still among the key players to address our past, present, and future problems in our environment. Microbial-based industries are expected to significantly catalyze our attempts at nation-building.

Microbiologists are everywhere. Seek a microbiologist. You can easily find one because they are in schools, in farms, in the industry, and in hospitals. Talk and interact with them and learn more about microbiology and what a microbiologist can do for you.

Printed in two parts on February 15 and February 22, 2007

Vicente Belizario Jr.

Epidemiology of Malaria in the Philippines

Malaria, caused by a parasite belonging to the genus *Plasmodium*, has long been a significant health problem in the Philippines. The recent unfortunate demise of the son of a known TV journalist and his companions after doing a story in the hinterlands of Palawan has put in the public limelight once again this deadly but preventable and curable disease.

Malaria is the world's most deadly tropical parasitic disease. It kills more people than any other communicable disease, except tuberculosis. In many developing countries, especially in Africa, malaria takes an enormous toll on lives, medical costs, and days of labor lost. The geographical areas affected by malaria have shrunk considerably over the past fifty years, but control is becoming more difficult due to activities like road-building, mining, logging, and new agricultural and irrigation projects, particularly in "frontier" areas like the Amazon. Disintegration of health services, armed conflicts and mass movements of refugees have also worsened the malaria situation.

Malaria is a public health problem today in more than 100 countries inhabited by a total of some 3.2 billion people. High-risk groups in endemic areas include pregnant women, nonimmune travelers, refugees, displaced persons, and laborers entering those areas.

Malaria causes 1.5 million to 2.7 million deaths each year, and the majority of mortalities occur among young children in Africa, especially in remote rural areas with poor access to health services. Its worldwide prevalence is estimated at 350 million to 500 million clinical cases each year. Around 60 percent of clinical malaria cases and 80 percent of all deaths due to malaria are in sub-Saharan Africa. In the Western Pacific Region, the World Health Organization lists ten countries where malaria is endemic, namely Cambodia, China, Lao People's Democratic Republic, Malaysia, Papua New Guinea, the Philippines, Republic of Korea, Solomon Islands, Vanuatu, and Vietnam.

In the Philippines, malaria still ranks as one of the ten leading causes of morbidity with sixty-five out of seventy-nine endemic provinces reported. The profile of malaria varies throughout the country. There is relatively high endemicity in the provinces of Palawan, Kalinga Apayao, and Ifugao. Other provinces have been reported to have generally low endemicity and sporadic malaria. It appears that in areas of low malaria endemicity, there is clustering of cases, resulting in pockets of high endemicity.

National Institutes of Health and College of Public Health, University of the Philippines, Manila, Philippines.
vbelizar@yahoo.com

A total of 11.3 million Filipinos, or 14.8 percent of the population consisting mostly of farmers, indigenous people, miners, forest product gatherers, and soldiers, are at risk of infection. In 2003, the incidence in the Philippines was reported at 0.5 per 1,000 population. Standardized rates of reported malaria cases in the Philippines have been stable with a gradual decreasing trend, from a high of 1.4 cases per 1,000 in 1990 to 0.5 cases per 1,000 in 2003, but still above the regional average. The Department of Health reported 286 deaths due to malaria in 2000.

More than 70 percent of malaria cases in the Philippines are caused by *Plasmodium falciparum*, less than 30 percent by *P. vivax,* and less than one percent by *P. malariae.*

In the Philippines, the principal malaria vector is *Anopheles minimus var. flavirostris*, a night-biter which breeds in slow-flowing, partly shaded streams that abound in the foothill areas. Occasionally, it has the ability to adapt to or utilize new habitats such as irrigation ditches, rice fields, pools, and wells. Its horizontal flight range has been reported to be about one to two kilometers.

Malaria can also be transmitted through blood transfusion from infected donors, and by contaminated needles and syringes. Blood from semi-immune donors without clinical symptoms may also contain malaria parasites. In congenital malaria, infected mothers transmit parasites to their child before or during birth.

Malaria is initially suspected in patients based on clinical symptomatology. Signs and symptoms include fever, chills, headaches, and muscle pains, all of which are non-specific and can also be found in other diseases like common viral infections such as the flu. Hence, early clinical findings are not reliable, thus necessitating the use of laboratory tests to confirm malaria. Microscopic diagnosis for malaria remains the gold standard for laboratory confirmation, where malarial parasites are identified in a drop of a patient's blood smeared on a slide. Other tests available include the more expensive antigen detection or rapid diagnostic tests, which provide immediate results but may have variable sensitivity.

In the Philippines, the treatment of choice of malaria has been chloroquine, until reports of the emergence of drug-resistant strains of *P. falciparum* were made. According to the 2005 world Malaria Report, the efficacy of chloroquine in the country, as expressed in percentage of treatment failure, was reported to be 42 percent, which is above the acceptable value of less than 25 percent. On the other hand, with the use of chloroquine combined with sulfadoxine-pyrimethamine, the current first-line drug regimen for *falciparum* malaria, higher efficacy has been shown with a lower treatment failure rate of 18 percent. This regimen appears to be more expensive, thus illustrating the importance of optimizing present treatment protocols for malaria.

Although the challenge for more effective malaria control in the country remains, there are efforts that show considerable promise in the fight against this disease. The Agusan del Sur Malaria Control and Prevention Project, a collaborative work among the Department of Health, Research Institute for Tropical Medicine, University of the

Philippines College of Public Health, and the local government units in Agusan del Sur, with support from the Australian Agency for International Development, has shown what can be described as a success story in the battle against malaria. The project combined strategies such as early case detection and treatment, surveillance with the use of a Malaria Information System, health education and promotion, use of insecticide-treated nets (ITNs), social mobilization, and institutional strengthening. More importantly, the project was able to develop and test a malaria control program that is self-sustaining, community-based, and currently managed by the local government units of the province. The project has demonstrated a reduction of malaria cases by 85 percent compared to baseline levels.

The challenges for malarial control in the coming decade include ensuring the quality of available antimalarial drugs in both the public and private sectors, increasing the coverage of quality microscopic diagnosis and rapid diagnostic tests, and increasing distribution and utilization of ITNs. In all these, evidence must be derived as basis for policy and practice. Moreover, national and empowered local government agencies and units should then make possible access to quality and community-based health services that will be crucial for early diagnosis and prompt treatment, thereby ensuring reductions in morbidity and prevention of mortality. Meeting the challenge would mean the lives of the young journalist and his companions were not in vain and thousands more won't have to be taken again.

December 15, 2005

Jose Enrico Lazaro

Malaria Drug Development in the Third World

Malaria is a serious and sometimes fatal blood disease caused by parasites of the genus *Plasmodium*. It kills about two million people a year, and gives 300-500 million others debilitating cycles of fevers and chills. It is also a poor man's disease. The common cure, chloroquine, costs less than a quarter of a can of soda. But chloroquine is losing its effect because of drug resistance. The newer alternatives can cost up to 750 times more.

But, with the high cost of drug development, a low-income target market, and uncertain patent protection, one asks, "Who will do it?" Few drug companies would, that is why public-private partnerships were founded. These institutions combine government funds, the World Bank (WB), and the World Health Organization (WHO), and the expertise of drug companies and universities. Aside from relaxed regulatory laws, forming partnerships is an option for countries to distribute the costs and the risks of developing new drugs.

Since the Philippines is in a part of the world with one of the highest species richness, and therefore, a large treasury of potential new drugs, we might ask: "At this time, how likely are partnerships to support a proposal to develop a new, natural antimalaria drug product from the Philippines?" Not farfetched, since the classic cure, quinine, and the newest cure, artemisinin, were both natural products from the South. The answer to this question, however, is "not very likely."

Why?

Multinational corporations, whose job it would be to develop the drugs, would accept that challenge if 1) they also had a market in the First World; 2) development costs were shared; or 3) their commitment would result in long-term benefits.

Companies realize that most of their sales growth in the next ten to twenty years will come from developing countries. The growth of Southern markets stimulates research, promising long-term benefits and public goodwill. Drugs under development in these companies include pyronaridine, sulfalene, mefloquine, and halofantrine, all synthetic.

Artemisinin is the only major natural product in the pipeline. Its source, a weed, has been in use for some 3,000 years, and the drug itself is not patented, a factor contributing to its relatively low price. In this case, the Chinese firm Kunming supplies the drug

National Institute of Molecular Biology and Biotechnology, University of the Philippines, Diliman, Quezon City, Philippines.
jaylazaro@hotmail.com

(already registered and marketed in China), whereas the pharmaceutical giant Aventis registers it in endemic countries at a relatively low cost.

The WHO supports the artemisinin project. WHO funding is channeled through the Tropical Diseases Research program and a large network of government and international entities. The WHO also funds the Medicines for Malaria Venture (MMV), an independent public-private company first incubated within TDR. The MMV receives funds from private foundations like the Gates Foundation, governments, and the World Bank. The MMV funds projects on a competitive basis. The best performers are then linked with drug companies. The top candidates so far are mostly new combinations of old drugs, and all proponents are from the North.

Northern governments support drug development through science agencies and the WHO. Partnerships sponsored include the Federal Malaria Vaccine Coordinating Committee (US); the Japanese Pharma, Ministry of Health and Welfare (JPMW), a collaboration of the Japanese government, the WHO, and twelve Japanese drug firms; the Global Initiative for Traditional Systems (GIFTS) of Health, between the University of Oxford and WHO/Tropical Disease Research Division (TDR) for the study of traditional herbal medicines; and the Malaria Consortium, which coordinates control programs, training, distribution, and policy research.

Northern support also comes through the World Bank (WB). Recognizing that malaria is a major contributor to poverty, and poverty a factor in security and business, the WB has within the last decade become the largest source of loans for health in low and middle-income countries as well as a major player in pharmaceuticals. Most of that support, however, goes to infrastructure and health reform, outside of the WB's contribution to the WHO.

Cash-strapped Southern governments do not generally support new drug research. Higher priorities include infrastructure, surveillance, and quality control. HIV/AIDS, tuberculosis, and childhood diseases are more problematic than malaria. Local companies also do not generally engage in new drug research since copying is more profitable. Even though the potential storehouse of natural drugs is large as a result of high biodiversity in developing countries, this does not mean that discovering a drug with the right mix of strength, safety, and affordability will be cheap.

Hence, global research funding for malaria is small. It comes to about $84 per fatality per year, compared to $3,800 for cancer fatality, and $1,000 for cardiovascular diseases.

Prospects

The potential market for antimalarials is the population at risk—about three billion people. New drugs, donated by drug companies or purchased by the government, are usually distributed through health centers. This is not just because of price: such control helps to avoid inordinate use and the consequent emergence of drug resistance.

Third World governments' power to push for new drugs, though limited by inefficient administration, low budgets, poor distribution infrastructure, corruption, lack of expertise in drug development and regulation, and uncertain political stability, is offset by a strong potential pulling power. Standards of living are going up, and the obstacles just enumerated would be given more attention if health budgets increase as economies continue to grow.

Another advantage that the developing countries can provide to reduce the cost of research and marketing is that the notions of "acceptable toxicity" and "purity of material" are different from those of the developed countries. Developing countries also provide lower salaries, lower risks of litigation, and larger numbers of patients. Even if the Northern markets should be small, the finished products need not be sold there, or they may be sold at a much higher price, provided that "imports" from the Third World are prevented.

New drug development does take place in the Third World, as with artemisinin. The Korean effort to develop and market an antischistosomiasis drug and the Cuban effort to develop a local biotechnology industry are other cases worth studying.

The Korean case is an example of successful copying. Praziquantel was discovered and patented by the German Bayer and Merck. The Koreans were subsequently able to develop and patent a new process at a negligible fraction ($14,000) of the Germans' multimillion-dollar research cost. The patent allowed the Koreans to: a) license the drug to countries where the product was allowed to compete with the Bayer/Merck version; b) outprice the Germans; and c) gain a major share of the market. Shin Poong, once the world's leading producer of praziquantel, owes its success during this period to this one drug.

The Cuban *Centro de Ingeneria Genética y Biotecnología* in Cuba is a case of Third World high technology. The center's mandate is to "develop concrete activities ... which can directly influence the epidemiology of each country and offer less costly solutions for diagnoses, treatment, drugs and medicines, without having to make purchases from large transnational companies." The project is entirely a Cuban initiative. Malaria, however, is not on their priority list.

Printed in two parts on August 25 and September 1, 2005

FRANCESCA DIANE LIU

Plasmo 'die' uhmm? False-Ipa-room?

I DO admit, the title of my article sounds like a Yugoslavian tongue twister whose meaning seems elusive to many; but in the field of medicine and parasitology, it is a word that signifies a complex microorganism causing a hematoprotozoan parasitic infection commonly known as malaria. *Plasmodium falciparum*, the main causative agent of malaria has been discovered for over a hundred years yet is still responsible for 1.5 to 2.7 million deaths annually. The disease continues to affect 40 percent of the world's global population and represents 2.3 percent of the world's disease burden, making it one of the most deadly parasitic diseases of man. The biological complexity of the parasite is made even more complex because of its infective ability and adaptation, which enables the parasite to involve a mammalian host during the completion of its life cycle. The infection results in life-threatening anemia for children and infants while pregnant women are susceptible to induced coma, and *primagravidae* due to the placental sequestration of the parasite. Deaths commonly result from severe anemia, mild acidosis, pulmonary edema and cerebral malaria.

RESISTANCE AND TRANSMISSION

The effort to control malarial infection today lies in the prevention against the spread of mosquito vector (i.e., through the use of nets, insecticides) and the application of antimalarial compounds for case management. Antimalarial compounds commonly used for the treatment include quinine, chloroquine, mefloquine, and sulfadoxinepyrimethamine. If you apply the classic quote "*What does not kill me can only make me stronger*" with the *Plasmodium*, then you get antimalarial drug resistance. Indeed, *P. falciparum* is now highly resistant to chloroquine in most malaria endemic areas. Sulfadoxinepyrimethamine and mefloquine resistance is currently widespread and developing rapidly. The only antimalarial compound reported to have no trace of plasmodium resistance is artemisinin. This compound is currently an essential component for treatment but if *Plasmodium* resistance to it develops, then we would be faced with untreatable malaria. Perhaps, the development of a newer method of control is difficult since the basic biology of the parasite and its unique mode of infection has not been fully investigated.

International Max Plank Research School for Infectious Diseases and Immunology, Germany.
fdiane_liu@hotmail.com

When I was young, I should have answered to the popular adult question, "What would you like to be when you grow up?" with "I would want to be like the *Plasmodium*, sir." If *P. falciparum* were human, it would have obtained its PhD degree and post doctorate in two-years time or less because of its intelligent niche and technique for propagation. *P. falciparum* completes its schizogony (asexual cycle) in the red blood cells of vertebrates (that includes us) and sporogony (sexual cycle) in its carrier, the anopheline mosquito. Immediately after the *Anopheles* bites, the *Plasmodium* parasite enters our body and evades the immune system by sequestering in the liver until it is mature for the next cycle. After prematuration, the parasite (now called a merozoite) invades the red blood cells causing structural deformation and reorientation of the cell membrane and sequesters there until it matures into male and female gametocytes, upon which the red blood cells will rupture releasing the gametocytes in the blood stream waiting to be ingested as blood meal by another *Anopheles* mosquito. What makes this parasite so clever is its ability to fool or deceive our immune system (which releases antibodies specific for an antigen due to an invasion of any foreign material). Not only is the infection "strain-specific" but rather "variant-specific." A single strain can produce different variants of antigenic proteins. In the erythrocytic stage, the merozoites infect the red blood cells and different parasitic proteins are formed in the RBC membrane to aid in the survival, serve as a marker, and aid in other adhesion processes. One important protein antigen is the *Plasmodium falciparum* Erythrocytic Membrane Protein1 (PfEMP1). This protein is highly polymorphic and has evolved to have sixty copies of its gene in each genome therefore enabling it to change. However, even if there is a change in DNA sequence, the activity or function of this PfEMP1 protein is still conserved! One example is its ability to bind to the endothelium through its Cysteine Rich Interdomain Region1. This primitive property enables the PfEMP1 to anchor itself to endothelial CD36 and avoid spleen-dependent killing.

The acute, specific, and polymorphic cleverness of this parasite is the primary reason that researchers are having difficulty in developing a vaccine. To make things more complicated, each of the different maturity stages of the parasite is marked by different proteins. Examples of these proteins are the circumsporosite for the hepatic sporocyte stage, the PfEMP proteins for the erythrocytic stage, and the Pfs25 protein for the oogenic stage.

Although the possibilities of an antimalarial vaccine seem remote due to the nature of the parasite, there are clever ways to overcome the barriers imposed by the *P. falciparum*. How does an amateur like me propose to meet the challenge?

The Challenge

The first challenge is choosing the stage of parasitic invasion. This problem is important because it determines which type of antigenic protein is to be targeted, since each stage has its individual mode of parasitemic effect. Vaccines directed against the liver stages are designed to prevent merozoite invasion of the red blood cells, thereby avoiding all clinical manifestations of the disease. Vaccines directed against the erythrocytic stage, however, are designed to minimize the severity of infection. Vaccines directed against the mosquito stage aims to block the transmission of the disease. One important factor to consider is the length of time the parasite is exposed versus the length of time it is sequestered. This is for the immune system to gain advantage in terms of early detection and attack before the parasite hides again.

The second challenge is the problem of polymorphism. Because of the primitive need of the parasite to survive inside the host and, with that, avoid being detected and eliminated by the immune system, its antigenic proteins need to change form. How to overcome? Select an antigenic protein from a specific stage (in lieu of the first challenge) that is the most conserved. How do you determine conservation? Align and analyze multiple protein sequences of the antigen (provided that the sequences are known), then determine the specific regions that are highly variable or highly conserved. Elucidate the secondary structure of the protein to determine the presence of alpha helices, beta sheets or coils. Better yet, if the 3D structure is known, analyze the sequences on the basis of the structure. Somehow the structure is related to the functional characteristic of the protein. Warning! Do not be fooled by the ease of words, this part of the challenge is highly tedious and requires lots of mental and visual power. (Trust me, I know. I have been through it.)

The third challenge is the problem of species and strain. Malaria is caused by other *Plasmodium* species aside from the *P. falciparum.* These species are *P. vivax, P. malariae,* and *P. ovale* (though *P. falciparum* and *P. vivax* are the species that contribute most to the malaria burden) and each type has its own strains. To propose a vaccine that would be effective for malaria, all species and strains should be taken into consideration. The difficulty of the challenge is attributed to the difference in the manner of infection of individual species. How do we propose to answer this problem? Why not formulate the vaccine in such a way that the antigenic proteins to be targeted are not specific for only one species but a combination of antigens representing each type, much like a cocktail drink?

My interest in malaria began when our professor, Dr. Eduardo Padlan, asked us to propose a targeting molecule against any type of disease for our MS 397B exam on "Molecular Targeting of Marine Natural Products." We were to find evidence for feasibility and data for proof for our chosen disease. I chose to study malaria and target the disease through the *Plasmodium falciparum* Erythrocyte Membrane Protein. My classmate, Charie, asked me, why not cancer? It got me into thinking: yes, why not cancer? Maybe because oncology is such a glamorous field of research and I chose to study a less glamorous field. Or maybe because I was infected with dengue a few months ago and the irony of life was unforgettable, but I figured, I am simply here for the challenge. I even considered pursuing malaria as my masteral thesis in Biochemistry, hoping to elucidate the structural and characteristic binding of the CIDR1 region to CD36 (of course, with the help of my mentors, Dr. Gisela Concepcion and Dr. Eduardo Padlan of the Marine Science Institute, UP Diliman). I set aside thoughts that the combination of immunology, molecular biology, biochemistry, and parasitology would impose a degree of difficulty that my mind cannot even comprehend, resulting in the downfall of my graduate studies. I guess, I believe in taking risks and following where the urge of science leads me.

Let me end by explaining my title.

"*Plasmo*"—The main causative agent of malaria will be impossible to eradicate causing even more people to "*die.*"

"*Uhmm?*" My answer is "*false*" because I believe that with the extent of knowledge and research on malaria that we have at present, hopes are bright for a successful discovery of a vaccine possible to counter malaria.

"*I*" advice to leave "*pa-room*" for faith, conviction and intensive research work.

Who knows, we might outsmart this little bugger yet.

Combine all the words together Plasmo-die-uhmm False-I-pa-room. Doesn't it sound like *Plasmodium falciparum*. So much for the Yugoslavian tongue twister.

Printed in two parts on September 8 and September 15, 2005

Luz Acosta

A New Tool for an Old Disease

We may be blessed with beautiful tropical islands and year-round warm weather, but unfortunately, the Philippines is also haven for many deadly tropical infectious diseases afflicting many of our fellow Filipinos. We have malaria, still the number one killer parasite that threatens the lives of young children in practically all provinces where there are mountain streams and malaria mosquitoes. And schistosomiasis, a blood fluke infection, ranks second to malaria in terms of public health importance. Most of us who are from the city have probably never heard of schistosomiasis, which is still affecting thousands in the remote rice-farming villages of Sorsogon, Leyte, Samar, Bohol, and Mindanao. Like malaria, schistosomiasis has probably been afflicting man since ancient times. There is archeological evidence of the disease from schistosome eggs found inside mummies of ancient Egypt.

It is now almost 100 years since the first case of schistosomiasis was accidentally discovered in the Philippines in 1906 among "Waray" (originating from Samar) inmates of the Bilibid prison in Manila. After a routine course of intestinal parasite screening, eggs were found in stools, later to be identified as *Schistosoma japonicum* eggs. Popularly called "snail fever" among the local folk in Leyte, the disease is caused by flukes that infect man and other mammals, including dogs and carabaos, feed on their blood, and reside as adult worms in the blood vessels of the liver and intestines.

The disease creeps in very slowly. Early symptoms such as paleness and malnutrition may be mistaken for ordinary symptoms of roundworm or hookworm infections common among kids in the rural and urban areas alike. One female worm can lay thousands of eggs, the majority lodged in tissues causing inflammation of the liver and intestines, disrupting intestinal blood flow, later to manifest as the "big protruding belly" syndrome in severe intestinal schistosomiasis.

The parasite initially develops into different forms inside the *Oncomelania quadrasi* snail. These are minute snails, smaller than a grain of rice, found thriving in swampy rice-farming ecosystems of endemic municipalities. The parasite's infective form can penetrate the bare skin of its targets, thus farmers and their families who are in frequent contact with water and rice paddies where the snails are found are the ones who are ready prey for these blood-feeding parasites.

Vicious as they may seem, schistosome worms are no match to the drug Praziquantel (Biltricide). Discovered by Bayer to treat animal parasites in 1975, Praziquantel was

Department of Immunology, Research Institute for Tropical Medicine, Department of Health, Alabang, Muntinlupa City, Philippines.
lpacosta@yahoo.com

shown to be very effective in killing all types of human schistosome worms in one single dose. While large-scale community-based treatment made a major impact in decreasing infection and disease worldwide, sad to say, it is clear that the control of schistosomiasis was never an easy task. Schistosomiasis remains a public health nuisance in many endemic villages in the country where the snails still abound. In fact, a new endemic focus was recently discovered north of Luzon in Cagayan Valley. Even if treated regularly, due to extensive occupational exposure and the presence of infected animals continuously contaminating the environment with eggs, many continue to get re-infected, thus requiring repeated treatments (yearly treatment), which is often very expensive for many resource-poor endemic communities.

The detection of infection relies on finding eggs in stools of infected patients. This technique, however, is not 100 percent accurate in detecting all infections. Even a trained microscopist will not find eggs in as many as 35 percent of cases after examining duplicate slides (stool smears) from a single collection of stool sample. Far better techniques (e.g., DNA tests or the use of antibody targeting specific antigens) are available than simply demonstrating the presence of eggs from a very small amount of fecal material. For schistosomiasis, a sensitive test that is simple and affordable, one that can be used in the field for community-based control programs, is a better alternative diagnostic tool.

In our effort to contribute to and uphold sustainable development in our country, the Research Institute for Tropical Medicine (RITM) in Alabang, Muntinlupa, currently headed by Dr. Remigio M. Olveda, continues to enhance tools in biotechnology to upgrade its research and development program. Mandated as the research arm of the Department of Health, the RITM focuses on carrying out high-quality, multi-disciplinary research activities to contribute to the fight against tropical and infectious diseases.

With research funds provided by the Department of Science and Technology through the Philippine Center for Health Research and Development (PCHRD), the RITM developed an immunodiagnostic "dipstick" kit, called the SJ-URIDIP, for diagnosis of *Schistosoma japonicum* infection. The technique makes use of a locally produced monoclonal antibody to specifically bind or target the *S. japonicum* circulating cathodic antigen (CCA). CCA is a highly glycosylated proteoglycan molecule largely secreted by cells lining the intestines of adult worms. It is just one of the many antigenic molecules secreted by the parasite which are released in the blood. The good thing is that CCA is also found in the urine and therefore, the use of urine instead of blood or stool for diagnosis makes the collection of samples easier for community-based screening for schistosomiasis.

Antibodies are not only for the immune system. These molecules are now widely used in developing diagnostic kits for various infectious and systemic diseases. Many of these techniques rely on the strong binding affinity of antibodies to its specific antigen, addressing problems of sensitivity and specificity in many detection systems. Monoclonal antibodies are produced by fusing antibody secreting B cells of mice with mutant myeloma cells to allow the continuous propagation of these cells in cell cultures. There is no need for growing them inside the body. From a single cell, these are propagated and grown in large amount, secreting one type of antibody, and therefore the use of the term "monoclonal" antibody.

The SJ-URIDIP kit was tested in the field and showed promising results. Urine samples collected from several endemic barangays in Samar and Leyte were tested for the presence of CCA in urine and compared with the results of multiple stool examinations for the presence of eggs in stools. We demonstrated that the URIDIP test is very sensitive in detecting moderate and heavy infections. However, it is not much better than a stool examination in detecting very light infections, but still it is better than the results of examining a single stool sample. We also collected urine from infected individuals after they received Praziquantel treatment. The test results showed that one month after treatment, almost all CCA-positive individuals became negative in the urine. Since other parasitic infections are very common in these populations, and were not killed with Praziquantel treatment, this shows us that the URIDIP test specifically targets the schistosome antigen and not other parasites. The ease of collecting a urine specimen for the test instead of stool or blood would result in high compliance for screening and treatment in many endemic communities nationwide.

Clearly, we now have in our hands a new tool for an old disease. But getting this to where it is needed is another challenge to hurdle. The Philippines is now lagging far behind its neighbors in Asia in the field of biotechnology. It is high time that we promote biotechnology and contribute to the nation's sustainable development. As for RITM's concerns, we still have other diseases to target—malaria, TB, rabies, dengue and many other deadly infections. Developmental research may be a long and painful path, but it is worth pursuing and it can result in significant benefits to our countrymen.

January 5, 2006

Eduardo Padlan

Mad-ewe → Mad-cow → Mad-you?

Fear of "mad-cow disease" and the related new variant Creutzfeldt-Jakob Disease (nvCJD) in humans persists and the fear is not unfounded. Although only a little over 150 cases of nvCJD have been confirmed worldwide (mostly in the UK), the disease is invariably fatal. No cure for mad-cow disease or nvCJD is in sight, and there is no vaccine to prevent it either.

The fear of mad-cow disease can have devastating economic consequences. As soon as a single case of a "mad" cow is discovered in a country, other countries immediately impose a ban on the importation of beef and beef products from that country—a ban that could last for years—and that often results in the slaughter of cattle herds.

Unlike other infectious diseases that are usually transmitted by viruses, bacteria or other microorganisms, mad-cow disease appears to be caused by an infectious protein—a prion. Prions are unusual proteins. In much the same way that water seeks its lowest level, a protein molecule likewise seeks a state in which its energy is at a minimum—where it is most stable—and there is usually only one such state. For prions, there is more than one stable state and one of those states causes disease.

A number of prion diseases are known to afflict humans and various other animals. Prion diseases are also called Transmissible Spongiform Encephalopathy (TSE) because the damage they cause to the brain makes the brain take on the appearance of a sponge. The prion disease of sheep and goats is called scrapie and its existence has been known for more than 250 years. Mad-cow disease, also called Bovine Spongiform Encephalopathy (BSE), is believed to have arisen from the practice (now banned) of adding sheep brain and other nervous tissue to cattle feed. Other animals, including cats, rats, mice and mink, among others, also get TSE when fed scrapie-infected feed.

Some of the known human prion diseases are Creutzfeldt-Jakob Disease (CJD), Gerstmann-Straeussler-Scheinker Syndrome (GSS), Fatal Familial Insomnia (FFI) and Kuru. The nvCJD is believed to be caused by eating BSE-infected beef. Surprisingly, there is not a single case of scrapie being transmitted to humans even though sheep and goats have been on man's diet since biblical times.

(My colleague at the UP Marine Science Institute, Dr. Gisela P. Concepcion, and I recently analyzed the fragments that probably result from the digestion of prion protein in our stomach. We found that one of the fragments corresponds closely to a synthetic

---

Formerly with the US National Institutes of Health and currently with The Marine Science Institute, University of the Philippines, Diliman, Quezon City, 1101, Philippines.
edpadlan@yahoo.com

prion peptide that has been shown to be infective. We also found that the prions of rats and mice are the most similar to human prion in that part of the molecule—closer than even the prions of sheep or cattle. We therefore proposed that an alternative route of transmission of scrapie to humans is via rodents, possibly through rodent droppings.)

Once a prion molecule has switched over to the disease state, it induces other prion molecules around it to assume that state. Soon, there are large aggregates (plaques) of the disease form and those aggregates cause the death of cells. The consequences of cell death are greatly pronounced in the brain because we do not regenerate brain cells—at least not to the same extent, or with the same facility, that we grow new cells in our other organs.

Kuru, the prion disease discovered among the Fore tribesmen of Papua New Guinea, is almost certainly transmitted through ritual cannibalism. The Fore believe that they could gain the wisdom of an elder by eating his brain—after the elder had died, of course. Well, they may have gained the wisdom of the elder, but they also got his prion disease. The incidence of kuru dramatically decreased and eventually stopped after the Fore abandoned the practice.

There seems to be a strong genetic component in prion diseases. For example, some sheep are resistant to scrapie, some are susceptible, and some are very susceptible. Resistance or susceptibility to the disease is linked to the presence of particular amino acid types at certain positions in the amino acid sequence of the prion molecule. Further, susceptibility to CJD, GSS and FFI is linked to mutations in the prion gene. Even further, the persons who had died of nvCJD in the UK were found to be homozygous for methionine at position 129 in their prion protein (that is, both copies of their prion gene—one from the father, the other from the mother—code for the amino acid, methionine, at position 129 in the prion sequence).

Curiously, the gene coding for the prion protein that has methionine at position 129 is quite common in the human population. It is possible that heterozygosity at position 129 (the two copies of the gene code for different amino acids at that position) confers protection against an as yet unidentified disease. There is already an example of such protection: people who suffer from sickle-cell anemia are homozygous for the amino acid, valine, at position 6 of their hemoglobin beta chain; individuals who are homozygous for the more usual glutamic acid at that position do not get sickle-cell anemia, but are susceptible to malaria; those who are heterozygous at position 6 do not get malaria, nor sickle-cell anemia.

At present, the normal function of the prion protein is not known. There are suggestions that prions may be involved in long-term memory in the sea slug, *Aplysia*. (Yes, prions are also found in more primitive organisms, including yeast!) In higher animals, the prion protein is highly conserved. Indeed, the prion proteins of rats and humans are more than 90 percent identical. Clearly, the prion protein has an important normal function. It is just unfortunate that it can cause disease, although apparently only in susceptible individuals. There is obviously a lot more to learn about prions and the diseases they cause.

Incidentally, there is no evidence that mad-cow disease was transmitted by humans to cattle, despite rumors that the mad-cow epidemic began when a mad person from here bit a cow while on a visit to the UK.

September 23, 2004

Sevilla Detera-Wadleigh

Is Mental Illness a Curse?

What do Van Gogh, a painter of masterpieces, Handel, the composer of the *Messiah*, Hemingway, the famous writer, and John Nash, the Nobel Prize winner whose life was immortalized in the movie *A Beautiful Mind*, have in common? Aside from being extraordinarily creative, did you know that they all suffered from mental illness?

Over a century ago, a German physician named Kraepelin classified two of the most severe, disabling forms of mental illness into what are now referred to as schizophrenia and bipolar disorder (manic-depressive illness). Schizophrenics suffer from symptoms that include hallucinations, delusions, social isolation, and cognitive defects. Bipolar disorder is a dysfunction of mood characterized by episodes of uncontrolled elation (mania), and swings to major depression. Physically, many psychiatric patients look fit but they display aberrant behaviors. Our society loosely refers to them as "crazy," "*loko-loko*," "*sira ang ulo*" or more derisively, "cursed." More than likely, a mentally ill person is ignored, poked fun at, institutionalized, homeless or forgotten. These diseases, more than any other, inflict intense psychological trauma and stigma on patients and their families.

Even in the face of manifold economic challenges, a developing country cannot ignore mental illness because of its adverse impact on families, society, and the health care system. In the US alone, in 2001, $85 billion was spent on mental health. A US survey published this year revealed an estimated lifetime prevalence of 1 percent for schizophrenia, 4 percent for bipolar disorder, and 17 percent for major depression.

Therefore, if left untreated, or if unresponsive to treatment, these diseases cause hundreds of millions of people worldwide to have difficulty functioning or to lead productive lives. In Southeast Asia, neuropsychiatric disorders account for about 10 percent of the causes of disability. Global Burden of Disease surveys in 1990 and 2000 ranked depression as the fourth leading cause of disability worldwide, accounting for the largest cause of nonfatal burden. It is striking that major depression is estimated to become the second leading cause of disease burden globally in 2020.

To reduce the burden that these disorders generate, strategies that aim to achieve both urgent and long-term goals are important. One is treatment-based. Multiple drugs, singly or in combination, that target the dopamine and serotonin signaling systems, have been in use for decades for the treatment of schizophrenia and depression, respectively. Lithium salt, discovered by an Australian psychiatrist to be an effective therapeutic

National Institutes of Health in Bethesda, Maryland, USA.
DeteraS@intra.nimh.nih.gov

agent for bipolar disorder, has been the mainstay for the treatment of mania. Timely intervention and management through effective pharmacotherapy with or without psychological care has led to either treatment or amelioration of symptoms enabling many patients to function.

Unfortunately, it is not uncommon for ill individuals to be undiagnosed, under-diagnosed, or simply untreated. If treated, a substantial number fall through the cracks due to lack of follow-up, maintenance, and compliance. In addition, a proportion of patients are resistant to various types of treatment. It is obvious that an improved and sustained implementation of effective measures is essential.

Research is a critical component of a long-term strategy that comprises multifaceted but interconnected goals: understanding the molecular basis of the disorders, investigating the relationship of *phenotype* (observable features) and *genotype* (combination of two alternate forms of DNA, one inherited from the mother, the other from the father), drug discovery, identifying environmental triggers, developing preventive measures, and pharmacogenetics.

Schizophrenia and bipolar disorder are complex diseases caused by both genetic and unknown environmental factors. The onset of disease commonly occurs at around puberty; penetrance is variable, i.e., not all individuals who carry defective risk genes manifest symptoms, and these diseases are most likely etiologically heterogeneous. Unlike simple Mendelian disorders which are caused by single genes that exert major effects, the overall risk for a complex disease is conferred by multiple genes of modest effects. Together, these complicated characteristics account for the difficulty in identifying the underlying disease-predisposing molecular factors.

Lacking reliable biological clues on the molecular causes of schizophrenia and bipolar disorder, a classic approach is to dissect the genetics of these diseases. The study involves multiple steps which include scanning of the whole human genome to identify chromosomal regions that contribute to disease-susceptibility, followed by a comprehensive and systematic search within these flagged locations for gene variations that correlate with disease. In other words, a map of the entire human DNA (the hereditary material) is screened by using DNA markers located at specific positions of chromosomes. This is analogous to a search for buried treasures in various parts of the Philippines guided only by street maps of provinces and instruments that can detect faint signals emitted by these treasures. Some faint signals could turn out to be false leads but areas that detect stronger signals are finely combed and investigated until the treasure is found.

Teasing out the disease-predisposing genes from the known 25,000 human genes remains a formidable research endeavor. Thus far, however, this approach has produced a few candidate susceptibility genes for schizophrenia and bipolar disorder. The next major step toward uncovering the molecular basis of these disorders is to establish the role of a candidate gene in pathogenesis and this requires investigations into biology and function.

It would be important to relate specific DNA variations with phenotype that could help refine diagnosis, predict therapeutic response, and tailor therapy to specific risk variants. Another outcome is to use the product of the risk-conferring genes as targets for drug discovery and to identify environmental factors that perturb their expression. Knowledge of molecular networks that regulate mood, cognition, thought, etc., would be immensely interesting and eventually beneficial in reducing the burden caused by these diseases.

For families who suffer from the adverse effects of mental illness and for the rest of the society, it is essential to know that new developments in research provide compelling and irreversible proof that these complex diseases, just like diabetes and hypertension, are caused by molecular defects, some of which are starting to be unveiled. Consequently, these families should shed the embarrassing notion about mental illness. MENTAL ILLNESS IS NOT A CURSE!

October 13, 2005

Judy D. Ribaya-Mercado

Vitamin A Studies in Rural Philippine Communities

Vitamin A deficiency is a public health problem in many nonindustrialized nations. In the Philippines, the 2003 National Nutrition Survey of the Food and Nutrition Research Institute-Department of Science and Technology (FNRI-DOST) showed that 40.1 percent of children between the ages of six months and five years, 36.0 percent of school-aged children, 17.5 percent of pregnant women, and 20.1 percent of lactating women have plasma vitamin A (retinol) levels that are below normal values. Vitamin A deficiency can result in anemia, reduced resistance to infection, impaired cellular differentiation, xerophthalmia, and ultimately, blindness, and death. The World Health Organization estimates that about 140 million children worldwide are subclinically vitamin A deficient, and that 250,000 to 500,000 vitamin A-deficient children become blind every year, with half of them dying within twelve months of losing their sight.

During the past ten years, I have been very fortunate to be able to obtain research grants to conduct vitamin A studies in rural Philippine communities. It all started in 1995 when a Request for Proposals (RFP) was sent out by the International Atomic Energy Agency (IAEA), one of the agencies within the United Nations Organization. Based in Vienna, Austria, the IAEA is known for its role in monitoring the safe usage of radioactive materials as an energy source, and more recently, for its role in searching for weapons of mass destruction in Iraq. What many do not realize is that the IAEA serves other functions as well such as conducting and/or supporting studies on the application of stable isotopes in health-related research. The IAEA has a Nutritional and Health-Related Environmental Studies Section in a Division of Human Health, and it was this section that sent out the RFP "to study the development and application of isotopic techniques in studies of vitamin A nutrition." A stipulation of the RFP was that researchers from developed and developing countries would have to team up, as the fieldwork should be done in the developing country and laboratory analyses in the developed nation. This was the opportunity I had dreamed about—to be able to conduct research studies in the Philippines. But who should I team up with? As an undergraduate student at UST, I studied Chemistry and it was only in graduate school when I focused on Biochemistry and Nutrition that I started paying attention to Public Health Nutrition studies. Dr. Florentino Solon of the Nutrition Center of the Philippines (NCP) came to mind. An internationally known public health nutritionist, he is a contributor of articles to the *Sight and Life Newsletter*, a publication based in Basel,

Jean Mayer USDA Human Nutrition Research Center at Tufts University, Boston, Massachusetts, USA.
judy.ribaya-mercado@tufts.edu

Switzerland dedicated to vitamin A studies and projects worldwide. We had never met, but after a flurry of e-mails, we became a team and settled into the task of writing a successful grant proposal to the IAEA. Thus begun my first research collaboration with the NCP which took me to the backroads of Barangay Santa Elena in Santo Tomas, Batangas and Barangay Hukay in Silang, Cavite—a journey to be repeated in the coming years with other vitamin A studies in other rural Philippine communities.

Before any human research study can be initiated, for the protection of study participants, the protocol undergoes review by an institutional review board. All our studies were rigorously reviewed and approved by not one, but two, review committees—the Ethical Review Board of the Philippine Council for Health Research and Development, and the Tufts University-New England Medical Center Human Investigation Review Committee.

The long-term objective of all our studies in the Philippines is the alleviation and eradication of vitamin A deficiency. The specific main objective in Santa Elena and Hukay was to determine whether ingestion of locally available fruits and vegetables that are rich in provitamin A carotenoids can improve the vitamin A status of malnourished schoolchildren. At that time, there was some controversy regarding the bioefficacy of plant foods for vitamin A use because other investigators have reported no improvement in vitamin A status in lactating women with increased green leafy vegetable intake. The assessment methods used by these investigators may not have the sensitivity to evaluate changes in vitamin A status in response to an intervention. By using stable-isotope-dilution methodology, our data showed an improvement in total-body stores of vitamin A in schoolchildren fed fruits and vegetables for ten weeks.

Stable isotopes are used as tracers in metabolic research studies and are safe to ingest. Unlike radioactive isotopes which undergo spontaneous disintegration and emit radiation, stable isotopes do not disintegrate (thus, are "stable"), do not emit radiation, and in fact are naturally present in small amounts in the body. Because of their low abundance in the body, small amounts of compounds tagged with ^{13}C or ^{2}H (deuterium) can be introduced into the body and used as tracers. For example, a vitamin A molecule in which four hydrogen atoms (^{1}H) are replaced by four deuterium atoms is a stable isotope of vitamin A (i.e., tetradeuterated vitamin A) and, when ingested, it mixes with the body's vitamin A pool. When there is sufficient vitamin A in the body, the ratio of deuterated to nondeuterated retinol in serum is low since there is plenty of vitamin A in the body to "dilute" the labeled vitamin A molecules ingested. By using a number of assumptions regarding the efficiency of absorption and storage of the tracer dose, the difference in retinol isotopic ratios in liver versus serum, the half-life of vitamin A turnover in the body, and the liver to body weight ratio, the serum retinol isotopic ratio can be used in a mathematical formula to provide a numerical estimate of total-body vitamin A stores and of liver vitamin A concentrations.

My second study with the NCP was conducted in Barangays Malabanan and Palsara in Balete, Batangas with funding from the US Department of Agriculture Foreign Agricultural Service. In this study, we aimed to determine the vitamin A requirements of elderly people.

Very few vitamin A studies have been done among the elderly; previous studies were mostly conducted in children, and pregnant and lactating women. Using the deuterated-retinol-dilution (DRD) technique, we found that elderly people are also vulnerable to vitamin A deficiency; 14.5 percent of the study participants had low liver vitamin A levels, even though none had serum vitamin A values that could be considered low. These data, together with our previous and subsequent data, confirm that stable-isotope-dilution methodology is a powerful tool for assessing vitamin A status. By determining the usual dietary vitamin A intakes of those with adequate versus those with low liver vitamin A, we estimated that vitamin A intakes of about 500 and 400 micrograms daily by elderly men and women, respectively, are advisable for maintaining adequate vitamin A reserves in the liver. As examples, 100 grams of malunggay can provide about 645 micrograms of vitamin A, and 100 grams of kamote leaves, about 323 micrograms.

My third study in the Philippines is being funded by the US Department of Agriculture National Research Initiative. In this study, we continue to explore factors influencing the bioefficacy of plant sources of vitamin A. Specifically, this study, the fieldwork of which was conducted among schoolchildren in Barangays Banawang and Atiliano Ricardo in Bagac, Bataan, aims to understand the influence of amounts of dietary fat on carotenoid bioavailability and on the bioconversion of plant provitamin A carotenoids into vitamin A. The children were fed standardized vegetable-containing meals three times daily, for five days a week, for nine weeks in their schools; the meals differed only in the amounts of dietary fat they contained. Data analyses are currently in progress and will provide other information such as the vitamin A requirements of this age group, interactions between vitamin A and iron, effect of carotene-rich meals on the prevalence of anemia, and influence of parasitic infections on vitamin A status and iron status.

These studies in the Philippines have contributed substantially to what is now known regarding the use of stable-isotope-dilution methodology (i.e., DRD, as well as a shortened procedure called three-day DRD) in vitamin A studies. The DRD procedure has become the "gold standard" for biochemical assessment of vitamin A status; it is useful for assessing a wide range of status, from deficiency to subtoxicity. It is the best method available for monitoring the effectiveness of vitamin A intervention programs.

In all our studies in these rural Philippine communities, we make every effort to design the protocol to be as noninvasive as possible (with only the minimum number of blood draws at baseline and at post-intervention), and to provide a benefit not only to the study participants, but also to the community. For example, during the screening phase, the study physician conducted a physical examination of all children whose parents gave their consent, regardless of whether the child would decide to participate in the study or not; those found to have abnormalities were treated or referred to the local clinic for follow-up. Tests for intestinal parasitic infections were done routinely, and those found to have worms were treated. Blood pressure measurements were available to the parents and the community. The studies among the children were designed to be food-intervention studies which provided nutritious meals and/or snacks for nine to ten weeks at the schools, and employed local residents to assist with meal preparation and clean-up. After the study, some schools were left with a renovated room (the study kitchen/dining area) with running water. The children were provided with school supplies. The elders were given a physical exam, including an electrocardiogram and other medical tests not part of the research agenda. As a result of the tests, several were identified to have tuberculosis and received treatment for it. Medical and nutritional advice was available to everyone.

Our studies in Santa Elena, Hukay, Malabanan, Palsara, Banawang, and Atiliano Ricardo have given the research work that I do an added significance that is both profound and special, and I am very thankful. It has been a privilege and a pleasure working in all these places.

April 27, 2006

Antonio R. Villanueva

Heart Attack of the Bone

What is a heart attack of the bone? Dr. Robert P. Nirschl coined the term "heart attack of the bone" as synonymous with avascular necrosis. Avascular necrosis is a condition which can occur in people as young as thirty years old and as old as sixty. The bone most commonly affected is the thigh bone (femur) and the head of the femur—the ball and socket part of the hip joint. It could happen with an injury, or sometimes it is hereditary or metabolic.

Avascular necrosis is the result of alterations in the blood supply to the femoral head causing it to become "shut out" and to die. This complication can occur in a variety of conditions such as a fracture of the femoral neck or following a dislocation of the hip. The other causes of avascular necrosis of the femoral head cannot be so clearly related to the "shut out" process of the blood supply but can be associated with the use of corticosteroids, systemic lupus *erythematosus* with or without corticosteroids, renal transplant patients taking corticosteroids or cytotoxic drugs, gout, alcoholism, and idiopathic disease (of unknown or obscure causes).

Another concept, which has limited scrutiny, is the experimental evidence of an abnormal number of microcracks (microfractures) in the hip bone due to impairment of bone remodeling which repairs microcracks accompanying the normal usage of the hip. This abnormality occurs when there is an imbalance between the rate of production and repair of microcracks, the former leading the latter—thus microcracks accumulate, causing spontaneous fracture. Such is the case of spontaneous fractures in athletes and *march* fractures in military recruits.

The other concept which needs further research is the impairment of the activation of the osteoprogenitor cell response, which normally causes the repair of microcracks and prevents it from reaching functional significance. This impairment is probably caused by the defective cracks' detection mechanism by bone cells called osteocytes. It is believed that when a microcrack occurs, the osteocytes tend to distance themselves from the injury, or perhaps the line of microcracks occurs in an area populated by dead osteocytes, and as a result, the microcrack is not repaired and the bone becomes "sick," not "dead." Could there be any abnormalities of bone cell function? Or does avascular necrosis reflect low turnover bone? Is there a dysfunction of bone remodeling? Is this a disease that results from some focal abnormality in the control of bone remodeling? If this overall concept

Bone and Mineral Division, Henry Ford Hospital in Detroit, Michigan, and the Harrington Arthritis Research Center in Phoenix, Arizona, USA.
avillanueva11@cox.net

is correct, it should be possible to describe in detail exactly how a robust hip becomes transformed over time into an old and fragile bone, in terms of varying combinations of the balance within remodeling sequences and the frequency of remodeling activation. Understanding the links between the remodeling system and microcracks, and each of the other levels of study, may give us a clue as to why one acquires a "heart attack of the bone."

Hip fracture is one of the most puzzling and frustrating orthopedic problems, and it disables many in the most productive part of their lives. It is not known what the true incidence in the Philippines is, but in the United States, estimates as high as 60,000 patients with femoral neck fractures and 20,000 necrotic heads per year have been reported. The percentage of necrosis of the femoral head following fracture of the neck of the femur in the adult is approximately 35 percent. Most people do not know that they have avascular necrosis for sometime until severe incapacitating pain starts suddenly. Such pain is brought on when the blood supply is cut off, or by walking, standing, or moving the affected bone, and relief usually comes only when one is resting.

Current research shows that there are tiny cracks of the hip that eventually cause the bone to collapse. These tiny cracks are invisible to x-rays and can be seen only under a high-power microscope. In my ongoing research of this phenomenon, these microcracks are seen along the subchondral bone, and some emanating from the articular cartilage. Some of them are seen in the trabeculae, a network of bone, inside the marrow cavity, commonly joined together in much the same way as beams of wood are joined together to form a scaffold.

How were these tiny cracks ever discovered? Over forty years ago, the late Harold M. Frost, M.D., a world-renowned bone expert, my mentor, and boss, discovered that these tiny cracks are seen in the rib of normal people. (Frost, HM: Presence of microcracks in vivo in bone. *Henry Ford Hosp. Med. Bull.* 8:25-35, 1960). Because it is difficult to show them in x-rays, Dr. Frost and I developed a neat and simple way to make them visible under the microscope. We used a simple stain or dye dissolved in alcohol (Frost, HM: Staining of Fresh, Undecalcified Thin Bone Sections. *Stain Technol* 34:135-46, 1959; Villanueva, AR, et al: Staining and Histomorphometry of Microcracks in the Human Femoral Head. *Biotechnic and Histochem* 69:81-88, 1994). We left the whole slab of bone suspended in it for seventy-two hours to a week, then we made ground sections using sandpaper, while running water is dripped gently on the bone to minimize friction and heat. After the bone grinding is over, the thickness of the bone would come to approximately 75 to 100 microns. Since our discovery of these microcracks, which is now accepted by the medical and scientific community, several articles have appeared in the literature. One of them is the author's research article indicated above.

What is next after we found these tiny cracks in the hip? Information gained from this discovery, I believe, would need further studies, using femoral heads removed following fractures of the femoral neck; arthritic femoral heads removed for replacement;

avascular necrosis; ethanol abuse; and osteoporosis. This would be a great research project for a graduate student working for his/her PhD, or for a young investigator trying to make something of himself/herself in the field of bone research.

November 16, 2006

Antonio R. Villanueva

Mineralized Bone: A Panoramic View

Mineralized bone is a condition where properties of the bone tissue are still in its natural state. Several of its inorganic and organic properties, including bone cells, are still intact. The bone is a living tissue. It is in constant repair and renewal throughout life, varying in characteristic ways with age, during growth, and in the presence of many diseases. This is referred to as bone remodeling. When a bone remodels, unknown factors or mysterious phenomena trigger the activation of "mother cells" or osteoprogenitor cells. These cells start to proliferate and divide into specialized cells called osteoclasts. Osteoclasts are "bone destroyers"—multinucleated giant cells, with a destructive lytic enzyme, capable of eroding or boring a hole in the bone. Eroding of bone takes approximately thirty days, after which another type of mononucleated cells called osteoblasts or "bone builders" takes over and starts replacing the bone that the osteoclasts removed. Immediately after the osteoclasts are gone, osteoblasts start laying down unmineralized bone (osteoid) centripetally, and keep doing it until the hole is all filled up. Some osteoblasts become trapped in the bone they formed and become osteocytes. Osteocytes live about twenty-five years and probably affect blood-bone exchange of mineral ions. Bone remodeling is the sum of the resorptive (osteoclastic) and formative (osteoblastic) activities occurring over time. Normally the length of time to complete the process of one metabolic unit takes approximately 90 to 120 days, depending on the age of the individual, and the presence of various diseases.

Since the middle part of the sixteenth century when bone technology was still in its infancy, experimenters have turned their intellect to devising methods of studying bone tissue. They encountered some challenge to their ingenuity because bone tissue is hard to cut into thin sections, and it is difficult to do simple chemical and physical tests on it. The problems can be partly resolved either by learning to remove the inorganic elements in the bone before sectioning or by devising simple ways of making fairly thin sections while retaining the minerals in their natural state. The removal of mineral salts by acids or chelating agents has been the classic procedure for many years, whereas the latter is at its beginning stage.

Early investigators would rub or grind bone sections between plates of old ground glass with water and pumice powder, sand or carborundum paper. These early methods

Bone and Mineral Division, Henry Ford Hospital in Detroit, Michigan, and the Harrington Arthritis Research Center in Phoenix, Arizona, USA.
avillanueva11@cox.net

for making sections of mineralized bone were tedious and difficult. Consequently, for many years bone has been one of the most neglected tissues in the pathology laboratory, and most pathology laboratories overlooked it entirely.

In recent years, pathologists have become concerned with questions about mineralized bone for diagnostic pathology on the light microscopic level. Initially, examination of bone was conducted mostly by pathologists whose real interest was in the ultrastructure of various cells and tissues. When the mechanisms of plastic embedment and the technical complexities of bone preparation were understood more clearly, several investigators, including some pathologists, became interested in light microscopic examination of this material. When all their observations had been collected, they found they had vast amounts of information of potentially great clinical value, including answers to a variety of important problems like skeletal growth, articular architecture, function, and repair. Now several other areas are ready for exploration: bone remodeling, bone biomechanics, fracture healing, and criteria for diagnosis of bone tumors and metabolic bone diseases.

Notable is the pioneering work of Harold M. Frost, MD, elucidating the dynamics of bone remodeling. In the early stages of this area, most of the research had been done by internists, orthopaedists, dentists and others from related disciplines, who had little knowledge of mineralized bone pathology. What is really surprising is that then, and even now, only a few pathologists have been truly cognizant of mineralized bone pathology. Today the use of mineralized bone has dramatically increased. It is useful not only in the study and evaluation of metabolic bone diseases, but also in space travel studies. A major concern among astronauts is the possible adverse effect of weightlessness on bone remodeling in space. Studies have shown that 25 percent of bone mass is lost after one year in space.

It is now clear that mineralized bone has a place in research, and in diagnostic pathology, particularly in metabolic bone disease. With tissue time markers, particularly tetracycline, we can obtain accurate measurements of rates, quantities, and locations of bone mineralization in relatively simple ways. Because of the geometrical process of making the organic matrix and subsequent calcification, bone has several properties analogous to the architectural and time-based proportion of growth rings in trees. Thus a tetracycline marker can identify where bone formation has occurred and on what particular day. Two sequential markers spaced several days apart can give us the actual amount of bone made within a time interval.

The information gained from these markers is inherently associated with the mineral salts rather than the organic part of bone. Decalcifying the tissue with chelating agents or acids to cut a section more readily wastes all such information. Before quantitative measuring techniques and analytical schemes based on these markers could be developed, it was necessary to find some method of making sections of bone that retain both the mineral and any incorporated marker. Such methods have since been developed and enjoy increasingly wide use as an integral part of the battery of techniques in medical centers and hospitals around the world. The methods have proven extremely useful and informative, particularly in follow-up studies (through serial biopsies) of disease and its treatment. They will permit doctors or scientists to look at the cells and study their properties relating structure, organization, and chemistry in terms of both composition and behavior.

May 31, 2007

Yasmyne Castillo-Ronquillo

Translational Research in Medicine

Medicine is a field of study that has the goal of establishing the best human care using epidemiologic, clinical, and basic researches. One example of epidemiologic research is the recent nationwide national blindness epidemiologic survey conducted by the Institute of Ophthalmology, University of the Philippines Manila, and funded by the Department of Health. This research has generated data which identified the most prevalent eye diseases. Thus, much-needed government resources can be allocated wisely for both preventable and remediable eye problems. An example of clinical research is the pharmacologic drug trial. The efficacy, safety and reliability of drugs are monitored by these multi-center, multi-physician, multi-patient, single-protocol studies. The more famous Western clinical drug trials have centered on cancer and cardiac problems. The herbal drug clinical trials have also been discussed frequently in Philippine newspapers and talk shows.

Clinical research data may be directly applicable to patient care since it leads to the use, disuse, recall, improvement, ban or endorsement of a drug, product, diagnostic procedure, surgical procedure, treatment option or health management protocol.

Basic medical research, as distinguished from clinical medical research, involves many basic sciences such as chemistry-biochemistry, physics-biophysics, engineering-bioengineering, atomic-molecular technology-biotechnology, pharmacology-pharmacy and pathology-immunopathology, to name a few. Basic medical research involves considerable time, and requires a multidisciplinary task force of basic scientists and immense financial resources. As an example, a technical procedure nuclear magnetic resonance imaging (MRI), started as a chemistry tool and is now at the forefront in medical diagnostic imaging. It took considerable time for this development to occur. Another example is the discovery of the genetic basis for cystic fibrosis which has led to a treatment breakthrough; again, after a considerable amount of time.

Translational research is a term recently established to refer to the collaboration of basic and clinical investigators to provide rapid clinical translation of novel basic and preclinical observations. In short, basic research or technologic experts at the cellular, molecular, histopathologic or imaging levels are engaged with the clinical staff of doctors

UP College of Medicine and Institute of Ophthalmology, National Institutes of Health, University of the Philippines Manila, and Sentro Oftalmologico Jose Rizal, Philippine General Hospital, University of the Philippines Manila, Philippines.
pinkyronquillo@hotmail.com

and investigators who are interested in applying new discoveries or information in the clinics. This type of research underscores the importance of identifying the most pressing or priority health problems in a country.

Most Western countries have data to prove that their most pressing problems are cancer, cardiovascular diseases, diabetes, and other age-related problems. Thus, their translational research agenda is focused on those identified diseases. The Philippines' demographic data show that infectious diseases, especially in the young, are claiming many lives in the rural areas. The problems of maturity and lifestyle diseases are slowly catching up in the metropolitan areas. What translational researches are necessary given that information? On which areas will we concentrate our health resources? We do not necessarily give up on basic research, in favor of clinical research. Translational research, properly focused, is an alternative.

Genetic studies allow us a glimpse into the underlying mechanisms of disease. While these studies do not necessarily lead immediately to a treatment, the data gathered increases the chances of discovering a treatment outside of the traditional drugs. We can wait for breakthroughs in the major centers of the world to help us solve our own problems, or we can actively participate in these researches. Participation in multi-center collaboration is one way to minimize the costs of an extensive genetic study.

Basic studies on the mechanisms underlying infectious diseases have been carried out. The staff of the UP Manila Institute of Public Health has done schistosomiasis, malaria, and parasite-borne infectious disease researches. The UP Manila College of Pharmacy has done studies on herbal medicine in collaboration with the UP College of Medicine Department of Pharmacology.

Translational research is the responsibility of any health organization. Working within the constraints of financial resources necessitates prioritizing allocation for results-oriented goals. Scientists and administrators of science funds must work together for the just use of local resources as well as for tapping nontraditional sources of research funds.

February 1, 2007

YASMYNE CASTILLO-RONQUILLO AND JOSEPH RANCHE

Genetic Research and Retinal Diseases

THE RETINA is that part of the eye on which light energy is converted into signals that are relayed to the visual centers in the brain. If you think of the eye as a camera, the retina is the film. Retinal degenerative disease is diagnosed mainly in the adult population. Some patients have incapacitatingly poor vision before they turn 40 while others live to their 70s with minimal visual problems. Retinitis pigmentosa, one of the degenerative retinal diseases, is caused by the destruction of the rods. The rods are the retinal cells that act as photoreceptors in light and dark adaptation. The cones, retinal photoreceptors that subserve color vision, are preserved. Retinitis pigmentosa has been discovered to have a genetic basis.

Genetics is the science that helps us understand how the biological nature of parents influences that of their children through the transmission of genes. Genes themselves are stretches of the long, chain-like molecules of DNA (deoxyribonucleic acid) that are embedded in the microscopic structures known as chromosomes within living cells. Each gene serves as a repository of information enabling a cell to produce one or more kinds of proteins that themselves perform specific biological tasks. A change in the chemical structure of a gene constitutes a mutation, which can alter the quantity or quality of proteins produced by cells and thereby result in disease. For example, retinitis pigmentosa is known to be caused by certain mutations in the genes for the proteins rhodopsin and arrestin—both of which play key roles in the light-sensing process in rods. What techniques are used to discover the mutations in these genes?

Linkage analysis is one technique that is used in the genetic studies of retinitis pigmentosa. Linkage is the tendency for genes and genetic markers to be inherited together because of their location near one another on the same chromosome. A genetic marker is simply a segment of DNA whose physical location on a chromosome is known and whose inheritance can be followed from one generation to the next. Markers are often used as tools for tracking the inheritance pattern of a gene that has not yet been identified.

Yasmyne Castillo-Ronquillo: UP College of Medicine and Institute of Ophthalmology, National Institutes of Health, University of the Philippines Manila, and Sentro Oftalmologico Jose Rizal, Philippine General Hospital, University of the Philippines Manila, Philippines.
pinkyronquillo@hotmail.com
Joseph Ranche: Institute of Ophthalmology, National Institutes of Health, University of the Philippines Manila, Philippines.

To study the genes involved in retinitis pigmentosa, DNA is needed from both affected and unaffected members of a family suspected to have a hereditary form of the disease. This DNA is subjected to a chemical process known as PCR (Polymerase Chain Reaction) which selectively replicates specific portions of the DNA to yield collections of DNA fragments of varying lengths. These DNA fragments are then separated according to size in a gel with the aid of electrical force, producing distinctive patterns (analogous to the lines of a barcode) that are subsequently studied by linkage analysis.

Linkage analysis is used for gene hunting and genetic testing through the use of LOD scores (Log of Odds ratio). The LOD score is used to decide whether an observed pattern of inheritance in a family arose by chance alone or as a consequence of genetic linkage. The LOD score is a statistical measure of the likelihood that two positions on a chromosome are near enough to each other so as to be consistently inherited together, in which case they are said to be linked. A LOD Score of 3 or higher strongly suggests that the suspected mutant gene and the known genetic marker being followed are in fact linked. A chromosomal region likely to contain the gene of interest can be defined around the position of a known genetic marker to which it is linked, and candidate genes can then be selected from the set of genes identified within the region as determined from the DNA sequence data of the Human Genome Project (HGP). (Completed in 2002 by the US National Institutes of Health, the HGP involved the participation of hundreds of scientists in many laboratories worldwide.)

Identifying the exact gene among the candidate genes requires many more steps. Once this gene is identified, gene mutation studies are performed. These consist of analyzing the DNA sequence of the mutant gene and determining its difference from the normal gene. With the use of linkage analysis and mutation screening techniques, discovery time for gene-disease association and gene-mutation identification is shortened.

Mutation screening refers to the identification of mutations that have altered the quantity or quality of proteins produced by cells, or have otherwise resulted in some form of biological change. The complete DNA sequence of a mutated gene may be determined for this purpose, but even just a part of this DNA sequence may suffice to disclose mutations. In the latter case, PCR can be used to produce a relatively short segment of the gene within which mutations are known to occur frequently; this strategy often leads to the identification of mutations without having to study the entire DNA sequence, thus saving on time and other resources.

We have reviewed the records of forty-four patients with retinitis pigmentosa at the Sentro Oftalmologico Jose Rizal, Philippine General Hospital, University of the Philippines Manila. The patients ranged from eight to seventy-two years old. They complained of chronic, progressive clouding of vision and night blindness. Eighty percent of them were documented to have retinal changes and visual impairment as defined by World Health Organization (WHO) standards. The genetic changes, if any, that have caused retinitis pigmentosa in these patients are puzzles that remain unsolved.

Several eye research centers in the United States, Europe and Asia are participating in an effort to find novel genes and gene mutations that cause retinitis pigmentosa. The ultimate goal of genetic retinal research is definitive treatment or at the very least, a better management of retinal diseases so that our patients will maintain good visual function as long as possible.

June 14, 2007

Part 6

Herbals and Other Natural Products as Health Remedies

Gisela Padilla-Concepcion

Nature: Still the Best Provider of Cures

We know that there is a high incidence of infectious diseases such as tuberculosis, malaria, and AIDS in our country, and drug resistance to currently used drugs is a major concern. I think that the long-term control of TB, malaria and AIDS should involve the use of local natural resources, local human samples, and local health scientists and health workers. The Philippines can provide two major sources of novel molecules that can be explored for developing new anti-infective therapies. These are the rich biodiversity of lower life forms and the large number of human disease samples. The Philippines could benefit directly from therapeutic products developed from these sources. Filipino patients would gain direct access to the treatment which would be affordable and suited to their particular form of disease.

The Philippines' terrestrial and marine biodiversity could yield novel compounds with potent anti-TB, antimalarial, and antiviral activity. Over the last fifty years, more than 80 percent of all antibiotics and anti-infective drugs were developed from compounds isolated from terrestrial plants and land-based microorganisms. Up to the present, marine organisms have been explored largely or only for cancer and other diseases that are important in developed countries. Because marine bioprospecting is a costly and difficult undertaking, it is pursued principally to yield high-value therapeutic products such as anticancer drugs.

Philippine marine biodiversity is the third highest in the world. It should now be explored as a source of new anti-infective compounds to help treat TB, malaria, and AIDS, aside from cancer. The Philippine government has passed a law to regulate scientific studies and commercial bioprospecting of biological and genetic materials, and this law is now being enforced with simpler implementing rules and regulations.

There are already thousands of marine natural products reported in the literature over the last thirty years, including some compounds isolated from Philippine marine organisms. These are compiled in several databases abroad. These compounds should be reviewed, requested from laboratories abroad and screened for anti-infective properties. Their chemical structures could be evaluated, modeled, and modified to fit and bind the known molecular targets associated with the TB, malaria, and HIV pathogens.

The Philippines has another rich resource useful for TB, malaria, and AIDS research—human disease samples. Clinical isolates could be a source of new strains or

The Marine Science Institute, University of the Philippines, Diliman, Quezon City, 1101, Philippines.
gpconcepcion@yahoo.com

forms of the pathogens, and thus a source of new antigens and molecular targets. The new antigens could be the basis for designing new vaccines. The new molecular targets would be used to find new small molecule drugs. Samples from patients who have the disease in various forms and stages, and samples from patients who have recovered or are asymptomatic, with or without drug treatment, are potential sources of new therapeutic molecules. Survivors must have produced some antibodies and other proteins, and probably some small molecules in their bodies to combat the disease. These molecules would be associated with the particular form of the disease, the strain of the pathogen, the genetic make-up of the patients, and would thus be most useful to the community or population from which they were obtained.

In order to collect human samples, the patient's informed consent should be obtained. The patients should also be provided monetary compensation. It is important that the research conducted should follow strict bioethical principles and should be performed under utmost caution, using the highest level containment facilities (P3 level). The health and safety of the researchers must be ensured. However, while diagnostic and therapeutic products can be developed using novel biomolecules obtained from the disease samples, the Philippines is not yet very competitive in this area due to limited capabilities in molecular biology and biotechnology. A thorough characterization of the new antigens and molecular targets would also require capabilities in the field of proteomics.

The small molecule or metabolite profile (metabolomics) of human samples obtained from various stages, forms, and treatment of the disease could indicate the particular physiological state of the individual and his response to the pathogen. A human survivor would have metabolites which could act as important signals (inhibitors or activators) of mechanisms which may be critical in overcoming the disease or keeping it at a residual level. Sensitive analytical methods such as mass spectrometry are now available for detecting such small molecules in biological samples over a period of time. Metabolomics could contribute to modeling metabolic pathways which are affected in disease progression and regression. This then could form the basis of finding new natural small molecule drugs. (It is interesting to note that many of the classical anti-cancer drugs used today are DNA topoisomerase 2 or topo 2 inhibitors isolated or developed from plants and microorganisms, and yet small molecule topo 2 inhibitors from humans, which probably exist in normal cells in order to regulate the structure and function of human topo 2, have not been found and explored as models of anti-cancer drugs.)

It is not surprising that Nature provides us the best source of bioactive small molecules. These molecules, which have specialized functions in the source organisms, are products of a natural selection process that has been occurring throughout evolution. These compounds could be used in their natural form or can be chemically modified to produce new effective drugs for human health.

Since small molecule drugs are still the more affordable drugs for TB, malaria, and AIDS, I believe that research in this field should be pursued in our country. If necessary, we should collaborate with foreign laboratories that can transfer to us more advanced technologies. In the past, I had proposed that the Philippine government set up a National TB (Natural Products) Drug Discovery and Development Program that it would support in a substantial way. To assist in the funding, foreign foundations would be invited to support the program. Foreign experts would be invited to help advance technology here. With the government's bioprospecting regulations in place, this program would be feasible and could bring in a lot of funding and training for research on natural products, drug discovery, and development. Further, the products developed could undergo preclinical and clinical trials in the Philippines.

The proposed initiative focuses on scientific and technological innovations. But the success of health prevention, diagnostic, and therapeutic programs in a country depends not only on the potential to find and develop new drugs but also on the participation of key individuals and agencies in effective scientific research and education programs, health care, and community service programs. The cooperation of various sectors of the health and science community in our country will translate into greater national self-reliance in dealing with public health problems, such as infectious diseases, in the long term.

December 22, 2005

Gisela Padilla-Concepcion

Marine Biodiversity and Marine Biotechnology: A Philippine Competitive Advantage

The field of marine biotechnology is now recognized worldwide as an emerging and promising area of scientific research and development. The marine environment is considered the world's last frontier in the search for novel drugs from nature for serious human diseases such as cancer, AIDS, and drug-resistant infections. Several leading universities in technologically advanced countries are engaged in marine bioprospecting with support from their governments. Their aim is to discover and characterize novel bioactive compounds and to develop and produce these as useful drugs through biotechnology. The ultimate challenge of marine biotechnology consists of utilizing and translating into economic value the rich resources of the seas.

The Philippines is identified as the third highest in marine biodiversity in the world, and as a marine biodiversity "hot spot," i.e., with threatened biodiversity. To protect, conserve, and promote wise utilization of our rich natural resources, the Philippine government enacted the Wildlife Act in 2002. A section of this legislation concerns the regulation of bioprospecting, which refers to the collection, study, and utilization of biological and genetic materials for commercial use. The Wildlife Act aims to ensure conservation of biodiversity and provide benefits to the government and local communities if useful products are developed from Philippine raw materials or natural resources.

Local communities, considered the custodians of the marine resources that surround them, can benefit from marine bioprospecting, drug discovery, and biotechnology in the long term. The Indigenous Peoples Rights Act (IPRA) recognizes and protects indigenous knowledge of medicinal and agricultural uses of Philippine natural resources. The Wildlife Act requires that benefits be shared with indigenous communities in the event that useful products are developed based on their indigenous knowledge. Farming and field-based, "low-tech" marine biotechnology for large-scale production of materials may provide work opportunities for these communities in the future.

The university should play a key role in transforming raw materials or natural resources to useful products by providing the appropriate milieu for scientists to conduct meaningful and productive research, e.g., an academic climate for interaction among

---

The Marine Science Institute, University of the Philippines, Diliman, Quezon City, 1101, Philippines.
gpconcepcion@yahoo.com

researchers and mentoring of students to become competent researchers; physical infrastructure and basic facilities and equipment to conduct research; an intellectual property office to protect the intellectual property or inventive rights of scientists; and a technology transfer office to find industry partners to develop results obtained from basic R&D for useful applications and products. Over the last twenty-five years, the UP Marine Science Institute (UPMSI) has played a leading role in documenting, conserving, improving, and propagating Philippine marine resources; establishing basic scientific information; and identifying and developing useful applications for them. Value is added to raw materials through scientific discovery, intervention, and invention, and scientific ingenuity originates and resides in the scientists working in a conducive, supportive environment in a research institute such as the UP Marine Science Institute.

Biodiversity and human health issues are, indeed, closely linked. Biodiversity is the source of chemical diversity, which has been the basis for discovering useful drugs and medicinal agents from Nature throughout history. Over 70 percent of all clinically used anti-infective drugs today are natural or Nature-derived compounds, and so are over 60 percent of all currently used anticancer drugs. The loss or depletion of biodiversity then would imply the loss of a valuable pharmacological resource. Thus, conservation of marine biodiversity is of great importance to biomedical research and human health. While terrestrial biodiversity has long been studied extensively and fewer new classes of terrestrial compounds are now being discovered, marine biodiversity remains largely unexplored. Based on the scientific literature in the last twenty years, marine invertebrate animals and microorganisms continue to yield novel compounds with unprecedented chemical structures and potent biological and pharmacological activities. Thus, the oceans are rightfully considered an untapped pharmacological gold mine.

At UPMSI, there exists a multidisciplinary team of well-trained scientists, e.g., marine biologists, ecologists, natural products and polymer chemists, biochemists, toxinologists, and molecular biologists, who together have shown the capability to study unique marine habitats and rehabilitate endangered ones, to identify marine invertebrate organisms that produce compounds that can serve as leads for new drugs, and to propagate marine plants that produce polymeric materials that can be useful in industry and medicine. With strong productive scientific collaborations with foreign laboratories, technology continues to be transferred to the UPMSI. Marine biotechnology at UPMSI is quickly advancing from low level to intermediate and advanced level. This, combined with our rich indigenous marine resources, could translate into a real competitive advantage for the Philippines. While other institutions are also active in marine biotechnology, the UPMSI is one of the country's pioneer institutions in the field. It is at the forefront of aquaculture, marine polymer, marine natural products, and marine biomedical research and could have one of the best chances to produce innovative useful marine products that would benefit Filipinos.

Our Marine Natural Products (MNP) Group at UPMSI has projects on the isolation and purification of bioactive marine compounds or "marine natural products" (MNPs) from marine invertebrates, e.g., sponges, sea squirts, and associated marine microorganisms (MMOs) that live in coral reefs. These samples have been collected from different biogeographic regions of the country, led by my colleague, Dr. Porfirio M. Aliño, who heads the UPMSI Marine Chemical Ecology and Biodiversity (MCEB) Group. Compounds with potential anticancer, antimicrobial, anti-TB, anti-platelet aggregating and immunosuppressant activity have been isolated and are being studied intensively. Several international publications have resulted from the work of the MNP Group and its collaborators in the last few years. We have also produced a number of patentable results for which we are filing patent applications in the name of UP, the collaborators and funding agencies.

We pioneered in the collection of marine microorganisms (MMOs) associated with marine invertebrates, realizing that these MMOs are almost completely unexplored and could therefore be a good source of novel bioactive compounds. An additional advantage is that while only a limited amount of sponges or sea squirts can be collected by scuba diving, MMOs may be cultured in the laboratory to produce the bioactive compounds. The MNP Group currently has a collection of over a thousand marine invertebrates and over a thousand MMOs in storage at UPMSI, maintains a "PharmaSeas" database to document these samples and, in addition to a natural products chemistry lab, runs a bioassay lab, a mammalian cell culture lab, a microbial culture lab, an immunology lab, and a small animal testing facility. Our team consists of about twenty researchers and thesis students. We interact closely with other research groups at UPMSI, other UP Diliman units, other local universities, local government research institutions and hospitals, local foundations and foreign universities. Our major funding sources are the Philippine and US governments and international foundations. But in spite of this support, we are still unable to pursue many of our promising bioactive marine samples because of limited research equipment, supplies, and manpower.

Our vision is to continue building multidisciplinary, coordinated research programs in the marine biomedical field. We would like to explore our vast marine biodiversity as a source of small molecule drugs for the treatment of serious diseases. We believe that this is one area of competitive advantage for our country. Our immediate goal is to contribute to the development of an anticancer therapy making use of marine compounds or MNPs isolated from Philippine marine organisms. Related projects undertaken to help attain this goal include the chemical modification and synthesis of MNPs to improve anticancer activity, linking of MNPs to antibodies that target cancer cells to improve specificity and reduce toxicity, attachment of MNPs to polymers to improve drug delivery, and the determination of molecular targets of MNPs to understand their mechanism of action. An end goal is to be able to demonstrate "proof of principle" or in vivo activity of the anticancer therapy in an animal tumor model.

Very significantly, in 1998, the first Commercial Research Agreement (CRA) signed in the country for marine bioprospecting was for the project of the UPMSI MNP Group with the University of Utah on marine drug discovery for cancer. The CRA allowed us to collect marine samples from the different biogeographic regions of the country. This CRA, between UP-Utah and the Department of Agriculture (DA), was the first and for a long time the only case of implementation of Executive Order 247 (access regulation on the collection of biological and genetic materials; predecessor of the Wildlife Act) in the country. While other foreign universities have discontinued collaborations with other Philippine universities because of difficulty in obtaining a CRA and complying with bioprospecting rules, we succeeded in having our CRA renewed in 2003.

The UPMSI MCEB and MNP Groups pioneered in conducting the public consultation and PIC (prior informed consent) process among local communities living near marine collection sites, which is required prior to the granting of a CRA. Further, the MNP Group has faithfully and satisfactorily complied with the government's reporting requirements and the execution of a proper Material Transfer Agreement (MTA) of samples to a third party. The UPMSI's CRA and bioprospecting activities have been cited by NGOs as a model of compliance with government regulations and of good bioprospecting practices. These actions indicate that UPMSI recognizes the role and importance of the various stakeholders in the marine drug discovery and development process. Together with government lawyers and technical staff, we worked hard to negotiate equitable terms of benefit sharing for the local communities, the Philippine government, UP and Utah, in the event that a marine compound or MNP from a Philippine marine organism is developed into a drug or used in a therapeutic preparation in the future.

Printed in two parts on September 30 and October 7, 2004

Irene Villaseñor

Nutraceuticals in Cancer Prevention and as Antioxidants

The term "nutraceutical" is taken from the words "nutrition" and "pharmaceutical." It is any substance that may be considered as food, or part of a food, and provides medical and health benefits. Other names are functional foods, designer foods, medical foods, pharmafoods, and FoSHU (food for specified health use).

According to the World Health Organization, at least 35 percent of cancer causes are diet-related. Proper nutrition could prevent 50-90 percent of all cancers. An Internet search lists some examples of nutraceuticals used in cancer prevention. Those with high anticancer activity are soybeans, garlic, ginger, cabbage, carrots, celery, parsley, parsnips, and licorice. Those with moderate levels are citrus fruits, onions, flax, broccoli, Brussel sprouts, cauliflower, tomatoes, peppers, turmeric, brown rice, and whole wheat, while oats, barley, rosemary, mint, thyme, oregano, sage, basil, cucumber, cantaloupe and berries have measurable levels.

The research group of Dr. Clara Y. Lim-Sylianco, one of our National Scientists, many years ago pinpointed the following nutraceuticals as antimutagens: garlic, bell pepper (green and red), *mongo*, onion, *sineguelas* (1), *ampalaya* fruits and leaves, avocado fruit, *duhat, kamatsile, kamias,* melon fruit, papaya fruit, radish, squash fruit, flowers and leaves, *suha* and tomato (2). Several vitamins and minerals were also found to be antimutagenic (3).

The chemical structures of antimutagens are diverse. Some sterols found to be antimutagenic in studies conducted at the University of the Philippines Diliman-Institute of Chemistry (UPD-IC), are spinasterol from squash flowers (4,5), b-sitosterol from peppermint (6) and an acyl glucosyl sterol from *ampalaya* fruits (7). Several sulfur-containing compounds were also isolated from onions and garlic. Nitrilosides are present in apricot kernels, apple seeds, grape seeds and grasses. Some antimutagens are bioflavonoids such as ellagic acid from berries, proanthocyanidins from grapes, catechins from green tea, quercetin, hesperidin, and rutin from citrus fruits, genistein and daidzein (both isoflavones) from soybeans, indole-3-carbinol and its oligomer from cruciferous vegetables. These bioflavonoids are also known as phytoestrogens.

Estrogen is, of course, the female hormone. It programs the breast and uterus for sexual reproduction, controls menstrual cycles, maintains the heart and healthy bones,

Institute of Chemistry, University of the Philippines, Diliman, Quezon City, 1101, Philippines
irene.villasenor@up.edu.ph or ivillase@yahoo.com

and promotes youthful cellular division and regrowth of cellular components. However, several harmful effects of estrogen are known such as weight gain, abnormal blood clot formation, uterine bleeding, increased risk for gallstones and fibroid tumors, and the ability to promote the proliferation of cells in the breast and uterus. Women who used estrogen therapy for six to eight years had a 40 percent higher risk of ovarian cancer while women who used estrogen for eleven or more years had a 70 percent higher risk of ovarian cancer.

Estrogen, in the form of estradiol, is metabolized into estrone, which is further metabolized into 2-hydroxyestrone, 4-hydroxyestrone and 16-hydroxyestrone. There is an increase in the levels of 16-hydroxyestrone in women with breast cancer and with familial high risk for breast cancer. It is genotoxic in breast epithelial cells and it promotes abnormally rapid cell growth and proliferation.

Xenoestrogens, on the other hand, are synthetic chemicals that can act like human estrogen. They are not easily broken down and can accumulate and be stored in the body's fat cells. They bind to estrogen receptors (ERs) and act as estrogen mimics. They bear little structural resemblance to natural estrogens. Some examples are DDT (1,1,1-trichloro-2,2-bis-(p-chlorophenyl)ethane, BHT (butylated hydroxytoluene), BHA (butylated hydroxyanisole) and DES (diethylstilbestrol).

Phytoestrogens are weak estrogens leading to lesser breast cell division. They significantly increase the 2:16-hydroxyestrone ratio and hence, decrease the risk of estrogen-related cancers. They act as antiestrogens by binding tightly to the ERs without activating transcription.

Nutraceuticals are also antioxidants. Antioxidants are compounds capable of inhibiting oxygen-mediated oxidation of diverse substrates and hence, inhibit formation of free radicals and free-radical chain propagation. Antioxidants prevent oxidation that is responsible for premature aging, wrinkling of skin, hardening of arteries, stiffening of joints, formation of cataracts, and development of cancer. Common examples of antioxidants are Vitamins A, C, E, and beta-carotene.

A diet high in Vitamin A and C and fiber is highly recommended for cancer prevention. Vitamin A-rich foods are cantaloupe, carrots, lettuce, mango, spinach, sweet potato, and squash. High in Vitamin C are broccoli, Brussel sprouts, cabbage, cantaloupe, cauliflower, chili peppers, melon, kiwi, orange, pineapple, spinach, strawberries, bell peppers, tomatoes, and watermelon. High in fiber are apple, banana, berries, Brussel sprouts, carrots, cherries, beans and peas, kiwi, orange, pear, prunes, spinach, and sweet potatoes.

Reactive oxygen species (ROS) are produced in the mitochondria, as a consequence of normal aerobic respiration; in peroxisomes, which are organelles responsible for degrading fatty acids and other molecules; in phagocytic cells, which destroy infected cells with ROS; by cytosolic enzymes in response to stimulation by growth factors; and by cytochrome P450 enzymes, involved in one of our primary defense systems. Exogenous sources of ROS are UV light, ionizing radiation, chemotherapeutics, and environmental toxins. Examples of ROS include: ONOO-, H_2O_2, RO·, O_2, ·NO, ·OH, $\cdot O_2$, $\cdot NO_2$, ROO·, $\cdot O_2$. The hydroperoxyl radical (·OH) is the most reactive ROS, followed by the alkoxyl radical (RO·) and peroxyl radical (ROO·), and the least reactive is the superoxide radical ($\cdot O_2$-).

Antioxidants have also been implicated in the prevention of skin photocarcinogenesis. When evaluated for their effectiveness against chemically induced skin tumors, four nutraceuticals, namely; *Beta vulgaris* L. (sugar beet), *Tetragonia expansa* Murr (spinach) (8), *Curcuma longa* L. (turmeric) and *Cucumis sativus* (cucumber) (9) decreased the percent skin tumor incidence by 30 percent to 70 percent, compared with 100 percent skin tumor incidence for the positive control. There was also a decrease in tumor multiplicity and a delay in tumor onset. These studies were conducted by BS Chemistry thesis students at UPD-IC.

These nutraceuticals may prevent cancer because they act as phytoestrogens and as antioxidants. The daily intake of fruits and vegetables is therefore highly recommended.

References

1. Sylianco, C.Y.L., et. al. (1986). *Philippine Journal of Science 115,* 293-98.
2. Sylianco, C.Y.L., et. al. (1986). *Philippine Journal of Science 115,* 23-30.
3. Sylianco, C.Y.L., et. al. (1981). *Asian Journal of Pharmaceutical Science 3,* 19-25.
4. Villaseñor, I.M., et. al. (1996). *Mutation Research 360,* 89-93.
5. Villaseñor, I.M.,et.al.(2000). *Teratogenesis, Carcinogenesis, and Mutagenesis 20*, 99-105.
6. Villaseñor, I.M., et. al. Phytotherapy Research (in press).
7. Guevara, A. (1988). PhD Dissertation, UP Diliman.
8. Villanueva, Ainstein Marie A. (2001). Nutraceuticals: Anti-Tumor Potential of Beets (*Beta vulgaris* L.) and Spinach (*Tetragonia expansa* Murr.). BS Chemistry Thesis, University of the Philippines, Diliman, Q.C.
9. Simon, Ma. Karenina B. (2001). Nutraceuticals: Anti-Tumor Potential of Cucumber (*Cucumis sativus* L.) and Turmeric (*Curcuma longa* L.). BS Chemistry Thesis, University of the Philippines, Diliman, Q.C.

January 13, 2005

Pacita Zara

Research and Development Model: The Philippine Medicinal Plants' Development

Fifty-billion-pesos-a-year domestic pharmaceutical industry! That's a huge market and herbal medicine barely made a dent with its share of 0.5 of one percent. According to Senator Juan Flavier, one-third of the population can fully afford Western medicine; one-third can partially afford them; and therefore, one-third should be allocated to herbal medicine for those who cannot afford high-cost Western drugs. The industry is highly competitive. Opportunities are tremendous. But why did it take so long for drugs from medicinal plants to be introduced into the market?

Nearly three decades ago, the government—through the then National Science and Development Board (NSDB), which subsequently became the National Science and Technology Authority (NSTA) and now the Department of Science and Technology (DOST), which created the Philippine Council of Health Research and Development (PCHRD) as the highest policy-making body for health research—gathered together a multidisciplinary/multiagency group of researchers on medicinal plants to analyze the situation and hurdle the roadblocks to the development of herbal medicine in the country. The experts realized the need for financial and human resources and above all investment in research and development (R&D). For the natural products to be highly competitive, the missing link is science and technology.

In their desire to provide low-cost quality drugs and to cushion the effects of the escalating cost of medicine in the market, the group initiated the National Integrated Program on Medicinal Plants (NIRPROMP). The program strategy was to develop a new order in the investigation into medicinal plants departing from the old method of initial phytochemical studies prior to pharmacologic tests by developing a more rapid but still highly scientific screening process to determine the efficacy and safety of medicinal plants and, thus, obtain a faster solution to the immediate and pressing needs of the country.

The Approach

1. Multidisciplinary—Experts (agriculturists, chemists, pharmacists, pharmacologists, clinicians) were needed to conduct R&D.

Philippine Council for Health Research and Development, Department of Science and Technology, Philippines.
gpconcepcion@yahoo.com

2. Multiagency/multicenter—The following agencies, both government and private, academe, research institutes, medical centers, hospitals, were involved:
   a. UP System (Colleges of Medicine, Pharmacy, Agriculture, Institute of Chemistry);
   b. University of Santo Tomas;
   c. Ateneo de Manila University/PIPAC; and
   d. Jose Reyes Memorial Medical Center

The Mission

The actual research was divided into two missions:

1. Mission I—Production of crude drug preparations against common ailments:
   a. Mission I a—Formulation of decoctions and infusions from crude plant parts. One-hundred-two plants were subjected to preclinical trials and rapid clinical screening involving toxicology, mutagenicity, and bioassay to establish efficacy and safety.
   b. Mission I b—Formulation, manufacture, quality control testings of pharmaceutical dosage forms (tablets, syrup, lotion, etc.) from plant materials and extractives.
      i. The dosage formulations were subjected to preclinical trials (LD50, subacute, subchronic and pharmaceutical studies);
      ii. Clinical trials (Phases I, II, III); and
      iii. Appropriate cultural management and propagation techniques.

2. Mission II—In-depth studies leading to the development of drugs from semipure and pure plant constituents. Investigations done on Mission I b will be conducted on the identified isolated and purified active constituents.

The Selection Process

The plants were prioritized based on the following process:

1. Identification of common ailments (top causes of mortality/morbidity) in the country;
2. Identification of common plants which address the diseases based on folkloric claims in cooperation with 1,280 herbolarios in 800 barangays in the country; and
3. Selection of plants based on availability, ease of propagation, cultural management, and conservation.

The Research and Development Process

The standard research process has been established by the NIRPROMP Team based at UP Manila. It is composed of three major components:

The agriculture component— for the supply of raw materials:
1. Identification of plant/plant parts and disease indication;
2. Collection/authentication of plants/establishment of gene bank;
3. Propagation;
4. Harvesting techniques, handling procedures, facilities, storage, labeling;
5. Postharvest techniques (proper sanitation and clean facilities).

The medical component—for the validation of formulations on human subjects:
1. Pharmacologic—effect/mechanism of action/safety of preparations;
2. Toxicologic—toxic/adverse effects of a given dose over a period of time (one day, 21, 90, 270 days);
3. Bioassay—dosage of drugs that exhibits effectivity/potency;
4. Mutagenicity—genetic effects of the formulations;
5. Heavy metals analysis—metal content and tolerable limits of the preparation;
   a. Dosage formulation—tablets, syrups, lotion, ointment, etc.
   b. Clinical trials:
      i. Phase I—trials on fifteen to thirty human subjects to determine the range of a relatively safe dosage;
      ii. Phase II—evaluation of the efficacy and determination of the most effective dose, including administering the drug among thirty to sixty subjects;
      iii. Phase III—confirmation and refinement of the formulation's efficacy, safety, and toxicity based on trials involving at least 200 subjects. Community participation is encouraged;
      iv. Phase IV—though not strictly required by the Bureau of Food and Drugs (BFAD) on medicinal plants, the NIRPROMP still confirmed the drug efficacy on a broader range of subjects through the multicenter approach involving hospitals and medical centers for drug monitoring.

The pharmaceutical component—for the determination of the appropriate formulations for each plant/plant parts and the shelf-life qualities.

The Technology Transfer and Utilization Process

Technology transfer

To date, 102 plants have been scientifically validated for safety and efficacy. Ten of these plants have undergone or are still under different stages of development under Mission I. These are *yerba buena, tsaang gubat, niyog-niyogan, bayabas, ulasimang bato, bawang, lagundi, sambong, akapulko,* and *ampalaya.* Studies on these plants have been completed—*lagundi, sambong,* and *akapulko.*

The technology of the production of *lagundi* tablets (300 mg), *sambong* (250 mg), *akapulko* lotion, and *lagundi* pediatric syrup was adopted by the Department of Health through its four regional herbal manufacturing plants in Davao, Cotabato, Tuguegarao, and Tacloban.

Private sector transferees were Pascual Laboratories, Gruppo Medica, and Lloyd Laboratories through a license agreement per product. Altermed, the subsidiary of Pascual Laboratories, is actively manufacturing and commercializing *lagundi* tablets (Ascof) and *sambong* tablets (Releaf) which bested 100 other products and inventions and winning silver medals at the 25th International Exhibition of Inventions, New Techniques and Products in Geneva, Switzerland in April 1997. Altermed is now earning millions of pesos from the sale of these products.

Technology utilization

Results of the NIRPROMP studies were documented in the following publications distributed to different communities, schools/colleges/universities, barangays, NGOs, medical centers, DOH, etc.: *Guidebook in the Proper Use of Medicinal Plants, National Trainers' Manual on the Use of Philippine Medicinal Plants, Philippine Materia Medica, Philippine National Formulary,* Brochures on the Use of Medicinal Plants, *Selection and Scientific Validation of Medicinal Plants for Primary Health Care, Manual on Bioassay Procedures, Monograph on Vitex Negundo (Lagundi)*, and *Agricultural Manual.*

The NIRPROMP Team conducted six months of technical manpower training programs and seminar/workshops to raise awareness of the country's health and drug situation on the validated safety and efficacy of medicinal plants as alternate/ complementary medicines.

Through NIRPROMP, the DepEd has encouraged the adaptation of the herbal medicine gardens in schools.

The NIRPROMP research results have been adopted in the development of the ASEAN Program on Herbal Medicine.

Above all, Executive Order 302 signed by President Arroyo on March 29, 2004 adopted the Philippine Pharmacopeia I as the official book of standards and reference for pharmaceutical products, which include crude plant drugs in the Philippines.

Sources

Philippine Council for Health Research and Development (PCHRD), Department of Science and Technology

UP Colleges of Medicine, Pharmacy and Agriculture, Institute of Chemistry

Philippine Institute of Traditional and Alternative Health Care (PITAHC), Department of Health

Printed in two parts on August 11 and August 18, 2005

Flor Lacanilao

Public Understanding of Science II: Research on Medicinal Plants

The Department of Health (DOH) has been promoting products of medicinal plants as alternative medicines, citing their benefits and effectiveness. One-third of our population who cannot afford Western drugs can depend only on herbal medicines. Their low costs compared with those of conventional medicines will help many Filipino families, announced the DOH secretary, and that medicinal plants have been a subject of extensive research by Filipino scientists.

According to the Department of Science and Technology, 102 plants have been "scientifically validated for safety and efficacy." Ten of these plants are under different stages of development, and studies have been completed on sambong, lagundi, and akapulko. From sambong and lagundi alone, a local maker of herbal drugs is reportedly earning millions of pesos.

But recent reviews of the scientific literature on herbal medicinal products have a warning: THEY ARE NOT FREE OF RISK. One such review was conducted by scientists from the Department of Complementary Medicine, Universities of Exeter and Plymouth, United Kingdom and published in the journal *Pharmacoepidemology and Drug Safety* in 2004. The study focused on the toxicity, interactions, and quality of herbal products. Toxicity data indicate that some herbal drugs "have the potential to cause serious adverse events and fatalities." They "affect pharmacokinetic and pharmacodynamic factors and thus cause herb-drug interactions." Contamination, adulteration, or substitution of botanical material has repeatedly put patients in danger, and that most often implicated are herbal drugs from Asia. The review concludes that "the widespread notion of herbal drugs being inherently safe is naive at best and dangerous at worst," and that more research and more information are required to ensure consumers' safety.

This paper examines the extent and quality of research done on sambong, lagundi, akapulko, ampalaya, coconut oil, and virgin coconut oil. Are there enough scientific studies to justify the booming market of their medicinal products?

Formerly with The Marine Science Institute, University of Philippines, Diliman, Quezon City, 1101, Philippines and the Southeast Asian Fisheries Development Center.
flor_lacanilao@yahoo.com

References were obtained through the Google Scholar, which provides ways to search for scholarly journals, books, and unpublished papers. Some of them have abstracts and a few have full texts. Sourced materials are from academic publishers, professional societies, universities, and other scholarly organizations.

The search for literature used scientific names as keywords for sambong (*Blumea balsamifera*), lagundi (*Vitex negundo*), akapulko (*Cassia alata*), and ampalaya (*Momordica charantia*). These local plants have different names in other countries, and their scientific names are used in international publications. For coconut, the keywords are coconut oil and virgin coconut oil. To focus on articles with them as the main subjects, the search was confined to titles of articles that contain the keywords.

Retrieved titles were examined to separate those published in international journals. These are journals covered in the Clinical Medicine and Life Sciences editions of *Current Contents* (CC). Clinical Medicine covers 1,229 journals and Life Sciences, 1,368. These are indexes of the Institute for Scientific Information (*www.isinet.com*). No Philippine medical journal has yet met the standards for inclusion in these indexes.

Results of the search for references consisted of only four titles for virgin coconut oil, five for sambong, 31 for lagundi, 63 for akapulko, 234 for ampalaya, and 291 for coconut oil. The total is 628 titles of which 245 (39 percent) were published in int. journals, with Filipino authors in the country contributing only four int. journal papers (see Table).

Summary of information from titles of published papers obtained through Google Scholar on six medicinal plants with commercial products

Keyword in title of article	*No. of titles retrieved*	*No. published in int. journals (%)*	*No. in int. journals with Filipino authors*	*Medical indications (in humans or animals)*
Blumea balsamifera (Sambong)	5	2 (40)	0	Blood clotting
Vitex negundo (Lagundi)	31	9 (29)	0	Anti-inflammatory, liver protection
Cassia alata (Akapulko)	63	20 (32)	1	Antimicrobial, other bioactives
Momordica charantia (Ampalaya)	234	115 (49)	2	Reduces sugar level
Coconut oil	291	97 (33)	0	Various health conditions
Virgin Coconut oil	4	2 (50)	1	Reduces cholesterol, moisturizer
Total	628	245 (39)	4	

Note: International journals are those covered in the Clinical Medicine and Life Sciences editions of the index *Current Contents*.

One journal paper with Filipino authors is on bioactivities of akapulko by IM Villasenor et al. from UP Diliman's Institute of Chemistry. It was published in *Phytotherapy Research* in 2002. Another journal paper by Filipino authors is on virgin coconut oil by AL Agero and VM Verallo-Rowell of the Makati Medical Center. Their paper, published in *Dermatitis* in 2004, shows that virgin coconut oil is effective and safe when used as a moisturizer. The other two journal papers by Filipinos are on antimutagenic properties of ampalaya by AP Guevara et al., also of UP Diliman's Institute of Chemistry. Their papers appeared in *Mutation Research* in 1990 and in *Phytochemistry*.

Ampalaya is featured in the largest number of int. journal papers, 115 (49 percent). Authors are predominantly scientists from India. Only two papers are authored by Filipino researchers. Most of the titles are medicinal, and are related to blood sugar level or show hypoglycemic effect in laboratory animals. One article published in 2003 reviewed ampalaya's efficacy and safety, concluding that adequately placebo-controlled trials were still needed to properly assess its safety and efficacy before it can be routinely recommended.

Coconut oil appear also in a large number of journal papers, 97 (33 percent), but none written by Filipino authors in the country. Almost all of the titles contain a wide variety of health-related and clinical terms—infant nutrition, organ function, metabolic effects, skin disease, STDs, atherosclerosis, platelet function, serum lipids, etc. And just about all are on laboratory or domestic animals. No review of literature is seen on the safety of coconut oil as medicine. None of the titles are on HIV or AIDS.

In sum, results show that out of 628 titles examined on the five local plants, 245 (39 percent) are scientific papers (int. journal papers). The four papers with Filipino authors are on akapulko, virgin coconut oil, and ampalaya. There are many retrieved titles with Filipino authors or published in the country; ten are on coconut oil. They are among the non-journal papers and can hardly be verified by international peers. Hence, as mentioned earlier, such papers are not counted in research evaluation, and they are not considered here.

The data show that there are no sufficient scientific studies to justify the development of medicinal products from sambong, lagundi, akapulko, and virgin coconut oil. Where are the results of the extensive research on medicinal plants by Filipino scientists, as claimed by the DOH? I think it is clear that the state of scientific research in the country, or elsewhere, on the four products does not support the government's campaign to use these products. The potential of ampalaya and coconut oil for developing drugs is indicated. But problems with toxicity and interactions with prescribed drugs require more research. Hence, considering all examined data, the answer to the question asked in the introduction is no, there are not enough research done on the efficacy and safety to support the booming market of medicinal and food supplements from these plants. And I will focus the rest of my discussion by asking a few more questions.

Does testimonial or anecdotal evidence on the effects of drugs derived from plants justify the government's promotion of their use? What is our health department doing to guard the consumers' safety in view of scientific findings that herbal drugs are not free of risk? Why did the few and unverified studies on the much-publicized virgin coconut oil win the approval of its products by the Bureau of Food and Drugs Administration? On the other hand, numerous studies on saw palmetto (from the berry of a dwarf palm found in the US) have repeatedly verified its viable treatment of enlarged prostate in men. It has been clinically proven to relieve the symptoms of medically diagnosed benign prostate enlargement in men. But it is yet to be evaluated by the US Food and Drug Administration.

Let me stress the recent findings of the scientific review on herbal drugs mentioned above. All herbal medicinal products (HMPs) contain a variety of pharmacologically active constituents, and users often combine HMPs with prescribed drugs. "Thus herb-drug interactions are a real possibility." Since they are natural products, some constituents can be good or harmful to body systems or organs. Each of these constituents has not been thoroughly studied to know at what amount and required time each is effective in giving the cure or harm to the body. The studies on our medicinal plants confirm that most of the claimed benefits from their products are anecdotal or testimonial. And literature reviews of the use of herbal products worldwide reveal numerous examples of liver, kidney, or other organ damage. More research is required on our plant medicinal products, particularly on the harmful constituents, to minimize the risk they pose to consumers' health. The excellent training of our medical doctors is shown by their reluctance to prescribe herbal drugs routinely.

Finally, on the question of promoting science in the country, which starts with the proper publication of results so that these can be taken seriously: Why hasn't the DOST made publication experience in int. journals a requirement for funding proposals? Mere printing of the results does not validate them. And they don't count in established international evaluation of research performance. Changing our funding practice will not only save money but will also justify increasing the R&D budget. It will improve the state of scientific research, and someday this may make herbal products effective and safe medicines. We have some able researchers (scientists) on natural products, including those who authored the four papers on medicinal plants mentioned earlier. They are the ones who should be given support and unlimited funds to be able to do enough science. The future of drug development from our natural products depends on them. A final note to our administrators of funding agencies: Yes, we are a poor country but "if you think research is costly, try disease."

February 2, 2006

Lunasin Natin ang Cancer! Lunasin—A Novel Cancer Preventive Substance from Seeds

BEN DE LUMEN

LUNASIN (from the Pilipino word *lunas* for cure) is a novel cancer preventive peptide (a small protein) that was first discovered in soy and now found in other seeds. Strictly speaking, lunas is not the proper word because lunasin has so far been found to be preventive not curative—but it sounds good anyway! Would iwasanin (from *iwasan*) be more appropriate? The serendipitous discovery of lunasin in our laboratory provides another explanation for the known anticancer properties of soy and other seed crops and opens up the opportunity to develop nutraceuticals and pharmaceuticals based on lunasin technology. It is not unusual that many profound discoveries were made serendipitously—without the intention of making the discoveries. Lunasin is such an example.

Many years ago, my laboratory initiated a project on enhancing the nutritional value of soy protein and other legumes through bioengineering. This requires increasing the level of methionine, the essential amino acid, which is most limiting in soy and other legumes, including mung bean. The strategy is quite straightforward: clone a gene coding for methionine-rich protein (MRP) and over-express the gene in soy or any target legume. We were faced with two choices for the source of the MRP gene—obtain it from other plants or from soy itself. I made the fateful decision to clone the MRP gene from soy itself, which eventually led to the discovery of lunasin as a cancer preventive agent. Had I not made this decision, I would not be writing about lunasin now—but who knows? At that time, there were two other groups that made the other choice—one group (Pioneer Hybrid International Inc.) cloned the MRP gene from Brazil nut which has a high methionine content and over-expressed it in soy; the other group (Dr. T.J. Higgins at Division of Plant Industry, Canberra, Australia) chose to clone the MRP gene from sunflower seed and over-express it in lupine, a cover crop used to feed sheep in Australia. The Brazil nut MRP in transgenic soy turned out to be allergenic to humans and led to the termination of the project. The Australian project is successful, in fact resulting in increased yield of wool for sheep fed on the transgenic lupines.

The choice to clone the MRP gene from soy is based on the hypothesis that there must be non-abundant MRPs in soy because most of the proteins in soy seed are

Department of Nutritional Sciences and Toxicology, University of California at Berkeley, CA, USA.
nitto@nature.berkeley.edu

notoriously low in methionine and therefore, there must be other MRPs that contribute to the overall methionine content of soy protein but they are non-abundant. The process of cloning the MRP gene turned out to be not easy, taking about four years, two graduate students and a postdoctoral scientist. One of the graduate students is a Filipina (now Dr. Jamie Revilleza from UPLB) and the postdoctoral scientist is another Filipino from UPLB (Dr. Alfredo Galvez). Dr. Revilleza contributed to the purification of the MRP from soy that eventually led to the cloning of the gene and Dr. Galvez took over the project and discovered the antimitotic effect of the lunasin gene when transfected into mammalian cells and the cancer preventive effect of the lunasin peptide. While it was not by design that the major contributors to lunasin discovery were both Filipinos, it is a source of pride to point this out.

Since its discovery in 1999, lunasin has been shown to be effective in preventing the transformation of normal cells into cancer cells caused by chemical carcinogens as well as cancer-causing agents from viruses such as the one known to cause cervical cancer in women. This is relevant to the Philippines, other Asian countries, and other developing countries all over the world where cervical cancer is the most prevalent cancer among women. Indeed, there is a plan to conduct clinical trials on cervical cancer in the Philippines when the proper time comes.

In the first animal model, lunasin was found to prevent skin cancer in mice when applied topically. It is likely then that the first commercial application of lunasin would be in skin cancer or at least in the formulation of over-the-counter "cosmeceuticals" to maintain a healthy skin. Topical application to prevent cervical cancer is also a likely application in the immediate future.

Since lunasin is a small protein, it is expected to be digested when we eat soy. We now have evidence that lunasin is protected from digestion by naturally present protease inhibitors in soy and other seeds.

The way lunasin works still needs to be elucidated further. However, accumulated evidence points to a unique and novel "epigenetic" mechanism—lunasin selectively kills cells that are being attacked by carcinogenic agents while leaving other cells unharmed by interfering with the unfolding of the chromosome that is necessary for cell proliferation. This unique epigenetic mechanism lends scientific legitimacy to lunasin and makes it different from currently studied soy chemopreventive agents such as phenolics and isoflavones that function as antioxidants and phytoestrogens.

Lunasin may play a key and fascinating role during seed development. Seed development in angiosperm is characterized by three major stages. First, there is rapid cell division and differentiation; second, cell division stops and the cells enlarge, accompanied by synthesis of DNA, lipids, proteins and carbohydrates; and, finally, the seed dries up. We believe that lunasin could be an effector molecule that allows the arrest of cell division to initiate the second stage. The hypothesis remains to be tested and proven.

Lunasin technology, because of its promise, is protected by numerous patents owned by the University of California at Berkeley, a necessary first step in its commercial development. Patents have been filed and granted in the US, Canada, Japan, European Union, and other countries. A start-up biotech company, FilGen BioSciences Inc., has been formed and granted a license to commercialize lunasin. It is noteworthy that the initial investors in the company are Filipino-Americans.

Lunasin's story is a fascinating one and continues to be so. Its discovery is serendipitous, its applications myriad, and its mechanism of action novel.

January 6, 2006

Elsie Jimenez

Protein Modification: To D or not to D...

With the proliferation of diet regimens, more people are turning to a "low carb (carbohydrate), high protein" diet. Have you ever wondered about the nature and quality of the protein in your diet? Are the constituent amino acids (substances that make up a protein) essential? You might have a ready answer to this question. Furthermore, are the constituent amino acids in the L- or in the D-form? Before we settle this question, let me tell you something about L- and D-amino acids in relation to my research.

Proteins and peptides ("small proteins") are chains of amino acids; they are synthesized from a set of twenty standard L-amino acids in the cell's protein machinery called the ribosome. Naturally occurring proteins and peptides contain predominantly standard L-amino acids which are directly encoded by genes. After synthesis in the ribosome, they undergo changes which are referred to as posttranslational modifications. A modified protein or peptide may contain one or more constituent amino acids which have been transformed into nonstandard amino acids, such as L-5-hydroxylysine and L-4-hydroxyproline which are present in the muscle protein collagen, and L-gamma-carboxyglutamate which is found in the blood-clotting protein prothrombin.

A remarkable posttranslational modification of a peptide is the rarely occurring conversion of a standard L-amino acid to the D-form—a process that transforms an L-amino acid into its mirror image, resulting in a different structure of the peptide. Conversion of L- to D-amino acid has been documented in peptides from different phyla. Among those are the neuroexcitatory peptides Achatin-I and Fulicin from the ganglia of African giant snail, a neuroexcitatory peptide called FFRF-amide from the muscle of a bivalve mollusc, an anti-pain peptide known as Dermorphin from the skin of the Argentinian tree frog, glucose regulatory hormones from crayfish, and a calcium channel blocker called omega-Agatoxin from the venom of the American funnel-web spider.

About a decade ago, I did my PhD dissertation on a sandwich program at the University of Utah in Salt Lake City, Utah, USA, under the sponsorship of the Department of Science and Technology-Engineering and Science Education Project (DOST-ESEP). During that one-year stint, I discovered a D-amino acid-containing peptide, the first one of its kind, in the venom of the marine cone snail *Conus radiatus. Conus radiatus* is known to feed on fish and uses its venom to immobilize its prey. Dr. Baldomero M. Olivera, distinguished professor of biology at the University of Utah, who was my PhD

University of the Philippines-Baguio, Philippines.
elsiecjimenez@yahoo.com

dissertation co-adviser, coined the term "contryphan" for the novel peptide because it was obtained from a cone snail and the D-amino acid found was D-tryptophan.

The discovery of a D-tryptophan was sort of serendipitous; D-amino acid was initially not included in my PhD dissertation proposal. Dr. Lourdes J. Cruz, then concurrently research professor at the University of Utah and professor at the Marine Science Institute, University of the Philippines Diliman, who was my PhD dissertation adviser, and Dr. Olivera encouraged me to pursue the study. Later, I informed DOST and requested my PhD dissertation committee at the Molecular Biology and Biotechnology Program (now the National Institute of Molecular Biology and Biotechnology) in UP Diliman for a change in my dissertation title and objectives. I got an approval, for which I am grateful. The decision to pursue the research on the D-amino-acid-containing peptide ("To D or not to D...," because I thought that I took the risk of getting a disapproval from my dissertation committee) is now bearing fruit. That discovery initially led to two accelerated publications in the prestigious ISI-accredited journals, the *Journal of Biological Chemistry and Biochemistry*.

Dr. Olivera took keen interest in pursuing further research on D-amino-acid-containing peptides. Since I finished my PhD in 1996, I have been shuttling between the University of the Philippines in Baguio and the University of Utah to undertake research on D-amino-acid-containing peptides and other posttranslationally modified peptides.

A challenge in assessing posttranslational modification is that the larger the peptide, the more difficult it is to ascertain the presence and/or determine the position of a modified amino acid. With the genetic recombinant technology being more widely used, there is a greater possibility that posttranslationally modified peptides will be missed when using the conventional gene cloning approach. Moreover, in contrast to other posttranslational modifications, conversion to a D-amino acid is not detectable by any of the standard methods used in protein chemistry, such as mass spectrometry, since both L- and D-amino acids have the same mass.

In small peptides, transforming one L-amino acid into the D-form can result in a significant difference in properties. In contryphan, which has only eight constituent amino acids, the presence of D-tryptophan is responsible for two different structures that give rise to two distinct peaks upon chromatography. In contrast, the corresponding peptide with L-tryptophan gives a very broad peak upon chromatography, and turns from white to a violet substance upon drying. Those were some of the experimental leads to the working hypothesis for a D-amino acid—an idea first suggested by Dr. William Gray, a researcher collaborator in the University of Utah.

As the peptide gets bigger, converting a single L- to a D-amino acid results in more subtle changes. An excitatory peptide belonging to a toxin family which blocks potassium channels in the central nervous system has been purified from the venom of *Conus radiatus*. This peptide was recently found to have a D-amino acid called D-phenylalanine. The determination of the presence of a D-phenylalanine in the 44th position of this peptide which has forty-six constituent amino acids was not an easy task. There was a need to cut the peptide separately by two enzymes. During the course of the experiments, Dr. Olga Buczek and Dr. Grzegorz Bulaj, who were my research collaborators, and I would find ourselves betting on which peptide fragment would contain the D-amino acid. As it turned out, no one won because we all made the wrong guesses. The peptide with a D-phenylalanine is strikingly different in biological activity as compared to the one with an L-phenylalanine. Using an in vitro assay, Dr. Doju Yoshikami, a research collaborator, confirmed that the naturally occurring D-phenylalanine-containing peptide is greater than 250-fold more potent than the peptide with L-phenylalanine. Thus, the posttranslational modification of an L-phenylalanine to the D-form is clearly functionally significant in the biology of *Conus radiatus*.

It has been documented by Dr. Pradip Bandyopadhyay and coworkers, University of Utah researchers, that there is a signal in the precursor protein or peptide for the modification of L-glutamate to L-gamma-carboxyglutamate, as found in both blood clotting proteins and *Conus* peptides. It is expected that a similar signal for the modification of an L- to the D-amino acid is also present. The need for an L-amino acid in a protein or peptide "to D or not to D..." must somehow be justified in terms of either functional or structural significance or both. So far, it is not yet known whether any single amino acid in proteins that we usually take in our diet is in the D-form.

(Note: The research on *Conus* peptides is done in accordance with a Commercial Research Agreement signed by the Department of Agriculture, University of the Philippines, and University of Utah.)

January 27, 2005

Mary Ann Endoma

Synthetic Organic Chemistry in the Philippines: Where Do We Go from Here?

In the Taxol success story, one cannot help but admire the ingenuity of the chemists who devised its reaction scheme. In reality though, all it took was some common sense and plain simplicity. Taxol is a very potent anticancer drug that is being marketed by Bristol-Meyers Squibb Co. as Paclitaxel. It was approved in 1994 by the US Food and Drug Administration for the treatment of breast cancer.

But just what is the Taxol success story? A compound called 10-deacetyl baccatin (10-DAB) was extracted from the leaves and needles of the European yew tree *Taxus baccata*. The active compound, however, was Taxol, which on the other hand, was isolated from the bark of the same tree. Removal of the bark resulted in the killing of the tree. However, close inspection of the structure of Taxol and 10-DAB revealed a close relationship between them such that 10-DAB was actually a precursor in the biosynthesis of Taxol by the plant. In six chemical steps and a whopping yield of 50 percent, 10-DAB was converted to Taxol.

Before the discovery of the structural relationship between 10-DAB and Taxol, several renowned synthetic organic chemists had set their eyes on Taxol as their target molecule for synthesis. Due to its structural complexity, it posed tremendous intellectual challenge and thus, it was a very attractive target. Also, due to its extreme importance in the treatment of a dreadful disease, it received considerable attention from health-related funding agencies. The synthetic approach was to prepare the Taxol molecule from commercially available starting materials. After more than ten years of continuous struggle and hard work, some researchers had succeeded in their efforts, but their success was hampered by a very low conversion or yield. There were just too many steps involved in the synthesis and the reactions used were too complicated for possible transfer from the laboratory to industry. Total synthesis, as their approach is called, was rendered impractical. On the other hand, the approach of starting the synthesis from another isolated product is called semisynthesis. The conversion of 10-DAB to Taxol is a semisynthesis.

This illustration serves to introduce to you a field called synthetic organic chemistry. "Synthesis" loosely translates into preparation and "organic" indicates compounds of natural origin and mainly not possessing a metal center in its structure. Synthetic organic

---

Institute of Chemistry, University of the Philippines, Diliman, Quezon City, 1101, Philippines.
maryannendoma@yahoo.com

chemistry involves the use of known reactions and protocols in the literature for the preparation of pharmaceutically important compounds. A group of pharmaceutically important compounds comes from the natural products group. Taxol itself is a natural product. Most of these natural products are isolated in very small amounts (micrograms). Synthesis, therefore, steps in for the production of biologically active natural products in greater quantities. These compounds of interest are to be produced in the laboratories with the use of commercially available reagents via established protocols. A major limitation in the choice of targets for synthesis is the complexity of the molecular structure. In general, large and more complex molecules demand more sophisticated techniques and equipment. In the Philippines, only simple structures are considered to be doable targets. For more complex structures, however, semisynthesis is the approach of choice.

So far, two approaches have been introduced to you: total synthesis and semisynthesis. A third approach in the preparation of compounds is called derivatization or analog formation. An analog is a compound, which has a slight structural difference from the parent compound. For example, technically, 10-DAB is considered an analog of the parent compound Taxol. Most analogs, however, are made from the parent compound. An active area of research in advanced countries like the US is the preparation of analogs. In some cases, it has been found that analogs exceed the potency of the parent compound.

In the long run, what is doable in the Philippines? Which of the three approaches is to be taken by a researcher who is interested in the preparation of organic compounds? These questions will be answered based on an assessment of about five years' worth of our continuous, dedicated, and extensive research work in a Philippine setting. A direct comparison with a research conducted in a US setting will be done to illustrate the point.

First, the only reactions considered doable here are those that are most elementary—simple mixing of reagents and reacting components, so to speak. One cannot expect to carry out advanced reactions. The more advanced the reactions, the more sophisticated the setup required. The main limitation is the availability of special equipment. In the US, a typical synthetic organic chemistry laboratory is equipped with the necessary equipment, provisions, and amenities for special reactions. Second, there is a very long waiting time for the reagents to come since they are imported from other countries. In the US, one week is considered a long time. Here, the average delivery time is about one to three months. Third, there is no access to important journals. One has to be aware of what solutions exist as reported in journals for certain laboratory problems. Not getting the necessary help may result in loss of motivation to continue with the work. Frustration sets in and hence, the researcher is left with no recourse but to give up. In the US, the journals are accessed online, "twenty-four-seven" (24 hours a day, seven days a week or 24/7).

Before this article begins to sound like the TV commercial ("*Walang ganyan sa* States..."—meaning traffic and bad roads plus the pollution), now let's look at the positive side of things. So why is it important to do synthesis in the Philippines? Because that is the way up. Synthesis is a major technological innovation. It adds value to indigenous materials. At the moment, for a beginning researcher, the road ahead looks impossible to pursue. But it is only a matter of time. Filipinos are known for their creativity and positive outlook.

To solve the three problems discussed earlier, our research group at the UP Diliman Institute of Chemistry (UPD-IC) started seeking help from more established laboratories. By joining forces with other institutions, we were able to pool together our meager resources. We found the spirit of sharing and cooperation in the Marine Science Institute in UP Diliman (UPD-MSI), the Philippine Institute of Pure and Applied Chemistry (PIPAC), the National Kidney and Transplant Institute (NKTI) and the Research Institute for Tropical Medicine (RITM). These institutions were able to provide us with some needed equipment and expertise. In return, we also shared our capability and technological know-how with them. We are pursuing a lot of collaborative work at the moment. One still has to deal with the waiting time for the reagents. Considerable advanced planning has to be done to make sure the researcher does not waste time waiting. Eventually, they also arrive. In regard to the journal requirement, some Filipino friends abroad help by providing needed journal articles. Most of them are willing to help out. Again, a spirit of self-giving and cooperation solves the problem.

Eventually though, the three problems mentioned have to be addressed in a more realistic and direct way. Meanwhile, we will make do with what we have right now. Again, this shows how resourceful and patient we are as a nation. Finally, how is it going in our pursuit of synthesis? So far so good. We have prepared simple compounds in the laboratory and we have repeated some literature syntheses as a proof of principle. We are now moving on to tackle more advanced targets. The progress is slow but the results are definitely encouraging, enough for us to move forward. We are working on projects that involve the two approaches—total synthesis and analog preparation. But our targets are definitely easier and simpler than Taxol. We believe we are headed in the right direction. Synthetic organic chemistry will hopefully make its way as part of our country's technological capabilities.

January 20, 2005

Part 7

Food, Agriculture, Aquaculture, and Biotechnology

Rafael Guerrero III

Foodfish of the Twenty-first Century

A fish that we commonly know and consume as tilapia is considered the "foodfish of the twenty-first century" by Dr. Kevin Fitzsimmons, president of the World Aquaculture Society.

The tilapia, a native of Africa, is produced as a foodfish in over 100 countries in culture systems that range from backyard ponds to superintensive farms. Global production for the fish in 2002 was more than 1.5 million metric tons, making it one of the top ten aquatic species farmed throughout the world.

Dubbed as the "aquatic chicken," the tilapia has good attributes that make it suitable for aquaculture. It matures early, breeds readily, and is a hardy fish. It also has a growing international market demand as a foodfish.

Historically, the introduction of the first tilapia species, the Mozambique tilapia, in the Philippines in 1950 was initially not well-accepted by the industry because of the lack of appropriate culture techniques. Growth of the fish in ponds was stunted with too much breeding and overpopulation. The small size and dark color of the fish did not also appeal to local consumers.

The coming of the Nile tilapia in the 1970s improved the acceptance of tilapia in the country because of its lighter color and faster growth compared to the Mozambique tilapia. The tendency of the tilapia to overpopulate in ponds, however, still had to be solved. The problem not only limited commercial production of the fish in the Philippines, but in other countries as well. A practical solution was needed for the "family planning" of tilapia in order that large-sized fish could be produced commercially.

The technological breakthrough for the control of tilapia reproduction in ponds for grow-out of the fish to market size was achieved in the mid-70s with the development of the hormone-induced sex reversal (HSR) method. In applying the method, oral treatment of "sexless" fry is done with synthetic male hormone to convert would-be female fish into faster-growing and nonreproducing males. Compared to other methods of producing all-male tilapia, HSR is the most practical, cost-effective and efficient. Proper use of the HSR method has been approved by the Food and Drug Administration of the United States.

In 1986, the Philippines was the world leader in tilapia production from aquaculture with its commercial culture of the Nile tilapia in freshwater ponds and cages, and

Philippine Council for Aquatic and Marine Research and Development, Department of Science and Technology, Philippines.
pcamrd@laguna.net

application of the HSR technology. Our country produced 135,996 metric tons of tilapia in 2003 and was second to China in world production.

With the declining supply and spiraling prices of marine fishes like the roundscad ("*galunggong*") caught from our seas, President Arroyo has declared the tilapia as the new "fish of the masses." A master plan for the country's tilapia industry has been prepared by the Department of Agriculture and is now being implemented in coordination with the private sector.

In the Medium-Term Philippine Development Plan (2004-2010), tilapia will play an important part in ensuring the food security of the country in terms of foodfish. The government projects an annual increase in our tilapia production by 25 percent. Improved tilapia strains will be mass-produced throughout the country to increase food supply, income, and livelihood opportunities. The major tilapia-producing areas are Central Luzon and Southern Tagalog.

Fifty percent of the tilapia production in the Philippines is from freshwater ponds, followed by 37 percent from cages in lakes and reservoirs and 7 percent from brackishwater ponds. The species cultured are the Nile tilapia in freshwater ponds/cages, and the Mozambique tilapia and its hybrids with other species in brackishwater ponds.

The future prospects for tilapia farming are bright globally. With an annual growth rate of 12 percent, global tilapia production is expected to exceed that of carps within this century. There is a growing demand for frozen whole and filleted tilapia in the United States which is the world's largest importer of the fish. Worldwide, the business of trading tilapia as a foodfish is already estimated to be more than $2 billion.

An interesting development in the industry is the utilization of the by-products of tilapia processing plants such as skin, scales, and trimmings. Tilapia skin can now be made into leather goods, clothing, and even snack foods. Scales, on the other hand, can be processed into gelatin for pharmaceuticals, and trimmings (including heads) can be prepared for soups. Even small-sized tilapia can now be transformed into what is called *tilanggit* similar to the much-in-demand boneless *danggit* of Cebu with simple and low-cost technology appropriate for rural families.

A project supported by the Netherlands known as "Fish for Every Family" is being implemented by Plan Philippines in partnership with the Philippine Council for Aquatic and Marine Research and Development to help improve the nutritional status of families in the depressed areas of Isabela, Pangasinan, Occidental Mindoro, Southern Leyte, and Western Samar. The project aims to develop backyard ponds for tilapia, each with an area of at least fifty square meters, for 400 family-cooperators.

As Israel Snir, an international tilapia integrator, put it: "Tilapia is a fish that can be produced for the people and by the people. It can be produced everywhere and consumed by everyone."

August 4, 2005

Paciente Cordero Jr.

Sea Vegetables ... *Gulay Dagat*

The sea, often referred to as man's ultimate frontier for survival, is populated with plants such as seaweeds and sea grasses. They play a significant role in the food chain and in maintaining the marine ecosystem. Studies in the early 1900s on the seaweeds of the Philippines included the identification of some useful species, but still little is known about the many species of seaweeds that are edible to man. Edible seaweeds, appropriately known as sea vegetables or *gulay dagat*, grow in the shallower, rocky parts of the sea. Seaweeds are colored green, brown, and red depending on the dominant pigment present in these lowly plants.

Several chemistry studies reveal that the sea vegetable species found in Philippine waters like green *Ulva lactuca (gam gamet*–Ilokano*), Caulerpa racemosa (lato*–Bisayan, *ar-arucip*–Ilokano*), C. peltata var. macrodisca (laba-laba*–Aklanon, Ilonggo*), Codium (pocpoclo*–Ilokano*)*, brown *Hydroclathrus clathratus (balbalulang*–Ilokano*), Sargassum (aragan*–Ilokano*), red Porphyra (gamet*–Ilokano*), Eucheuma (ruprupuuc*–Ilokano, *guso*–Bisayan*), Gracilaria (guraman*–Ilokano, *gulamang dagat*–Tagalog, *gulaman*–Bisaya*), Hypnea (culot ti pusa*–Ilokano*), Laurencia (al alacon*–Ilokano*), Gelidiella (gulaman*–Bisayan*)* and *Acanthropora (culot)*, are rich and cheap sources of protein, carbohydrates, iodine, and Vitamins A, B, and C

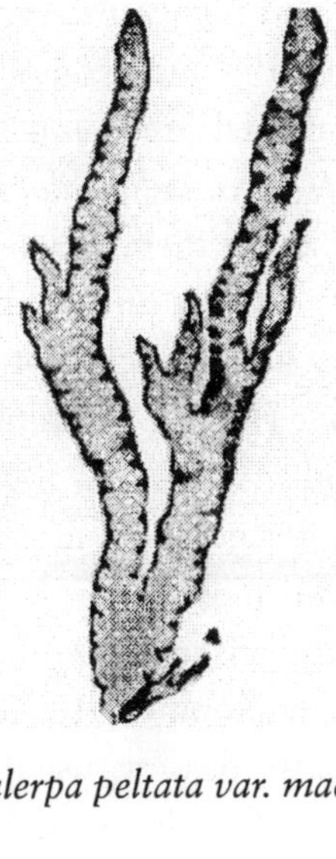

Caulerpa peltata var. macrodisa

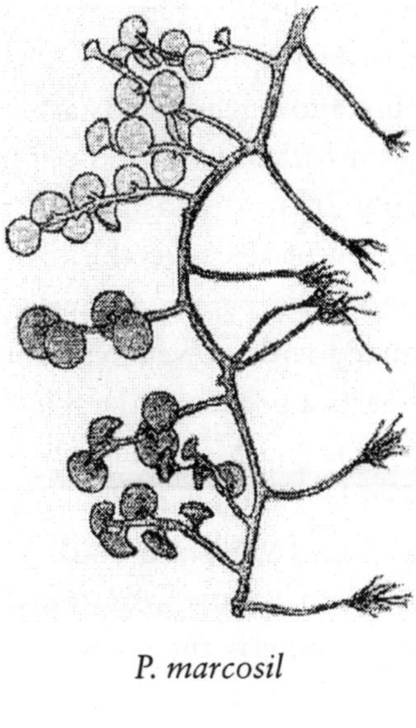

P. marcosil

National Research Council of the Philippines, Department of Science and Technology, Philippines.
drpacjr@dost.gov.ph

Aside from being among the most nutrient-rich plants on earth, sea vegetables are an excellent source of complex carbohydrates with varied commercial, pharmaceutical, industrial, and food uses. We have 1) agar (dessert gel ingredient), extracted from *gulamang-dagat*; 2) carrageenan, a hydrocolloid gel principally from *guso*, is processed into varied food and used in dairy products manufacture and in medical and cosmetic products; 3) alginates, found in brown seaweeds, are gelling ingredients that are used in both food and non-food products (e.g., dental impressions); and 4) gelatin or *gulaman*, obtained by boiling red edible seaweed such as *Gelidiella*.

Sea vegetables are popular among Ilocanos, especially those in northeastern Luzon. Public markets in Ilocos Norte and Cagayan sell about a dozen kinds or species of fresh and dried green, brown, and red sea vegetables. The popularity and acceptability of sea vegetables as part of meals have a long history; in fact, it is a culinary tradition passed on to the present generation. *Gamet* is the most popular and highly prized sea vegetable in Ilocandia. This is attributed to its known nutritive value, the limited area where it is found, and the difficulty of collecting it. Specifically, the three *Porphyra* species *(Porphyra crispata, P. marcosii* and *P. suborbiculata)*, mainly grow in and are gathered from Burgos and Pagudpud, Ilocos Norte and Santa Ana, Cagayan. Chemical analysis of *gamet* shows that it contains 20 percent digestible protein or more, Vitamin A at 36,000 to 50,000 IU per gram, much higher than chicken eggs (850 IU) or chicken liver (10,000 IU), among others.

Among the Ilocanos, their common food preparations of sea vegetables after blanching with lukewarm water, are in the form of a salad or mixed with vegetable or fish stews. Specifically, for *gamet* salad, after washing and blanching, the sea vegetable is mixed with crushed ripe tomatoes, sliced green mango, and sliced onion, complemented with native Ilokano *bagoong* to taste. Otherwise, blanched green *pocpoclo, gamgamet* and *lato*; brown *balbalulang*; and red *gulamang dagat, guso, culot, alalacon* and *culot ti pusa* are simply eaten fresh as a salad by adding or soaking them in vinegar. A modified salad preparation of sea vegetables is done by adding a sweet and sour mixture with a dash of ground black pepper. Young and tender leaves and branches of select species of aragan are mixed with stewed vegetables, fish or meat. The sea vegetable adds nutritive value and imparts a peculiar palatable taste to the viand.

In the Visayas and Mindanao, the more popular sea vegetables, mostly eaten fresh as salad, are green *lato* and *laba-laba*, and red *gulamang dagat* and *guso*. In Lapu-Lapu City, *guso* dipped in aged vinegar is sold in the market as appetizer or *pulutan* by Cebuanos. Similar salad preparations are found in Southern Philippines. In Biliran province, *Gelidiella* or *gulaman* is the source of homemade gelatin dessert, usually flavored with lemon juice. To them, a good quality gelatin extract is firm in texture upon standing; otherwise, it is soft and wiggles when shaken. Indeed, except for *gamet*, majority of the sea vegetables are found in most marine waters of the country (Pacific Coast, China Sea, and inland waters). Fisherfolk and inhabitants along coastal areas would easily recognize these seaweeds because of their dominant green, brown, and red colorations.

In a collaborative project designed to develop sea vegetable recipes, funded by the National Research Council of the Philippines (NRCP-DOST) and the Food and Nutrition Research Institute (FNRI-DOST), a panel of food tasters was organized to do sensory evaluation and determine the acceptability of eighty sea vegetable recipes. The acceptability of the recipes was evaluated using the Hedonic Scale of 9 to 1, representing a high to low acceptability range. The study utilized forty-three sea vegetable species, of which twelve were green, seven brown and twenty-two red, among the eighty species. Of these, fifty-two recipes had a high acceptability rate ranging from 8.8 to 6.0. The most acceptable among the recipes, at 8.8 level, was fish balls soup with *gulamang-dagat (Gracilaria coronipifolia)*. Other recipes registering within the high 8 level of acceptability included fish balls soup and *bangus relyeno* with red seaweed *Gracilaria arcerata; okoy tagonton; bangus sinigang* with *Gracilaria salicornia, Gelidiella acerosa,* and *Laurencia papillosa* used separately; fish *lumpia* with *Laurencia okamurai* and *Gracilaria coronipifolia*; and fried fresh *lumpia* with *Hypnea charoides*. There were ten recipes in the 8 scale bracket, twenty-nine in the 7 scale, and fifteen in the 6 scale attesting to the acceptability of sea vegetables as food for human consumption.

The Philippines is beset with a cyclical problem of food security, especially to feed the marginalized Filipinos. It is a reality that land-based food resources need to be supplemented by tapping the resources of the sea, especially edible seaweeds. Necessarily, conservation and controlled harvesting of seaweeds should be practiced. The tropical location of the country finds it endowed with the presence of favorable ecological factors, making the country rich in marine biodiversity. Records show that there are about "2,000 reef fish species, 800 species of seaweeds, sixteen species of sea grass, twenty-three species of cetaceans, and five species of sea turtles, making the Philippines one of the world's richest countries in terms of coastal and marine species." Among the edible seaweeds growing in the country, a high percentage belong to the red group or Class *Rhodophyceae*.

However, convincing more Filipinos to accept and serve sea vegetable meals to enrich their food intake appears to be a lonely battle. Hopefully, not for long, as we continue to grapple with and source out nutrient-rich and cheap food. The need to develop a scheme wherein part of the farmed-produced *guso* seaweeds for export are going to be processed as protein-rich food for the disadvantaged segment of the Filipino population should be a primary concern. Seaweeds, dominantly *Eucheuma* and *Kappaphycus*, in the dried, semiprocessed, and processed state (carrageenan and pet foods), rank third among the Philippines' fishery export commodities. Suffice it to say that the Philippines is ranked the world's number three, behind China and Japan, in seaweed production.

In 2002 alone, the Philippine seaweed industry reported 895 metric tons of seaweeds produced, worth $131 million in exports. Roughly 70 percent of Philippine *guso* production comes from Southern Philippines (Tawi-Tawi, Sulu, Zamboanga, Basilan), while 30 percent comes from the Visayas and Southern Luzon. The Department of Agriculture envisioned seaweed production to hit 1,000 metric tons in 2004. The proceeds were perceived to help increase the country's coffers. The real challenge for policymakers and decision makers is to seriously craft appropriate measures that will tap, conserve, and encourage the eating of sea vegetables among Filipinos. Such a strategy holds the key to arresting the chronic malnutrition problem and helping solve the scarcity of food supply through a managed development of one of the country's promising marine resources—sea vegetables.

January 6, 2005

S. Suzanne
Mingoa-Licuanan

Gentle Giants

I first encountered a giant clam when I was fifteen years old. I was in a church, and the clam shell was a baptismal font. It was just a shell, not a live clam, and it could fit maybe three babies, I thought. It was the biggest shell I'd seen so far. I tucked away that memory, without an inkling that my work would eventually involve such a creature. Ten years later, I left for postgraduate studies in Australia, to study the giant clam.

The giant clam is a bivalve mollusc, its soft body housed between two shells. Scientists have now grouped giant clams with the Family Cardiidae, Subfamily Tridacninae, which has two living genera *Hippopus* and *Tridacna* (which literally translates to "eaten at three bites"). This subfamily is quite small, with only nine species presently found in the Indo-West Pacific region. Of these nine, the Philippines has seven species.

Unlike other bivalves, the giant clam sits upright on the seabed, its brilliant fleshy mantle grinning at the sun through the water column. The world record size for the largest species *Tridacna gigas* is 1.37 meters. Although literally brainless, the giant clam has a nervous system that controls all bodily functions. The mantle has two "holes," or siphonal orifices, that differ in size. The large orifice, or incurrent siphon, sucks in food-laden seawater, which gets filtered through the clam's large white gills. As a matter of course, the clam may gently expel the water through the other smaller orifice, or excurrent siphon. Occasionally, a hungry fish pecking at the clam's mantle may get a strong jet of water from the excurrent siphon. In the mantle of *Tridacna*, there are thousands of minute "eyes," each with hundreds of lens cells. Giant clams are basically shy—at the mere passing of a shadow over the mantle, they opt to hide within their valves. Incredible as it may seem, these gentle giants have no means to defend themselves, having no lethal weapon to speak of. But then again, there must be something in being big.

Scientists have hypothesized that the giant clam has evolved into a form different from other bivalves, to allow its symbiotic zooxanthellae living in its mantle maximum exposure to solar radiation. The term "zooxanthellae," fondly shortened to "zoox" by scientists, refers to endosymbiotic dinoflagellates—plants that are single-celled, yellow-brown in color, and live within a host. The evolution of the giant clam begins sixty-five million years ago, from small ancestral cockle shells (not unlike the modern day cockle *Cerastoderma edule*) about the size of the *rambutan* fruit. After some fortuitous event,

The Marine Science Institute, University of the Philippines, Diliman, Quezon City, 1101, Philippines.
suzanne@upmsi.ph

this ancestor may have filtered out from the seawater its first zoox, most likely, from damaged larvae of corals and sea anemones. The ancestral cockle's phagocytic cells (whose function is to engulf anything foreign) may have ingested these zoox, which, over time, have found a suitable habitat in the cockle's mantle. The "marriage" between giant clam and zoox has prospered largely because it fulfills the needs of both partners in terms of nutrition and security. As plants, zoox are photosynthetic and produces by-products such as simple sugars and proteins that become "food" for the clam. In turn, the clam provides a safe haven in its mantle for zoox to reside and proliferate in high densities. The clam's dissolved nitrogenous wastes also serve as nutrients for the zoox.

How fast do giant clams grow? Growth rates will differ with species, but as a rule, the large species grow faster (like *Tridacna gigas* with average growth rate of 10 cm per year) than the small ones (like *Tridacna crocea* averaging at 2 cm per year). *Tridacna gigas* grows as fast as the common edible "slipper-shaped" oyster, locally known as *talabang tsinelas* in the Philippines. Giant clam growth is fastest during the juvenile stage, and slows down a bit during adolescence, and even more so when it becomes an adult. Giant clams naturally live long. Today, nothing can be farther from the truth.

Any fisher will tell you how delicious giant clams are. Large and conspicuous as they are on the reef, clams easily make a fisher's meal. Large species are highly priced in foreign markets. It took less than twenty years for man to bring the largest species, *Tridacna gigas*, to the level it is now—virtually extinct. But unlike fifty years ago, when commercial fishers would go their merry way harvesting clams from the reef, an export ban has been in effect since 1985 against the commercial trade of giant clams. Supposedly, there is a simultaneous ban against harvesting, but there are local markets south of the Philippines where wild clams, after being shucked out of their shells, are skewered and sold. In central Philippines, they are offered as exotic dishes in restaurants. Northward, along the west coast, they are sold whole in small piles.

But all is not lost for these gentle giants. If giant clams can be propagated in culture, and released on the reefs, then clams may grow in the wild, reproduce, and repopulate our seas. In addition, clams are released only in areas where protection is guaranteed by legislation, specific institutions or groups, or owners of adjacent coastal properties. This has been the strategy taken by The Marine Science Institute (MSI), one of the very few institutions in the Philippines working on giant clams. Prof. Edgardo D. Gomez, known to us as the "Father of Giant Clams," has been the moving force behind the culture and reseeding efforts of MSI, to return giant clams back to the reefs where they belong.

December 16, 2004

Marco Nemesio Montaño

Gelatin, *Gulaman*, "JellyAce," *atbp.*

In fiestas and parties, the common gel dessert is always served. Some of these gels are soft and wiggle when shaken, while some are hard and brittle. Some gels need refrigeration or cooling to prevent them from melting. Others can withstand the heat. These dessert gels are served with or without fruit.

Traditionally, the ingredient in making gel dessert is the animal-derived gelatin, to which water, sugar and flavoring are added. Gelatin is a protein derived from the bones, skin, and other collagenous materials of farmed animals. To form gels, gelatin is dissolved in hot water and cooled to set. Without cooling, the gels will not form well or the solution remains a liquid.

Recently, there have been objections to the use of gelatin that brought its price down in the world market, aside from oversupply. The controversial hoof-and-mouth disease and mad cow disease made the consumers wary of any products that come from farmed animals. Since gelatin is an animal protein from cattle or hogs, vegetarians and Muslims also have objections to its use in food and medicines. The use of fish gelatin has been introduced in Western countries with fewer objections than those derived from four-legged farmed animals. Fish gelatin has been traditionally produced in China, Japan, and Korea but not in large quantities. (One might notice the gel formed when leftover *sinigang* soup is refrigerated.)

Other commonly used replacements as gelling ingredient are seaweed-based. Unlike gelatin, agars, alginates, and carrageenans are complex carbohydrates. This article is written as an introduction to the use of seaweeds and their extracts in food, medicines, agriculture and other technical products.

Agar, as a dessert gel ingredient known locally as *gulaman*, is usually sold as wafer bars in the Philippines. It is also sold in powder and wafer strip forms. One has to boil the water to dissolve the wafer before allowing it to set as a gel. Because the gelling temperature is high, refrigeration is not necessary to obtain the gel. The agar gels are brittle and hard. Those containing relatively high sucrose have brilliant appearance. Some agar extracts produce harder gels when sucrose is added to the solution. Gels from purified agar do not have moist surfaces; hence, they are used in the culture of microorganisms and in tissue culture. Similar to carrageenans and alginates, they have pharmaceutical and technical applications too.

The Marine Science Institute, University of the Philippines, Diliman, Quezon City, 1101, Philippines.
coke.montano@gmail.com

The most common raw materials for agar production in our country are red seaweeds commonly called *gulamang dagat*. These marine plants grow in brackish to highly saline waters. There are other kinds of red seaweeds that are harvested and are also exported for agar production.

Another extract from *gulamang dagat* is called agarose. Agarose is a highly purified neutral agar that is used in biotechnology and in medical applications. Among the seaweed gels, agarose is high-priced. Unfortunately, we do not have an agarose industry in the Philippines.

Alginates are gelling ingredients obtained from brown seaweeds. Unlike gelatin, agar and carrageenan gels, calcium alginate gels are thermally stable, meaning they do not melt when heated. Hence, they are used mostly in refabricated fruits. Also, in synthetic crab sticks, meat ingredients are held together by alginates. Most of the time, alginates are used to increase the viscosity of solutions. Before the use of carrageenan in our local ice creams, alginates were used to make the texture creamier and improve mouth-feel. Talking about nonfood applications, one might also be aware that dentists use alginates in making the dental impression of a set of teeth. To print on textiles, one has to make paints viscous so that the paint will not smudge. This is done by adding alginates to increase the paint's viscosity. Unfortunately again, we do not have an alginate industry in the Philippines; however, we are exporting the ground dried brown seaweed as ingredient in animal feed and soil ameliorants.

One of the country's competitive edges is the farming of red seaweeds that produce carrageenan. Carrageenans have increasing applications not only in food but also in other products in the pharmaceutical and cosmetic industries. The ever famous "JellyAce" is a carrageenan gel mixed with another gum to modify the texture. The famous "Alsa Gulaman" is not agar but carrageenan gel. When mixed with other ingredients, carrageenan gels can be made to produce properties similar to gelatin gel. Aside from being a gelling agent, it is also used as an emulsifier and stabilizing agent. Filipinos have been taking carrageenan from consuming *guso*. For decades, we have taken evaporated filled milk that contains carrageenan as emulsifying agent. Nowadays, carrageenan has found its way to meat products like hams, sausages, canned meat, and other similar items, making them moist and tender. Some toothpastes and personal care products have carrageenan in them too.

The country produces two types of carrageenan: refined and semi-refined. Terms like Traditionally Refined Carrageenan (TRC) refers to the process in which the seaweed extract goes through a filtration stage to remove the cellulose found in the plant, thereby making the gel clear. Philippine Natural Grade (PNG), Semi-Refined Carrageenan (SRC), and Processed Eucheuma Seaweed (PES) seaweed flour all refer to carrageenan with cellulose, sometimes called Alternatively Refined Carrageenan (ARC); thus, the gel produced is not clear. However, if the application does not require clear gels, the semi-refined carrageenan can perform the function of the refined one. Because the process of production costs a third of that of the refined process, the Philippines has a number of processing plants producing this kind of carrageenan. Because of the world demand for carrageenan, the Philippines exports the final product and the seaweed raw material itself. We have been exporting the kappa-carrageenan and *iota*-carrageenan-bearing seaweeds to meet 70 to 90 percent of the world demand. Just for 2001, the exports reached almost $130 million. Seaweed farming for carrageenan production is mainly done in southern Philippines, particularly in Tawi-Tawi and Zamboanga. With the expansion of seaweed farming and increased production all over the Philippines, the exports would likely increase and positively affect coastal people who are mostly marginalized.

Seaweeds are just some of the contributions that enrich our marine biodiversity. Their products have an impact on our daily lives, particularly in our food and nutrition and in the economic upliftment of our country. They are just one clue to the rich resources in our archipelagic environment. With such information, Filipinos must realize that the sea is a blessing; thus, its environment should be protected and its products should be developed with sustainability for the present and future generations.

September 16, 2004

Rafael Guerrero III

Cultured Earthworms as Human Food

To most people, the mere mention of worms is enough to make them shriek or curdle in fear and detestation.

Are all worms really that abominable?

Not so, if you know and believe in the science of soil biology, the study of living organisms in the soil.

Yes, the "good news" is that not all worms are "bad" like the ones in your rose garden (by the way, they are really caterpillars or larval forms of insects), the parasitic ones that probably infested you such as the *bulate* (a nematode) at an early stage of your life, or the maggots (fly larvae) that scavenge on deal animals and dung.

This piece will be about the "good worms," particularly earthworms.

Before Charles Darwin, the "Father of Evolution," published his classic paper in 1881 on "The Formation of Vegetable Mould through the Action of Earthworms," earthworms were regarded as pests in the garden and were only good for the birds.

After studying the terrestrial annelids for forty years, he concluded: "It may be doubted whether there are any other animals which have played so important a part in the history of the world as have these lowly organized creatures."

Even before Darwin, the great Greek philosopher Aristotle already recognized the important role that earthworms played in promoting soil fertility. He referred to them as "the intestines of the earth."

Today, it is established and well-acknowledged that the more than 4,000 earthworm species (the Philippines has over 400) in the world comprise the most important macrofauna in the soil. They hasten the conversion of organic matter (mainly plant residues) into humus and improve soil structure and fertility. The burrowing habit of earthworms also enhances water drainage and aeration of the soil.

By digesting organic matter, earthworms produce casts that contain slow-release plant nutrients (nitrates and phosphates), increase the adsorptive capacity of the soil for minerals such as calcium, magnesium, and potassium, increase the water-holding and buffer capacities of soils, support a more varied microbial population, and reduce the toxicity of natural and man-made substances, according to soil ecologists.

Vermicompost, the compost produced through the action of earthworms (e.g., *Eisenia fetida* and *Eudrilus eugeniae*), is better than thermophilic compost (produced through anaerobic decomposition) because it is fully stabilized (it won't heat), has a granular

Philippine Council for Aquatic and Marine Research and Development of the Department of Science and Technology, Philippines.
pcamrd@laguna.net

texture and low carbon-to-nitrogen ratios, and contains plant growth regulators such as auxins, cytokinins and gibberellins, says Dr. Clive Edwards of Ohio State University (OSU) who is considered the world's leading authority on earthworms.

The application of vermicompost in the field at 2.5-10 tons/ha has suppressed the parasitic nematode infestation of tomatoes, pepper, strawberries, and grapes, reports Dr. Norman Arancon, a postdoctoral researcher of OSU. In greenhouse studies, the use of vermicompost at 20 percent of potting mix significantly reduced insect damage of cabbage.

Vermimeal, processed earthworm biomass, is an excellent substitute for Peruvian fishmeal which is imported by developing and developed countries and used as an animal protein source for animal and fish feeds.

A recent study conducted in Negros Occidental, with the support of the Technology Incubation and Commercialization Program of the Department of Science and Technology (DOST), showed that with sugarcane tops and chicken manure as substrates for the "African nightcrawler," the vermimeal production of 5.5 g/day/sq.m is economically viable if the value of the resulting vermicompost as a by-product is accounted for.

Earthworms have been used in traditional Chinese medicine for more than 2,000 years. Over forty kinds of prescriptions for medicines from earthworms are known in the ancient medical journals of China. It is claimed that such remedies have antipyretic, anesthetic, antihistamine, antihypertensive, and anti-inflammatory properties.

In modern medicine, clinical trials are said to have been conducted in assessing the efficacy of recently discovered pharmaceuticals from earthworms such as an enzyme termed lumbrokinase that dissolves thrombi (blood clots) in the blood vessels of humans. An antibacterial peptide in the *E. fetida* and lectins in the *E. eugeniae* have also been found recently. In fact, there is a commercial product from China that is sold in local drugstores with the brand name "Plasmin." Marketed as a food supplement, it is said to contain natural enzymes from earthworms for dissolving blood clots.

There is ongoing research collaboration between Drs. Clive Edwards and Norman Arancon of Ohio State University and this writer of the PCAMRD-DOST for studies on producing plant growth media, soil amendments, animal feed protein, and pharmaceutical materials from animal wastes using innovative earthworm-based technologies with support from the United States Department of Agriculture.

The Philippines hosted the International Symposium-Workshop on Vermi Technologies for Developing Countries in Los Baños, Laguna on November 16-18, 2005. Dubbed as ISWVT 2005, the meeting brought together researchers and industry stakeholders from different regions of the world to assess the current status of vermicompost, vermimeal, and vermiceutical technologies, and their prospects for developing countries.

With all these and much more, can anyone still say that earthworms are only for the birds?

May 4, 2006

Rafael Guerrero III

Earthworms Are Not Only for the Birds

In a television reality show, a man was shown eating a lot of live earthworms with gusto. This was repulsive to many viewers. Are earthworms really edible? Are they safe to eat?

Earthworms are terrestrial invertebrates that have been used as human food for ages. They are in fact considered a delicacy by the natives of Africa, South America, Japan, China, Papua New Guinea, and New Zealand.

Today, you will still find canned earthworms and dishes of mushrooms and bread with earthworms consumed in Western Europe. Restaurants in the Chinese provinces of Guandong and Fujian serve earthworm soup as a traditional food and mix earthworms with meat to make it tastier, respectively. How nutritious are earthworms as food? Studies have shown that the protein content of earthworm meal (processed worm biomass) is 54.6 percent higher than that of eggs or soybean. Worm meal is also rich in vitamins, particularly thiamine and riboflavine. In terms of minerals, its iron content and copper, manganese, and zinc contents are ten times and six times higher, respectively, than those of fish and soybean.

The body fluids and tissues of earthworms contain bioactive compounds such as enzymes (e.g., fibrinolytic, lumbrokinase, and collagenase). Pharmacologically active substances like triglycerides, squalene, and free fatty acids (omega-6 EPA) have also been identified in earthworms in addition to endorphin, encephalin, a hypoglycemic factor, and monoamines. With such health-giving benefits, the worms can be a functional food like fish, milk, and soybean.

To be acceptable as human food, earthworms have to be cultured under strict sanitary and biosafety conditions. The earthworm recommended for production in the tropics is the West African nightcrawler (*Eudrilus eugeniae*) for being a composting species (can live without soil), prolific breeder, and fast grower.

In the culture of *E. eugeniae* for human food, the use of soil and animal wastes should be avoided to ensure noncontamination of the product with pathogenic microorganisms and heavy metals. Commercial farming of the species in the Philippines has been achieved using grass and *kakawate* (*Gliricidia sepium*) leaves as substrates. The mass culture of earthworms is economically feasible with the production cost of ₱10-15 per kilogram.

Philippine Council for Aquatic and Marine Research and Development of the Department of Science and Technology, Philippines.
pcamrd@laguna.net

In preparing the earthworms for human food, they should be washed in clean running water to remove foreign materials on their body surface and then depurated to devoid their gut content by placing them in a container with moist tissue paper for at least twelve hours. The worms are blanched in 60 degrees Celsius water to kill them.

Earthworms cooked as "meatballs" were found to be as acceptable as pork meat cooked in the same way by ten members of a taste panel in a thesis conducted at a local university. The recipe for "Surprise Worm Poplette" is: Ingredients—one-and-a-half cups of ground earthworms, one-fourth cup of bread crumbs soaked in water, two eggs, two pieces of minced garlic, one tablespoon of kinchay, one-half teaspoon of salt and one-fourth teaspoon of pepper. Procedure—mix bread crumbs thoroughly with the ground earthworms; add the beaten eggs to the mixture; add grated cheese (optional), garlic, kinchay, salt and pepper; shape the mixture into small balls rolled in flour; and fry in hot oil for about ten minutes or until golden brown.

October 20, 2005

Domy Adriano

Organic Farming: Principles, Practices, Benefits, and Drawbacks

Healthy soil promotes healthy food which, in turn, promotes healthy people. This is the basic tenet of organic farming which has become popular in Europe, North America, Oceania, and parts of Asia.

Organic farming is a form of agriculture that largely prohibits the use of synthetic fertilizers and pesticides, plant growth regulators, and livestock feed additives. It is not synonymous with sustainable agriculture but it is a farming system that promotes sustainable land use. Organic production can be defined as an ecological production management system that promotes and enhances biodiversity, biogeochemical cycles, soil and plant quality, and soil biological activity. It is based on minimal use of off-farm inputs and on management practices that restore and sustain ecological, holistic harmony on the farm. More specifically it requires replacing commercial fertilizers with compost, traditional pesticides (i.e., insecticides, fungicides, and herbicides) with certain approved eco-friendly products, and prohibits giving livestock antibiotics and growth promoting substances.

The foundation of organic farming is healthy soil or the so-called pedosphere—the crossroad of the biosphere, hydrosphere, atmosphere, and the lithosphere. As such organic farms should have soils that contain ample organic matter that promotes desirable physical, biological, and chemical properties to produce optimum crop yields and quality, and provide a healthy environment for soil organisms. This translates to the soil being fertile relative to the essential major nutrients (nitrogen, phosphorus, and potassium), secondary nutrients (calcium, magnesium, and sulfur) and the micronutrients (zinc, manganese, copper, iron, boron, and molybdenum). The soil should have desirable water–air relations relative to the crops under cultivation. With the exception of paddy rice, crops prefer a well-drained, aerated soil.

By having ample organic matter, a balanced, productive soil could be achieved. Excessive tillage, wind or water erosion, and irrigation and drainage typical in conventional agriculture may reduce the amounts of organic matter and nutrients in soil. Hence pro-organic farmers and scientists maintain that conventional farming could deplete the soil of its organic matter content, rendering the soil less productive and more

Formerly with the University of Georgia, USA.
domyc@uga.edu

susceptible to compaction (poor soil structure), flooding (as soil with lower organic matter holds less water), erosion, and eventually, infertility.

Conventional farming relies heavily on the use of commercial fertilizers and pesticides, and at times recycled municipal sewage sludge. Certain phosphate rocks used in fertilizer manufacturing and sewage sludge, especially those from urban areas contain toxic heavy metals that are persistent, eventually accumulating in the soil. Pesticides could have metals, chlorinated hydrocarbons, or organo-phosphates—some of them known to be carcinogenic. These undesirable residues, if not biodegraded or broken down in the soil, could be absorbed by the plant roots and transferred in the food chain—first by accumulating in crop tissues, then being incorporated in feed, and finally ending up in poultry, eggs, and milk. Therefore, organic farming regulates practices that would damage the soil due to persistent and potentially toxic chemical residues.

Organic farming began in the 1920s in Europe, mainly in Germany, spreading to Great Britain in the 1940s as an alternative to conventional agriculture. Jerome Rodale popularized organic gardening in the US in the 1940s after realizing that resource-depleting conventional agriculture may not be sustainable. In 1972, the International Federation of Organic Agriculture Movement (IFOAM) was founded in Versailles, France whose aim was to disseminate information on the principles and practices of organic agriculture across cultures and continents. In the early 1990s, the European Union adopted formal standards regulating organic production and in 2002 the US Department of Agriculture (USDA) established production standards, under the National Organic Program (NOP), which controls the commercial use of the term organic (as in the "USDA Organic" label).

Organic regulations generally require that organic foods are raised without chemical fertilizers, synthetic pesticides, growth hormones, antibiotics, GMOs (genetically modified organisms), or sewage sludge. The use of synthetic substances is not allowed in organic farming unless they are on the National List of Allowed and Prohibited Substances. Organic farming employs various methods that enhance soil structure and fertility such as crop rotation, tillage practices, cover crops, and natural products. In short, organic farmers recycle animal and crop wastes; employ botanical, biological, and nonsynthetic pest control; and use synthetic materials that can be broken down rather quickly by reaction with oxygen, water, or sunlight. They also employ practices that avoid or minimize air, soil and water pollution. Hence biosolids like sewage sludge that generally contain inorganic (like heavy metals) and organic contaminants (like hydrocarbons including certain pesticides) are prohibited. Toxic heavy metals like mercury, cadmium, and lead from biosolids can be transferred in the food chain.

Organic farming became one of the fastest growing segments of US agriculture during the 1990s and is still growing fast. Since the USDA formalized national production and processing rules in 2002, organic food has been the fastest-growing segment of the retail food industry. Even Wal-Mart, the world's largest retail store chain, has jumped fairly recently into the organic bandwagon.

Organic food sales have been estimated to have increased nearly 20 percent since 1990 in the US, with consumer sales reaching $13.8 billion in 2005 (versus $13.7 billion in 2004 for Europe). However, organic foods constitute only about 2 percent of all food sales. The organic sales in the US are dominated by fruits and vegetables ($5.37 billion), dairy ($2.14 billion), beverages ($1.94 billion), and bread and cereal ($1.36 billion) sectors. The sale expansion is dominated by poultry and eggs.

What is fueling the dramatic expansion of organic farming in Europe and especially in the US? It is simply the profitability for the farmers and the consumers' appetite for organic foods. In Europe, consumer motivations relate to human health, environment, taste, and animal welfare. The motivating factors have evolved in Europe—mostly the ecological benefits (more natural, kinder to animals, better to conservation of natural resources) of organic production in the 1970s to health (nutrition and food safety) and taste (freshness) today.

In a world survey in 2005 involving 21,261 people from thirty-eight countries, the foremost motivating factors in descending order are: healthier for adults, healthier for children, eco-friendly, and more humane treatment of animals. On the other hand, the leading factors discouraging consumers are: high prices, inaccessibility of organic shops, and distrust of producers complying with regulations. In general, surveys in the US indicate consumers' preference to purchase organic foods because of the perceived health and nutrition benefits from them. In one survey, the main reasons consumers buy organic products were to avoid pesticides (70 percent), for freshness (68 percent), for health and nutrition (67 percent), and to avoid GMOs (55 percent).

In a number of surveys in Europe and the US, it was concluded that pesticide residues were 2.9 to 4.8 times more prevalent in conventional than in organic products and that the levels (in parts per billion) of these residues were generally higher in the former. In a US survey comparing residue data from 194 samples of organic produce and 26,000 conventional produce, it was concluded that organic samples contain only about one-sixth the amount of residues in nonorganic samples. However, when the frequency of occurrence and magnitude of concentrations are translated into risk, the doses are well below the accepted reference or tolerance values established by the WHO, European Union or the US.

More recent findings indicate that switching children to organic fruits and vegetables results in the rapid disappearance of trace metabolites of organo-phosphate pesticides from the children's urine. The US Environmental Protection Agency "Guidelines for Carcinogen Risk Assessment" in 2005 indicates that children receive about 50 percent of their lifetime risks of having cancer during their first two years of life. Children, younger than five years, are the most sensitive segment of the population to exposure to hazardous substances. This has spurred the growth of the baby food industry in the US.

Additionally, studies in the US and Europe indicate that organic products more often contain lower levels of nitrate than nonorganic products. In one survey, the ratio of nitrate levels in nonorganic to organic foods ranged from 97 to 81.9 percent. Nitrate is another health concern as once ingested by human it can be converted to nitrite, which may eventually become carcinogenic. Preliminary results from the Organic Center for Education and Promotion in the US indicate that organic produce may have higher levels of antioxidants, polyphenolic compounds, and other plant secondary metabolites, which may provide protection against cancer and age-related diseases.

It is clear from numerous studies and surveys in the US and Europe that there are qualitative differences for fruits and vegetables—that organic products possess fewer pesticide residues and lower nitrate levels than do conventional produce. This gives the perception that organic foods may be less risky than conventional foods simply because organic farming relies on far fewer pesticides and lower nitrogen fertilizers which result in fewer and lower pesticide residue and nitrate levels.

Because more extensive data on organically produced crop and animal products across continents are needed, it is rather premature to conclude whether the more often favorable observations on organic foods being more nutritious and safer than nonorganic foods are for real.

There are several perceptive reasons why consumers prefer organic foods over conventionally raised products—better nutrition and safety, freshness, eco-friendliness, and more humane treatment of animals. However, it is the notion that pesticide residues and nitrate from commercial fertilizers could cause cancer, especially in young children, that scares many consumers. A wide variety of pesticides have been reported to reduce the immune system's ability to combat infectious agents—a phenomenon of greater concern in developing countries where infectious diseases account for the majority of deaths and where at least two billion people living and working in farming areas are exposed chronically to pesticides. The recent food crises involving mad cow disease and foot-and-mouth disease have lessened consumers' confidence in food safety in general but more so with conventionally raised foods.

And now with drastic climate change (higher temperature, severe drought or flooding, and more violent storms and hurricanes) evidently caused by global warming, citizens of the world are becoming more serious in joining the "green" movement to curb the emissions of greenhouse gases, primarily carbon dioxide. The motives in organic farming are consistent with those in the green movement.

Organic farming is a major practice that can promote the green approaches to conserve energy and natural resources, and limit the emission of greenhouse gases. By being less dependent on synthetic fertilizers and pesticides, less tillage, and relying on compost, organic agriculture promises to facilitate sustainable agriculture and conserve natural resources. Thus organic agriculture, by all measures, has become synonymous with green agriculture.

Composting recycles plant and animal residues that can serve as slow-release fertilizer. More importantly, mixing it with the soil promotes carbon sequestration that minimizes the emission of greenhouse gases. In a long-term field study at the Rodale Farming Systems Institute in Pennsylvania, the carbon levels in organic farm soils increased by 15 to 28 percent while there was little or no change in the nonorganic system.

In spite of all the hype on organic farming, there is a dark side to it. Nobel Laureate Norman Borlaug has argued that organic agriculture would not be able to feed the burgeoning world population, now at over 6.5 billion. He contends that the only way modern agriculture can feed humans is to employ modern crop cultivars that would require the use of commercial fertilizers and pesticides, sometimes at high rates, to ensure high yields. In organic livestock production, bacteria and other pathogenic organisms may oftentimes be present in edible portions due to the absence of antibiotics. However, these potential drawbacks are not enough to outweigh the perceived benefits from organic products.

It has become evident that the consumers' appetite for organic foods, especially fruits and vegetables, and poultry and eggs, has fueled the rapidly expanding organic food sales, even if they have to pay premium prices. The popularity of farmers' markets in the US has grown concurrently with organic production and consumers' interest in local, fresh, and organically produced foods. The economic incentive has attracted many farmers to shift to organic farming.

Organic products sell at about 10 to 40 percent higher than conventional products in grocery stores in the US. Organic prices are usually lower in farmers' markets than in groceries. In developing countries, the organic and conventional price differential could be as high as 100 percent or more, giving the notion that organic foods are intended only for affluent consumers.

Organic agriculture has rapidly developed worldwide during the last few years and is now being practiced in about 120 countries. Its share of agricultural land and farms continues to grow. Because obtaining organic certification has become quite cumbersome, it is reasonable to assume that uncertified organic farming is being practiced in even more countries.

A recent survey indicates that more than 31 million hectares worldwide are currently operated organically by some 623,175 farms. Currently, the countries with the greatest organic areas are Australia (12.1 million hectares), China (3.5 million hectares), Argentina (2.8 million hectares), Italy (0.95 million hectares), and the US (0.89 million hectares). However, a great proportion of the organic land in Australia and Argentina is devoted to pasture.

In Europe, around 6.5 million hectares were managed by some 165,000 farms in 2004, mostly in the European Union; whereas in North America, almost 1.4 million hectares were located in around 12,000 farms (8,035 farms in the US alone), representing a mere 0.30 percent of the total agricultural land. In Asia, the organic area totals about

4.1 million hectares, scattered over 130,000 farms. The biggest organic producers in Asia are China, India, and Russia. Organic regulations are already in place in Japan, Korea, India, the Philippines, Taiwan, and Thailand.

In terms of share of the overall food sales, EU countries, most notably Switzerland, Denmark, Austria, Germany, and Great Britain, boast of over 2 percent of the respective national sales, while it is only around 2 percent in the US. However, the expected growth rate in organic sales is highest in the US at around 20 percent annually.

In developing countries, the apparent, most motivating incentive for small farmers to adopt organic farming is primarily economic-driven. Manufacturing commercial fertilizers like nitrogen and phosphate fertilizers requires huge amounts of energy to manufacture and farm machinery also consume large amounts of fossil fuel. For example, a typical American organic farmer uses just 63 percent of the energy required in conventional farming systems. In the farming heartland in America, pesticides, nitrates, and other agro-chemicals have contaminated ground water and streams, compelling many households to purify their drinking water—another environmental cost.

Evidently, the organic madness has proliferated across continents. While the organic movement has somewhat slowed down in Europe, it has accelerated at an unprecedented pace in North and South America, and parts of Asia. Until only recently, organic farming has been largely limited to the temperate Asian countries of Japan and Korea. Today, organic farming in India is undergoing explosive growth. In addition to economic impetus, the Indian government is now supporting organic methods to reverse the trend in soil deterioration that intensive tillage and agro-chemicals have induced.

Filipino farmers have a similar sentiment—that reverting to organic farming could alleviate them from poverty and malnutrition while also conserving the natural resources by using homemade composts. The main incentive motivating Filipino farmers to go organic is more economic than environmental. Premium price is not enough motivational incentive. Farmers are more motivated by their desire for lower production cost, more sustainable yield, and better access to production capital.

In South America, the Brazilian minister of agriculture has pledged to increase the share of organic farming in its total agricultural output, from the current three percent to about 20 percent in the next six years. Indeed, the Food and Agriculture Organization (FAO) espouses the benefits organic farming can provide in reducing hunger and poverty in the world—about four million people go hungry every year worldwide.

The main challenge that farmers face in developing countries is educational awareness, especially in the rural areas of the Philippines, Thailand, China, India, Brazil, and other countries where farming output and socioeconomic status are largely marginal. They need to form cooperatives or unions, in unison with scientists and government specialists, in order to have hands-on knowledge of the prevailing methods and technologies on organic farming—on composting, crop diversity, and other practices that organic farming requires. Of particular emphasis would be on composting technology.

Organic farming tends to be more labor-intensive which favors small operators in developing countries. By shifting to organic farming, small farmers in the Philippines have reported substantial reduction in expenses due to minimal usage of commercial fertilizers and availability of cheap labor. For example, an organic rice farmer in Mindanao has claimed that her expenses on farm input, labor, and land preparation on her 1.74-hectare farm drastically dropped from about ₱15,000 to about ₱2,500 per cropping (approximate currency exchange at original print time: $1 = ₱48). Another organic rice farmer also experienced even more savings—from about ₱35,000 to only about ₱10,000 per cropping on his five-hectare farm after adopting organic methods. And yet another farmer claimed spending only about ₱288 for chemical fertilizers for his organic farm, drastically down from the previous fertilizer expenditure of about ₱7,680 per season.

But many Filipino farmers are discouraged to shift to the organic system due to the stringent and time-consuming certification process. It normally takes a three-year waiting period of transition from conventional to organic farming before a certification can be issued. It takes from three to six months to complete the inspection of organic farms before a certificate, which roughly costs ₱14,300, can be issued which is valid only for one year. Thus the expensive certificates are deemed designed primarily for organized and big farmer groups.

Because crop yield reductions have been observed during the transition period, farmers cannot count on big profits right away. However, a few years after transition, farmers can expect steady profits. Observations in Japan confirm that applications of composts have enhanced the root growth of five crops, resulting in higher yields. Worldwide observations indicate that on the long term, repeated applications of livestock manure and crop residues to soil bring about a gradual improvement in soil productivity and crop performance.

The most crucial item needed by organic farmers is the organic materials for composting—where to get and how to compost them. Livestock manures are considered the best source of compost, especially chicken and cattle manures. These livestock, when raised commercially, are fed high-energy rations under confined housing conditions. As such, manures accumulate, making collection rather easy. It is therefore advisable to locate large organic farms nearby large livestock operations as a ready source of organic materials. Because livestock manures have favorable carbon and nitrogen content and ratios for composting, they are unquestionably the best composting materials. However, organic arsenic, a toxic substance from an antibiotic used in poultry (sold under the brand name "Roxarsone"), can accumulate in the soil from poultry litter application and move into the food chain. Therefore, livestock manures should still be chemically analyzed for contaminants before they are utilized in composting.

To sustain my organic gardening hobby—growing Filipino crops (*sitao*, *calabasa*, *upo*, *patola*, *talong*, *gabi*, and chayote)—I have used chicken manure both from broiler and hen (laying eggs) operations for my composting. Manure from hens has higher nutrient content and more favorable carbon and nitrogen ratios than broiler litter. The latter is mixed with wood chips and hay applied as bedding material, consequently widening the carbon/nitrogen ratio which slows down the decomposition process. In this case, I add some nitrogen fertilizer like ammonium nitrate to facilitate composting, which organic farming rules allow. I have never used fungal or microbial inoculants, but others do. It is best to protect the manures from rain, as much as possible, to minimize the odor and flies and to also minimize nutrient losses.

In rural areas where livestock manures are not available, there are other organic materials that can be substituted. Crop and other plant residues—like banana, rice straw, and other farm by-products, as well as tree and shrub cuttings and lawn/garden residues—can be good sources of compost.

Composting procedures are available as extension bulletins that farmers can easily obtain, often free of charge. A general composting method entails layering plant residues alternately with livestock manure at a rate of 10:1 (by weight) to make a pile of about 1.5 meters high. Urea or ammonium nitrate is added to the compost pile and sometimes rock phosphate. A microbial inoculant can be added to hasten decomposition. In the Philippines, a decomposer fungus (*Trichoderma reesi*) is combined with a nitrogen-fixing bacterium *Azotobacter*, to make an inoculant for the compost—producing compost in a shorter period and with higher nitrogen content.

Before application, livestock manures need to be thoroughly composted to decompose the organic materials and to kill pathogenic microorganisms. The elevated temperature in the compost pile would kill the pathogens. Never apply raw cow manure especially on green leafy vegetables to avoid contamination with deadly *E. coli*. These microorganisms adhere strongly to the plant surface that even chlorinated washes cannot completely remove. There was a deadly food poisoning outbreak caused by *E. coli* contamination of spinach from an organic farm in Salinas Valley in Central California last year. The bacteria were spread when feral pigs intruded into the farm after trampling cow manure in a nearby pasture. The contaminated spinach was bagged and distributed in several states, causing human fatalities and illness to many more.

In summary, the popularity of organic farming has been fueled by several motivational incentives, the scope of which depends on the cultural, sociopolitical, economic, and environmental values among nations of the world. In developed countries, nutrition and food safety, followed by freshness and environment, appear to be the main drivers of the organic movement. However, in developing countries, it is more economic than anything else. In these countries, organic foods are perceived to be intended for the affluent. But there are common driving factors worldwide: health protection from cancer and infectious diseases, and the "green" appeal of organic farming at a time when there are widespread concerns about the effects of global warming.

Printed in three parts on June 21, 28, and July 5, 2007

Benigno Peczon

The Continuing Bt Corn War

In 2001, President Arroyo issued a National Biotechnology Policy: "(The Philippines) shall promote the safe and responsible use of modern biotechnology and its products as one of several means to achieve and sustain food security, equitable access to health services, sustainable and safe environment, and industry development." In April 2002, the Department of Agriculture released Administrative Order No. 8, which provided rules and regulations for the importation and release into the environment of plants and plant products derived from the use of modern biotechnology. In December 2002, the Bureau of Plant Industry approved the commercial propagation of Bt corn MON810.

In 2002, about 120 hectares of Bt corn were planted. Because of the significantly increased yields obtained, by mid-2004, 12,000 hectares had been planted on various islands. In the planting season ending in May 2004, one farmer obtained 10.2 tons of shelled corn per hectare in his village, where people have planted over 300 hectares of corn. One advantage this farmer mentioned is the capability of planting corn three times a year. Previously, no more than two crops could be planted per year because moisture patterns saw massive infestations of the third crop by borers.

The Philippines annually imports about one million metric tons of corn or corn substitutes. With the potential of Bt corn to significantly increase yields, for the first time in many years, there is the strong likelihood that the Philippines will not only stop importing corn, but could also divert more of this corn to increasing value-added produce in the form of increased production of chicken, pork, and other agricultural output.

This potential to improve food security and to directly address malnutrition that still exists in the country can be adversely affected by FUD (Fear, Uncertainty, and Doubt) tactics. Media printed the pronouncements of Norwegian scientist Terje Traavik on February 22, 2004 at the first Meeting of the Parties to the Convention on Biological Diversity in Kuala Lumpur, Malaysia. Traavik was reported to have said that thirty-eight people in Kalyong, a village in Mindanao, carried increased levels of three different target antibodies, which he considered as evidence of an immune reaction to the Bt toxin built into the Bt corn to combat pests. Traavik was also reported to have said that "there's no literature or scientific studies that could prove GMO (Genetically Modified Organism) safeness to human health and environment." During an early March visit to Mindanao, media reports said that when pressed as to whether the traces of Bt toxin in the blood samples were the result of the natives' exposure to Bt corn, Traavik was quoted as saying, "It's difficult to completely conclude if the Bt toxin, indeed, came from the Bt corn. We

Biotechnology Coalition of the Philippines Inc., Philippines.
peczon@gmail.com

need to conduct further researches on that. However, this alarming finding should serve as early warning that it could be harmful to our health."

Within days of the pronouncements, Agriculture Secretary Luis Lorenzo Jr. called for an investigation. Dr. Artemio Salazar, director of the Department of Agriculture's corn program, was reported to have explained that the claims of Traavik are not in consonance with the tenets of modern science. Nina Gloriani Barzaga, MD, PhD, professor of medical microbiology and microbial immunology at the University of the Philippines' College of Public Health, was quoted as saying that Traavik should show pertinent scientific data supporting his claims before publicly releasing them and causing undue panic.

On March 23, 2004, a widely circulated newspaper reported that fourteen scientists and professors, mainly from United States, challenged Traavik to publish on the Internet the full details of his study so that it could be completely discussed by other experts in the field.

Government healthcare professionals in the area where the alleged ill effects of exposure to Bt corn occurred expressed the opinion that the reported illnesses are unlikely due to Bt corn. A private sector team involving physicians, which subsequently visited the place, confirmed the conclusions of the government health care professionals. Thus far (mid-May 3004), Traavik has failed to provide the raw data supporting his findings.

From the large number of reports from many newspapers regarding the pros and cons of GMOs, and the results of an independent survey showing popular print media as one of the major sources of information regarding these crops, it is clear that media plays a huge role in bringing down unwarranted barriers to international trade and in promoting science-based understanding and implementation of agricultural biotechnology.

Although media can be attracted by FUD newsmakers, it can also be utilized by responsible government officials and scientists to address relevant issues as quickly as possible to dispel doubts. Knowledge dissemination through daily broadsheets can augment the ongoing information campaign by entities such as the Biotechnology Information Center, the Biotechnology Coalition of the Philippines Inc., the Philippine National Academy of Science and Technology, the Department of Agriculture's Biotechnology Program Implementation Unit and other groups.

Another potential problem for the adoption of GMOs is stringent government regulations. To comply with the requirements of the Cartagena Protocol on Biosafety, the Department of Environment and Natural Resources (DENR) is crafting a National Biosafety Framework. In a March 2004 national workshop on the development of the Philippine Biosafety Framework, DENR Secretary Elisea Gozun pushed for the ratification by the Senate of the Cartagena Protocol, saying that the Protocol "creates an enabling environment for the environmentally sound application of biotechnology, making it possible to derive maximum benefits from the potential that biotechnology has to offer, while minimizing the possible risks to the environment and to human health."

While acknowledging the positive impact of modern biotechnology on food security, she noted its potential impact on human health and environment. "The use of biotechnology and its products must therefore be practiced in a safe and sustainable manner that minimizes the possibility of these adverse effects," Gozun said. She added, "We must follow precautions and take into account socioeconomic and cultural considerations in making biosafety decisions."

Some sectors want the Philippine National Biosafety Framework to address the biosafety issue apart from socioeconomic and cultural issues. It is clear that some countries have released GMOs into the environment many years (at least six years) before the Philippines did. Lessons learned by these early adopters could help later adopters like the Philippines and even later potential adopters, by shedding additional light through information sharing. Regional as well international cooperation would help pave the way for rational measures to resonably meet the stakeholders' needs.

While GMOs may provide advantages for producers and the public, FUD tactics and stringent government regulations can slow down their adoption.

July 15, 2004

Benigno Peczon

Updates on Biotechnology

The annual meeting of BIO (Biotechnology Industry Organization), a global organization with over 1,000 corporate members from thirty-four countries, was held on June 6 to 9, 2004 in San Francisco, California. Issues and my comments follow:

Public Perception of Biosciences: The Key Toward Moving Forward

Acceleration of scientific development leads inexorably to an ever-widening gap between experts and society at large, including public institutions, the media, and nongovernment organizations (NGOs). Policymakers are more likely to be influenced by the general public's "gut" reaction to a given issue than they are in focusing on the long-term interest of these very same people. Closing the gap between fact and fiction and helping the public to deal with science-based, ethically sound progress is crucial to the steady evolution of mankind.

Prof. David McConnel, of Trinity College, Dublin who represented the European Action on Global Life Sciences (EAGLES), described the organization's objective to benefit all, not just developed countries, by establishing and developing a platform for outstanding scientists from the Developing Countries and Emerging Economies (DECs) in the life sciences and biotechnology to:

- Give voice in Europe through various consultations and dialogues to the needs and competencies of the DECs;
- Apply European competencies for the benefit of the poor and needy; and
- Strengthen the global responsibilities of European Union programs in research, education, innovation, application and implementation.

A few years ago, Dr. Patrick Moore, one of the founders of Greenpeace and who served as its president at one time, felt the need to stand for something rather than against something. He considers the current Greenpeace assault on biotechnology a direct result of the collapse of the Berlin Wall when leftists moved in and began to use Greenpeace language. He describes environmental extremism as an abandonment of science and technology. Indeed, he believes Greenpeace is now defined as being anti-human, anti-science and technology, anti-trade and globalization, anti-business, and anti-civilization.

Biotechnology Coalition of the Philippines Inc., Philippines.
peczon@gmail.com

To help address current concerns, Dr. Moore refers to the Sensible Environmentalist. The concerns of such an environmentalist are renewable energy and materials, voluntary population management, development of logically consistent global analysis, increased wood usage, learning to be better gardeners of the Earth, controlling urban sprawl, and reversing deforestation in the tropics.

Perspectives on the Current and Future States of the Biotechnology Industry

A discussion featuring top executives of Genentech, Chiron, Biogen, Idec, and Amgen revealed that the biotechnology healthcare industry has spent a lot of time, effort, and money understanding pathways leading to diseases. The knowledge gained is shared with the public through publication in scientific journals. The main obstacles to progress are too stringent regulatory agency standards and financing.

Panelists from Genentech, Affymetrix, IBM, and Invitrogen provided several insights regarding future developments in biotechnology. Traditional biology will morph into systems biology. Heretofore, research has consisted of examining phenomena and then testing hypotheses. In the actual world, processes occur in a complex, ever-changing system—thus, the birth of systems biology that integrates huge amounts of data to develop models and generate hypotheses. The idea that one gene may be correlated with a specific disease may no longer hold. With vast amounts of data generated by traditional biology coupled with the massive capability that information technology can put to bear on genomics, the best gains in maintaining health lie ahead. The gains will likely be in predictive capability.

Communicating the Wonders of Biotechnology

At the lunchtime plenary session, Brooke Shields, star of the movie *Blue Lagoon* and model of Calvin Klein apparel, stole the show. After obtaining a degree from Princeton, she wanted what any woman would want—a baby. But she had some difficulty with her uterus. She and her husband just could not produce a baby. Enter biotechnology. Using molecules whose names she said she had difficulty pronouncing, she was able to convert one of her previously frozen eggs into what she happily called her "popsicle baby." Ms. Shields proved to be a very effective spokeswoman for biotechnology.

Agricultural Biotechnology

On the second day of BIO2004, at any given time from 9:30 a.m. to 5:30 p.m., there were seventeen simultaneous sessions covering biodefense, business development, drugs, global health, ethics, finance, industry and the environment, intellectual property rights and legal issues, policy, regulatory issues, science, technology transfer and licensing, and food and agriculture. Some topics in the agricultural arena follow:

Biobanking

There is enormous biodiversity worldwide which remains largely unexplored. Biobanking is an approach to conservation and sustainable utilization.

A scientist at the Missouri Botanical Garden (*www.mobot.org*) described how a staff of 150 people, fifty of whom have doctorates, have systematically classified and kept more than 5.5 million plant species in Missouri. Some small amounts of samples are available for R&D work.

The research leader of the USDA, ARS, Plant Genetic Resources Conservation Unit (*www.ars-grin.gov/npgs/*) described US-taxpayer-paid services provided by this government unit. The collection focuses principally on crops, i.e., clover, sorghum, peanut, eggplant, okra, watermelon, chili pepper, cowpea, mungbean, etc. The collection of seeds and tissues from a large number of genera comes from 178 different countries. Plant materials are kept at 40C and -180C. Seeds are replanted before degradation commences. Of 72,758 accessions, 87.2 percent are backed up. Anybody in the world can request and get a small amount of sample from this facility.

While these two agencies willingly share genetic resources, many small countries raise enormous barriers to the development of existing genetic stock without putting up resources to catalogue, let alone maintain, their genetic diversity. A balance must be struck to encourage developers to invest resources while ensuring that the interests of indigenous people are protected.

Biopharming

The production of beneficial proteins and peptides through the use of biotechnology in agricultural crops provides the enabling technology for delivering human health benefits. Although the benefits from the technology can be great, there is potential risk as well. Risk assessments are crucial and the state of Iowa has focused a major initiative on the development of risk assessments on this area.

A farmer from that state grows corn containing molecules of pharmaceutical value. To successfully produce these molecules in commercial quantities, there must be certainty that the crop used to produce them is adequately sequestered. He showed photographs of his farm where the transgenic corn is grown in the middle of soybeans. He described measures to ensure that all the corn is destroyed after harvest. The next step in the production of the desired molecules is extraction and isolation. This is done using modern chemical methods. Growing the crop and producing the desired molecules has been made possible by cooperation among regulators, the industry, educational institutions, and farmers. A panel evaluates risks, an advisory board examines the work, and a legal firm maintains a risk management system. All throughout the project, there is communication, particularly with regulatory agencies. Nonetheless, because this manner of producing drug molecules is so novel, some aspects of the whole process are still under evaluation. If successful, this manner of producing molecules of medical value by producing them in crops will provide inexpensive, uniform, quality drugs.

Using Biotech to Improve Food and Crops: Pacific Rim Challenges in International Trade

NGOs and the media often present unbalanced and inaccurate characterizations of crops improved through biotechnology that do not acknowledge the benefits these crops bring to consumers, farmers, and the environment. Such mischaracterizations can lead to the development of imaginary barriers to the adoption of agricultural biotechnology products.

Australia has huge potential, being a major exporter of wheat and canola. With its constraints (fragile, low nutrition soils and limited water), it must adopt the world's best practice and technology. Genetically modified cotton and canola have been assessed to offer distinct advantages. The regulatory system of Australia's federal government has a transparent, clearly defined process in line with world standards in evaluating process. Unfortunately, the state system thus far has been characterized by lack of a defined process to evaluate key issues driving the moratoria, varying levels of transparency, inconsistent treatment of evidence, and inconsistent policy across the states.

A spokesman for the International Biotechnology Regulatory Services who was present at the meeting of the parties for implementation of the Cartagena Protocol in February 2004 in Kuala Lumpur bewailed the fact that the agricultural biotechnology proponents at that meeting were badly outnumbered by people who were against agricultural biotechnology. For advocates of GMOs, there must be involvement in all events which affect agricultural biotechnology. He said the Cartagena Protocol is flawed in that it does not emphasize the benefits that biotechnology offers and makes no distinction as to whether developers of the technology are huge multinational companies or the public sector. In fact, a substantial portion of the development of innovative research in agricultural biotechnology is done through publicly funded research. The public research community must mobilize and act as a major stakeholder in the international decision making process that seriously affects its capability to deliver goods to the public. Time to get involved is of the essence.

Quoting verbatim from newspaper articles, I described the fear, uncertainty, and doubt (FUD) tactics of Norwegian Terje Traavik. Responsible media, however, published the government and private sectors' response to Traavik's data-poor pronouncement. The lesson is that responsible media practitioners are an asset in counteracting FUD tactics. Again quoting from newspaper articles, I described how the failure of government institutions to arrive at a unified front may slow down more widespread adoption of GMOs in the Philippines.

Take-home Message

Biotechnology has indeed taken off. Informed, principled, energetic, and reflective men and women are on the move growing the next sunrise industry. There are hurdles, i.e., in public acceptance, ethical issues, meeting development costs versus final consumer costs, intellectual property issues, evolving regulatory guidelines, and addressing the Cartagena Protocol. My gut feeling is that these men and women are passionate enough about their craft that they will effectively channel their collective efforts to bring better health and more and better food to mankind in a world more fit to live in than the present one.

March 31, 2005

Ma. Antonia Arroyo

Biotechnology: Building a Future for the Country

The Potential of Biotechnology

There are many problems that face Filipino businesses when they try to compete on a global scale. The biggest of these is that our advantages are easily imitated (command of the English language, for one). As a result, businesses are forced to compete on price by sacrificing wages to keep costs low. Put together low salary and a high inflation rate, and you'll understand why our best and brightest are fleeing abroad at the rate of 2,700 people per day.

But there exists a field, from which an entire industry can be born, where the Philippines is sitting on a gold mine... and we don't know it. This is biotechnology.

What is biotechnology? Biotechnology is essentially harnessing the machinery of living things to make products or provide services, from making dairy products to the manipulation of DNA and proteins.

Thus broadly defined, the biotechnology sector is currently estimated to account for more than a third of the world's Gross Domestic Product (GDP). Medical biotech is the largest sector worldwide, ranging from 12 to 14 percent of GDP in key markets. Health care and medical research spending in the US is projected to rise to 17 percent within the next decade. The size of other American industries transformed by biotechnology was estimated at $400 billion for chemicals, $800 billion for the food sector, and more than $1 trillion for biomaterials.

As of the end of November 2004, the US biotech industry had a market cap of $372 billion, raised $18.4 billion in private equity financing, and completed over 139 partnerships. The most interesting part is that this industry still has more room to grow, and many more discoveries are still to be made. This is where the Philippines can carve a niche for itself on the world market, and the best time to act is now.

Should we wait for more stories of lost opportunity due to the lack of coordination between the scientific and business community? Erythromycin, a common antibiotic, was discovered by a Filipino, Dr. Abelardo Aguilar, in our soil. This was the earliest documented Philippine medical biotechnology discovery.

The first Philippine agri-biotech breakthrough came from the work of Dr. Juan Salcedo, who discovered that milled rice mixed with Vitamin B1 was a good cure for

Hybridigm Consulting Inc., Philippines.
m.arroyo@hybridigm-consulting.com

beriberi (a disease caused by Vitamin B1 deficiency which can damage the heart and the nervous system).

In 1985, Dr. Baldomero Olivera and Dr. Lourdes Cruz isolated a toxin from the venom of a fish-hunting snail which became the basis for the drug Prialt Ziconotide. Recently concluded clinical trials have shown that the drug is 1,000 times more effective than morphine, produces no side effects, and is non-addictive. Elan Pharmaceuticals, which developed and commercialized the drug, recently enjoyed a $8.2-billion market cap gain, thanks to Prialt Ziconotide.

Think of how much better off the country could have been if someone had financially backed and advised these scientists properly on patents and the like.

A Wave of Success: Global Biotech Industries

In Asia, the dominant countries in biotech are Japan, China, India, and Singapore. Japan has four distinct types of biotech players—pharmaceutical companies, breweries and food makers with fermentation expertise, electronics firms, and local biotech start-ups.

The first three have funded and/or partnered with biotech firms, and the rewards for entrepreneurs have been very fruitful, since the Japanese pharmaceutical market has recently surpassed Europe as the second largest worldwide. It is valued at about $60 billion and accounts for more than 15 percent of global R&D expenditure.

How are the Japanese succeeding? Japanese pharmaceutical companies are actively acquiring technology via alliances with US and UK biotech firms. In addition, the government boosted funds for biotech and nanotech research, and this has resulted in the increase in biotech start-ups.

In Japan, we see the results of cooperation between the public (the government and the academe) and private sectors (entrepreneurs, investors, and large companies).

China and India present very different pictures of biotech industries. China's biotech sector has high potential but is still at the birthing stage, whereas India has major pharmaceutical companies but these will have to face up to the sharp transition from manufacturing generics to discovering and developing drugs.

The Chinese pharmaceutical market is valued at only $4 billion, but is expected to grow annually at a 15 percent rate. That leaves a lot of room for investors to join in. The country has 136 biotech firms, mostly traditional medicine therapeutics and enzyme manufacturers.

Meanwhile, India's biotechnology sector is the fastest growing in the world. For the last three years, it has grown consistently by 37 percent per year. The biotech industry is composed of ninety-six companies and is projected to clear $1.8 billion by the end of this year.

India has many competitive advantages, most notably its low cost innovation and manufacturing base—as low as 10 percent of those in developed markets—and its research infrastructure and biotech incubators.

Singapore is the most successful of all the relative newcomers to biotech in the ASEAN region. Although Malaysia, Taiwan, and Korea all offer attractive investment conditions for their emerging biotech industries, Singapore's trump card is the aggressive public funding by the country's paternalistic Economic Development Board (EDB).

Recent initiatives by the EDB emphasize biomedical research geared toward commercialization instead of publication, promoting start-up formation, and laying the groundwork to make Singapore a regulatory haven for stem cell research.

Singapore's biggest problem is its continued shortage of entrepreneurial scientists and managers. It is in this area that the Singapore government will meet its greatest challenge—it is difficult to inculcate creativity and entrepreneurial spirit in a population that has become used to an "autocratic democracy." These qualities are difficult to impose, even on a willing populace.

How can we be confident of the ability to compete in this field on a global scale? It is because the Philippines has been identified as one of the world's most biologically rich countries. Of the 500 known coral species worldwide, 488 are found in our seas. The coral reefs are home to over a third of the 2,300 fish species reported in the country. There is literally an ocean of opportunity in natural products, nutraceuticals, and potentially therapeutic natural compounds.

Add to that our brilliant scientists and our culture of entrepreneurship, and we can see that biotechnology is a lucrative high-technology industry where the Philippines can add value, compete and carve a niche for itself in the global marketplace.

January 19, 2006

Vic Ilag

Biotechnology in the Philippines—Impressions of an Accidental Tourist

One of the things that disappointed me during my visit in 2001 is that after being away from home for sixteen years, the rate of progress in biotech in the Philippines has been very slow. Inspite of the buzz about biotech in the Philippines, my personal impression is that its biotech industry is at least fifteen years behind other countries and the gap is widening further unless something gets done.

Misconceptions of Biotech and Academic Purity

One of the most interesting and revealing conversations I had was during a meeting with several faculty members at UP Los Baños. I was surprised when I was told that the faculty is not allowed to set up companies based on their discoveries or that they should quit their posts to pursue them. Apparently, this has been a policy to maintain academic purity. To counter this mindset, it should be pointed out that academic excellence is not compromised. If one looks at the top universities in the US, the faculty members, including several Nobel laureates, are free to participate in setting up companies based on their discoveries. The number of biotech companies associated with a professor is an unofficial measure of stature. Walter Gilbert, 1980 Nobel Prize laureate for his work on DNA sequencing and Harvard University professor, is co-founder of top-tier biotech companies. Furthermore, Genentech, one of the premiere biotech companies, has a higher publication rate when compared to other academic institutions.

Another argument for allowing faculty to set up companies based on their discoveries is that they have the expertise, and the technology is better transferred if the faculty is part of the company, and more importantly, they have the motivation and passion. Of course, a limit needs to be set so that the faculty member is not distracted from his or her major objective of teaching and scholarly work. In addition, this creates additional economic incentives for the faculty member and can discourage any temptations for academic corruption. Identifying and respecting each academic's strengths may lead to better management of his time and effort. Those who prefer mainly to teach should be justly rewarded as well for taking on the teaching credits that generate their income.

Patrys Ltd., Melbourne, Australia
vilag2001@yahoo.com

The academic environment should be mindful of and respect these multi-faceted roles the faculty takes. If innovation is the driver of the biotech industry and the source of talent and ideas is the university, it is common sense that we need to promote this activity within the university.

Lost Opportunities

One famous missed opportunity is the toxin from the *Conus* snails stemming from the research of Baldomero Olivera of the University of Utah and Lourdes Cruz of UP MSI. Ziconotide was being developed as a painkiller by a company called Neurex in the US. Neurex was eventually acquired by Elan Pharmaceuticals for a whopping $700 million. Of course, every misinformed opportunist took to the papers and streets crying exploitation by rich countries. However, we need to dig into the details and possibly consider that it was simply our fault. At that time, the medical value of the toxin was not obvious until Neurex stumbled on its potential use. What is disconcerting is that inspite of this potential today, the government has not yet invested more research into searching other *Conus* toxins. The reality is that we may have a lot of natural resources both in land and sea, but these resources need to be developed. A better understanding of intellectual property law needs to be communicated to lawmakers, government officials, and especially to militant and vocal nongovernment organizations. Nobody has claim on a living organism whether plant, animal, or microorganism. A patent only covers the isolated form of the material from these sources or a defined utility for an organism.

Recommendations in Creating a Biotech Cluster: A Brainstorm

Like any industry, the key is to establish an interactive community of suppliers and consumers directed on some sort of activity called a cluster. For biotech in particular, I see three key components that need to be addressed and developed—science education, finance, and legal structures.

The government should not get too involved and try to attempt to target research that needs to be commercialized. Let the markets decide which companies will be the winners.

The government should promote the industry by continuously supporting research at the university level. Research should be done at the universities and specialized research institutes should be slowly phased out. The lifeblood of biotech is innovation and this is best achieved in university research labs and by the training of future scientists. Boosting salaries of the faculty and modifying policies that will make it easier for scientists to be entrepreneurs would accelerate the development of this industry.

Economic incentives should be promoted to attract foreign venture capital firms and to entice even local companies to shift to biotech as a way of diversifying their investments. Once the money comes in, companies are set up, jobs are created, and the economic standard of living is raised. This effect should be directly felt by the academic community and then spread to the local communities. The government should recognize this multiplier effect and not worry about getting direct revenues from inventions.

One of the advantages of the Philippines is that there are a lot of Filipino scientists abroad, including respected experts in their fields, who can pitch in and help out. The government can help recruit these people and seek their advice. They are patriotic and respected and simply need to be organized to get them onboard

One should realize that creating a biotech cluster is a major endeavor and will definitely take time, money, and a lot of patience. I would say this is a ten- to thirty-year project. Although it is going to be a tough journey, I am optimistic because we have one of the greatest resources on earth—our people.

July 29, 2004

Benigno Peczon

Perspectives on Biotechnology: Proceedings of the Third Asian Biotechnology Conference

When public and private institutions work hand in hand with government regulators, modern biotechnology can effectively address concerns regarding health, agriculture and the environment in a socially acceptable fashion. But with a caveat. The biotech train is not out of the woods yet. There are many stones left unturned. There is still much to be done. This became apparent at the Third Asian Biotechnology Conference held at the Diamond Hotel in Manila, which saw the participation of 220 people from Australia, India, Indonesia, Italy, Myanmar, the Philippines, Singapore, and the United States.

With the theme "Biotechnology Opportunities for Developing Countries," the formal conference started off with Dr. Edgar DaSilva, former Director of Life Sciences at UNESCO, reading a backgrounder on Asian Biotechnology Conferences written by Dr. Nagesh Kumar, Director-General of Research and Information Systems for Developing Countries (RIS, New Delhi, India) The first two Asian biotechnology conferences were held under the aegis of RIS in India. At the end of the backgrounder, Dr. DaSilva unveiled and distributed Vol. 9, No.1 of RIS' *Asian Biotechnology and Development Review*, entitled "Bioethics in Asia."

In his welcome message, Dr. William Padolina, Deputy Director-General of the International Rice Research Institute (Los Baños, Laguna, Philippines), said that although we are making headway in biotechnology, we must not forget that we are confronted by legal, ethical, political, and commercial issues. These demand a high level of accountability and transparency.

Dr. Hubert Gijzen, Director of the UNESCO Office in Jakarta, gave an idea of UNESCO's interest in biotechnology. In her keynote speech, Dr. Estrella Alabastro, Secretary of the Department of Science and Technology (DOST, Taguig, Philippines) spoke on the growing importance of biotechnology all over the world. She pointed out that although the United States is presently the acknowledged leader in biotechnology, and that in 2005 the European Union (EU) had the highest ever increase in biotechnology revenue growth, the Asia-Pacific region showed the highest percentage revenue growth in biotechnology. Citing a UN report, she said biotechnology offers unparalleled growth opportunities. With regard to success factors, she pointed out that information

---

Biotechnology Coalition of the Philippines Inc. Philippines.
peczon@gmail.com

dissemination and partnerships at many levels are crucial for success in biotechnology endeavors.

Enabling policies are crucial for biotechnology initiatives to take off. Dr. Gurinder Shahi (University of Southern California, USA), session chairman on "World Perspectives on Biocommercialization," gave an overview of global trends and opportunities on biobusiness, a term he and his colleagues coined. He pointed out that the keys to success in biobusiness are smart people, smart ideas, smart money, and smart alliances. Although Asia is well placed in becoming a major player for value creation, the most difficult part of the success equation seems to be finding smart money, i.e., funding of start-ups before the rest recognize the potential.

Dr. S. R. Rao, adviser to the Department of Biotechnology of India's Ministry of Science and Technology, New Delhi, India, who works day-to-day on creating value through biotechnology in India, presented an evaluation of Indian policies and implementation strategies. He said that for policies to be effective, they must be backed up with coherent implementation strategies.

Atty. Peter Tabor, an international economist of the US Department of Agriculture (USDA) , Foreign Agricultural Service (USDA, Washington, D.C., USA), described the oversight of biotech crops in the United States. This oversight is based principally on the concept of substantial equivalence.

Dr. Saturnina Halos (Biotechnology Advisory Team, Department of Agriculture) presented the Philippine government's policies which guide the science-based approach in the utilization of biotech crops in the Philippines. Crucial to the Philippine system is the role of its Committee on Biosafety. The policies in place have permitted the Philippines to become the first country in Asia to commercially grow a biotech crop for food and feed, Bt. corn.

Dr. Patricio Faylon, Executive Director of the Philippine Council for Agriculture, Forestry and Natural Resources Research and Development of the DOST, presented enabling mechanisms for the promotion of agricultural biotechnology in the Philippines.

For all its potential, biotechnology must be utilized in a safe and responsible manner. Dr. Edgar DaSilva, session chairman on "Essential Issues," pointed out the need for sustainable development emerging from the use of ecologically sound, economically viable, socially just, and culturally appropriate policies and practices. Although "biosafety" is not defined in the text of the Cartagena Protocol on Biosafety, Dr. Wendy Craig, of the Biosafety Unit of the International Center for Genetic Bioengineering and Biotechnology (ICGEB, Italy), presented a comprehensive set of questions to help assess the safety/risk of biotech crops.

In addition, by giving a summary overview of biosafety policies around the world, Dr. Craig demonstrated how GMO labeling is likely to be the next battleground in the arena of international trade, and further, how ICGEB is providing assistance to developing countries in this and many other biosafety issues.

A program to assist developing countries develop regulations and policies which are practical, workable, appropriate, and science-based called Program for Biosafety Systems (PBS) involves the International Food Policy Research Institute, the Danforth Center, Michigan Sate University, and Western Michigan University. Dr. Hector Quemada (PBS, Washington, D.C. USA) presented ongoing activities and challenges facing this program. In particular, one component—the biotechnology/biodiversity grants—generates data that should be useful for risk assessments in developing countries, but the design and interpretation of these experiments, as well as their impacts on regulatory policy, should be thoughtfully and judiciously applied.

Pollution from oil spills, distillery wastes, landfill leachates, etc. must be addressed. Dr. Ernesto del Rosario of the Institute of Chemistry, University of the Philippines in Los Baños (UPLB), and his colleagues at the UPLB described potential solutions using biotechnology to address these problems.

Medical biotechnology took center stage in the session chaired by Dr. Jaime Montoya, Executive Director of the Philippine Council for Health Research and Development, DOST. A huge percentage of the global gross national product (GNP) is spent on healthcare. Citing 2001 figures, Dr. Gurinder Shahi presented numbers showing that globally, healthcare accounts for 9.1 percent of GNP. Biotechnology claims an increasing percentage of this sector. But the growth is not in medical biotechnology alone. An optimist, who sees a half-full glass instead of a half-empty glass, Dr. Shahi sees potential in many aspects of biotechnology. He sees the spinning off of agricultural biotechnology into pharmaceuticals, nutraceuticals, cosmeceuticals, specialty foods, industrial feedstocks, etc. To maximize potential, he described a game plan whereby the objective is to create "summit opportunities" (high-value products) from "cloud opportunities" (ideas which have not yet been converted to commercially viable products) rather than compete based on price alone in ubiquitous well-known commodities.

In the hunt for converting cloud opportunities into summit opportunities are Dr. Nina Gloriani Barzaga (College of Public Health, UP Manila) and her colleagues at the University of the Philippines and Nagoya City University Graduate School of Medical Sciences who are into the creation of plant-made pharmaceuticals. Dr. Barzaga described the group's ongoing quest to make edible vaccines to control *Salmonella typhimurium*, HIV, and rabies. Also in the same hunt for summit opportunities is the group of Dr. Filipinas F. Natividad (Research and Biotechnology Division, St. Luke's Hospital, Quezon City, Philippines). Dr. Natividad quoted a 2002 report from the University of Toronto which lists molecular diagnostics as number one in approaches to improve health. She discussed the utility, advantages and disadvantages, and the regulatory approval process for molecular diagnostics and shared her group's experience while working on a polymerase chain reaction-based test for the dengue virus. She also pointed out the huge growth potential in the use of molecular diagnostics in pharmacogenetics.

Two major issues, intellectual property rights and ethics, were covered in the session on "Enabling Issues" run by Dr. Reynaldo Ebora, Executive Director of the Philippine Council for Advanced Science and Technology Research and Development.

Dr. Sachin Chaturvedi brought out major concerns regarding plant variety protection, patents, protection of traditional knowledge and folklore, access to biodiversity, and benefit sharing. With respect to patents, there is concern that patents granted may be too broad. In some instances, patents which cover research tools wield an extremely large influence on downstream products. With respect to seeds, Dr. Chaturvedi reported that seed firms in India find it difficult to access relevant genes as their sequences have been patented. Since licensing fees are not regulated and can be incredibly expensive, the combined lack of access to desired gene sequences and high licensing fees place constraints on wider utilization of the technology. With respect to biodiversity and benefit-sharing, groups such as the Like-Minded Groups of Mega Diverse Countries (LMMC) may need to pursue legally binding instruments at appropriate international fora. The issues raised by Dr. Chaturvedi's presentation suggest that this area may need to be discussed in more detail at the Fourth Asian Biotechnology Conference, or at some earlier forum, if possible.

In his presentation on Ethics, Bishop Jesus Varela (Bishop Emeritus of Sorsogon, El Retiro, Sorsogon, Philippines) said that the Second Vatican Council affirmed that man "judges rightly by his intellect (that) he surpasses the material universe, for he shares in the light of the divine mind." Moreover, Bishop Varela said that "for every human need, the ever creative mind would always come up with the appropriate technology to adequately respond." These statements are in support of the belief that human intervention in the work of God's creation is not a violation of God's Mandate. This does not give Man carte blanche to do anything he wants. Bishop Varela advocates the use of "The Principle of Double Effect" and the "Precautionary Principle." According to the former principle, when both beneficial and harmful effects are foreseen, the following must hold:

1. The act must be morally good or at least neutral.
2. The intention is to achieve the good effect and avoid as much as possible (best efforts) the harmful effect.
3. The foreseen benefits must be at least equal to, if not greater than, the foreseen harm.
4. The beneficial effect must flow directly and immediately from the action and not as a result of the bad effect.

According to the Precautionary Principle, Bishop Valera said, "Prudent policies, based on the Precautionary Principle, require that decisions be based on a comparison of risks and benefits."

The session on "Advancement and Prospects in Agri-Biotechnology in Asia: Status and Priorities" was chaired by Dr. George Fuller, Executive Director of

CropLife Asia (Bangkok, Thailand). In the overview lead paper, Dr. Randy Hautea (Global Coordinator, ISAAA, SEAsia Center, c/o IRRI, Los Baños, Laguna, Philippines) described the global trends in agri-biotechnology. In 2005, 220,000 million acres were planted to biotech crops, an 11 percent increase over the 2004 acreage. Sixty-three countries produced or performed research on 57 biotech crops. In 2005, regulatory approvals on biotech crops were tallied as follows: USA, 73; Canada, 67; Japan, 48; Philippines, 30; China, 18; and EU, 18.

In the future, Dr. Hautea said that biotech crops introduced will be more tolerant of adverse agronomic conditions, will be safer as food, will have improved nutritional value, enhanced taste, better fiber content, and longer shelf lives, and will serve as platforms for plant-made products and renewable energy. The challenges enumerated for agri-biotechnology include:

1. Continuing responsible stewardship in risk assessment;
2. Improving communication with society; and
3. Ensuring that biotech crops in conjunction with conventional technologies contribute to sustainable agriculture, alleviation of poverty and a safer environment.

The eggplant is an important crop in India, requiring 510,000 hectares to meet local demand. Losses of up to 50 to 70 percent require from 25 to 70 insecticidal sprays for effective control of lepidopterans. Dr. Bharat Char, of the Mahyco Research Center of the Maharashtra Hybrid Seeds Co. Ltd. of Jalna, India, described completed and ongoing work to meet regulatory requirements in ongoing efforts to bring Bt eggplant to commercial production. Dr. Jerry Flint, director for technical development in the Asia-Pacific of Monsanto (Greater St. Louis, Missouri, USA), described the history of Bt corn development and factors leading to its successful commercial deployment in the Philippines. Noteworthy are the results of an independent study which shows that Filipino farmers retain 52 percent of the value created. In the Philippines, 40 percent of children from six months to five years old and 20 percent of pregnant and lactating mothers are deficient in vitamin A. Predominantly rice eaters, Filipinos would gain health benefits from consuming rice with built-in beta-carotene, the precursor to vitamin A. Dr. Rhodora R. Aldemita, chief science research specialist of the Philippine Rice Research Institute in the Science City of Muñoz, Nueva Ecija, who recently completed a postdoctoral stint at the Albert Ludwigs University in Freiburg, Germany, described the pioneering work of Ingo Potrykus and Peter Beyer in adding beta-carotene into rice. She also described ongoing activities in meeting intellectual property concerns and increasing the beta-carotene content of local rice varieties. The good news is that by 2011, Filipino farmers may expect to be able to plant the so-called Golden Rice and moreover, be able to replant the seeds. Drought is a major constraint in agriculture in many parts of the world. Dr. Kiran K. Sharma, of the Genetic Transformation Laboratory of the International Crop

Research Institute for the Semi-Arid Tropics (ICRISAT, Andra Pradesh, India), described completed and ongoing research at ICRISAT to alleviate poor production in drought-susceptible areas.

Dr. Dennis Gonsalves, Director of the Agricultural Research Service, USDA, Hilo, Hawaii, USA, related his success with papaya resistant to the papaya ring spot virus (PRSV) in Hawaii and his experience in transferring the technology to Thailand. In the district of Puna, on the big island of Hilo, in 1994, up to 95 percent of the papaya stand was affected by the PRSV, creating a big problem for papaya growers. Fortunately, Dr. Gonsalves and his colleagues had been working on creating transgenic papaya resistant to PRSV since the mid-80s. From their accumulated knowledge, by 1999, the decimated papaya plantations had been successfully replanted with healthy papayas which were resistant to the PRSV. In 1997, Dr. Gonsalves brought transgenic papaya to Thailand. Field trials conducted in 1999 showed excellent resistance to the PRSV. Then the project got bogged down in meeting regulatory requirements. By 2001, Thai regulators placed a moratorium on "on farm" testing of transgenic plants. This moratorium was lifted briefly in 2004 but in August of that year, Greenpeace made a massive effort against transgenic papaya, stalling the project up to the present. The lesson learned is that even with a willingness to meet regulatory requirements, a project can be stalled. The story of Dr. Gonsalves brings us back full circle to the message of Science and Technology Secretary Estrella Alabastro that there is a need for dialogue with all stakeholders to see the completion of a project and the reminder of Dr. William Padolina that there are many issues to address with heightened vigilance. Everyone must pull together to see to it that the promise of biotechnology comes into fruition in a manner that meets the needs of all stakeholders.

* With comments provided by some presentors; a version of this article was published in Biotechnology Vol. 2, No. 6 of *BioLife*, a publication of the Biotechnology Coalition of the Philippines in cooperation with J. Burgos Media Services, Inc. (Quezon City, Philippines).

Printed in two parts on December 14 and December 21, 2006

Ameurfina Dumlao-Santos

A Crash Course in GMOs

Picture this: you take a bite of boiled corn on the cob, only to find out that it has been genetically modified with Bt, a toxin that kills corn borers. Will you die just like these insects? Should you write your last will and testament?

This Bt corn is an example of a GMO—a genetically modified organism—so-called because a gene from an outside source, the soil microorganism *Bacillus thuringiensis*, has been deliberately added to the plant. This added gene makes a substance that is poisonous to a particular type of insect that includes corn borers. Man does this because he is sick and tired of these corn borers that extensively destroy his crops.

For thousands of years, man has been randomly mixing genes by crossbreeding two plants or two animals, hoping to get desirable traits in the offspring. This method is nonspecific and unpredictable; the offspring in fact may carry both desirable and undesirable traits of the parents.

Thanks to the discovery in the 1970s of tools needed for genetic engineering, we can now choose and transfer the gene we want into the genetic material of the recipient organism, which then becomes a GMO or a transgenic organism. The transfer is specific, and the desired trait is predicted.

In medicine, drugs may be produced in bulk by microorganisms. Take for example, insulin, a protein needed by diabetics to lower their blood sugar. Before the advent of GMOs, insulin was extracted from thousands of pancreata (plural of pancreas) of pigs and cows. Human insulin was isolated from the pancreata of cadavers. Now, we don't have to rely on these pancreata that are better put to use for organ transplants because the gene for human insulin has been introduced into bacteria.

And since these microbes multiply very fast, these millions of bacteria produce lots of human insulin. Whereas one pancreas would yield only a few milligrams of insulin, a few liters of the bacterial culture would produce much more. Vaccines, hormones, and other therapeutics are now produced by microbes and other organisms. We have ongoing researches at the National Institute of Molecular Biology and Biotechnology (NIMBB) at the University of the Philippines in Diliman where we produce vaccines and therapeutic antibodies from GMOs.

In "gene pharming," pharmaceuticals are produced from transgenic farm animals. ("Transgenic" means that the organism possesses genes from different species.)

National Institute of Molecular Biology and Biotechnology, University of the Philippines, Diliman, Quezon City, 1101, Philippines.
ameurfinas@gmail.com

Therapeutic proteins like human blood clotting factors may now be found in transgenic milk. Hemophiliacs eventually will no longer have to rely on blood transfusion from which diseases like AIDS may be contracted.

The benefits to food and agriculture are vast. Given the rates of Vitamin A deficiency and rice consumption of Filipinos, it is practical that Vitamin A rice, which is developed by IRRI, be made available. In UP Los Baños, papayas with a delayed ripening trait are grown for better storage and transport. This institute is also researching on the transgenic coconut with improved components.

The environment likewise benefits from GMOs. Pest-resistant crops mean reduced use of chemical insecticides that damage the biodiversity within the locality. It also means reduced accumulation of cancer-causing chemicals in the ecosystem. Herbicide-tolerant plants mean environment-friendly herbicides may be used to destroy weeds so that farmers do not have to till the soil, thus preventing soil erosion. There are also genetically modified bacteria that can clean up pollutants like oil spills.

So far, the only approved GM crop for commercial propagation is Bt corn, after six years of research and field trial testing in the country. However, there continues to be a raging controversy over the planting of Bt corn. The latest involves villagers residing near the Bt corn fields who were stricken ill. Bt corn pollen was allegedly the cause. However, there is no scientific evidence to substantiate this claim to date.

We are assured of the safety of GM foods by our government agencies. The Bureau of Plant Industry of the Department of Agriculture has stringent requirements for the importation of GM plant and plant products.

But in dealing with the GMO debate, it is human logic that will resolve issues on human technology. This Bt substance, a protein, exists in an inactivated form; it is activated only in the alkaline medium of the insect's gut. Once activated, the toxin binds to specific receptors in the insect's gut and bores holes through the membrane. The toxin is thus known as gut poison. But our stomachs are acidic; thus, the toxin cannot be activated. Furthermore, we do not have receptors to bind to the toxin. So what happens to the Bt substance in the corn that we eat? This will be treated by our digestive systems just like any protein, and will be broken down into tiny pieces. Eating Bt corn is no different from eating non-Bt corn.

Is there any perceived risk for GMOs? Like anything, of course, there is. There are concerns especially regarding the long-term effects such as "superbugs," or insects that have evolved to be resistant to insecticides. Constant vigilance and proper awareness of this issue are the best weapons against these. The bottom line is, do the benefits from GMOs outweigh the risks? Unless we want to continue living the problems of food shortage, crop destruction, medical deficit, and environmental decay, the answer is yes.

July 8, 2004

Virginia Monje

Cloning for the Curious

In the movie *The 6th Day*, Arnold Schwarzenegger was shocked to find a man exactly like himself—a clone. Today, we don't have to be shocked to learn that certain plants and animals can be duplicated down to their minute components. Various plants and animal species have been replicated. All this is due to the amazing work of genetic engineers.

A cell needs a complete set of genes to produce the whole organism. Like computer programs, genes are a set of instructions that produce every characteristic of the species, orchestrating its growth and development with precision. Because of genes, ilang-ilang trees keep their peculiar scent from tree to tree; zebras are covered by dark stripes; monkeys have tails and we don't.

Genes are made of DNA (DeoxyriboNucleic Acid), which is written out in a language with only four letters or bases: A, C, G, and T, the sequence of which directs the making of proteins. A sequence of instructions will read something like this: GAGATCTTAGCCTA. Yes, it looks like a meaningless group of garbled letters, but to living cells of a species, they mean everything. Based on the blueprint of DNA in genes, the cells create the species' form and function defined by proteins.

Science has taught us how to manipulate genes. As pieces of DNA, genes are chemicals that are not alive. Therefore, to clone the gene, it should be placed in a live host cell like a bacterium or plant or animal cell to make the gene work or we say "to express" the gene. For instance, at the National Institute of Molecular Biology and Biotechnology (NIMBB) at the University of the Philippines (UP) Diliman, we use insect cells and yeast cells to express genes for antitumor antibodies. This will eventually lead to antibody therapies for breast cancer, a disease that afflicts many women, particularly Filipinas. Scientists at the International Rice Research Institute in Los Baños cloned several genes in "golden rice" that produces Vitamin A. In the future, we can look forward to coastal areas planted with salinity-resistant rice produced by PhilRice in Muñoz, Nueva Ecija—all these using gene cloning technology.

Cloning, or making exact copies of a gene, a cell, or an organism, comes from the Greek word *klon*, meaning twig, a tiny branch which is genetically identical to the parent plant. To clone a gene means producing several copies of a gene, letting the genes work, and generating many gene products.

Formerly with the National Institute of Molecular Biology and Biotechnology, University of the Philippines, Diliman, Querzon City, 1101, Philippines.
virginia.monje@gmail.com

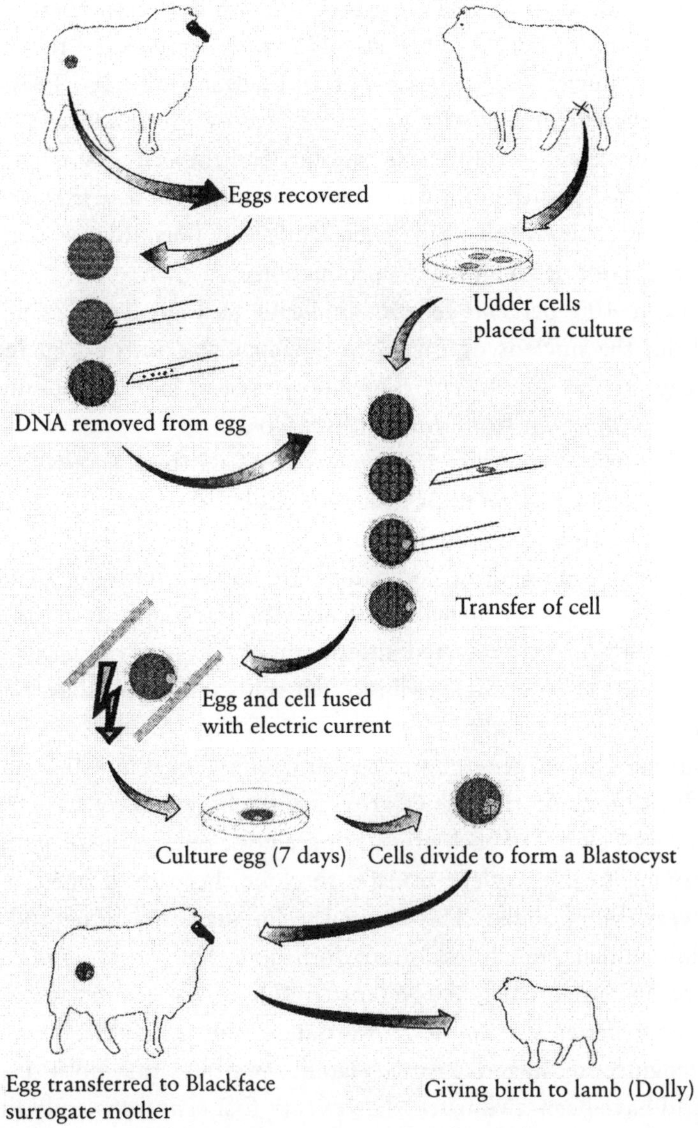

Diagram of the nuclear transfer procedure that produced Dolly (Source: Ian Wilmut, Keith Campbell, and Colin Tudge, 2000. *The Second Creation.* Headline Book Publishing, London. *www.headline.co.uk*)

Clones can be natural. Identical twins are clones of a single zygote (fertilized egg). When a bacterium is cultured in a laboratory dish, it grows into a colony of cells, each of them being a clone of the original cell. Is it possible to select all the qualities of an "elite" livestock and clone its entire set of DNA (or genome)? Nowadays, it is possible! One of the ways is by "nuclear transfer," like the way Dolly was made.

The team of Ian Wilmut started with an unfertilized egg from a ewe (pronounced "yoo"), a female sheep. Let's call her Ewe No. 1. They removed the nucleus of the egg, making the egg "enucleated" or devoid of all nuclear materials. They took the nucleus from a mammary cell of another ewe, Ewe No. 2. With an electric shock, the nucleus fused with the enucleated egg (nuclear transfer) and the egg was activated. The resulting embryo was cultured and then implanted into the uterus of a surrogate mother ewe, Ewe No. 3, which carried it to full term. Then Dolly was born. She was a striking image of Ewe No. 2—the result of an all-female production.

Was Dolly a 100 percent genetic duplicate of Ewe No. 2? No, not really. She was made from the nucleus of Ewe No. 2 which was introduced into the cytoplasm (enucleated egg) of Ewe No 1. The recipient cytoplasm had mitochondrial genes of Ewe No. 1 which were different from those of Ewe No 2.

Even with some successes in cloning genes in pigs, cows and goats, genetic modification in commercial livestock has not really taken off. "Normal breeding," says Harry Griffin, director of Roslin Institute, the home of Dolly, "is more effective and easier." But some leading drug companies are into "pharming"—designing lactating farm animals as factories for pharmaceuticals. Animals like goat or sheep carry genes for therapeutic proteins that are secreted into their milk.

Twins are made from a single embryo that splits into two. They have identical nuclei and cytoplasm. During gestation, the twins are located in different places in the same womb, which may affect gene expression differently. They are born in succession, and that affects them, too. As the babies grow up, they are subject to different experiences. They may be DNA clones of each other, but are not clones of their mother.

Those hoping to use nuclear transfer to clone their loved ones, whether human or pet, will be very disappointed. This cloning technique is not a method for making exact copies of adult animals, much less of grown people. The reason is that the physique and personality of an individual is a combined result of his genetic makeup as well as his upbringing and experience. Memory, for instance, is not clonable.

Genetic engineering has made such significant strides that both the academe and the business world have been awestruck. The twenty-first century is truly the "Biotechnology Century" with an avalanche of advances in technology. A rundown of the latest projects is astounding: attempts to coax bacteria to produce magnets, or marine organisms to manufacture unusual materials, and even an ambitious venture to clone the extinct woolly mammoth! With these possibilities, we can only imagine how cloning will revolutionize the way we live.

Despite the claims of success in human cloning, the technology is far from polished. It took 276 errors before Dolly was made. Recently, the journal *Nature* reports that when scientists at Tokyo University created female mice from all-female parents, the success rate of the expensive cloning experiment was only two out of 598 mice.

The issue on patents and regulations has also generated its share of controversy. Who owns the genes and to what extent scientists are allowed to manipulate genes are just two of the many questions that we as a society will have to deal with.

However, speaking as a wife and mother, I don't think cloning will replace the time-honored way of reproduction. There is no courtship, no romance in cloning. So finding exact copies of Arnold Schwarzenegger happens only in the movies.

July 22, 2004

Celia Aurora Torres-Villanueva

DNA Vaccines: When Less Is More

Formerly, when religion was strong and science weak, men mistook magic for medicine; now, when science is strong and religion weak, men mistake medicine for magic.—Thomas Szasz, M.D.

"Take two shots of DNA and call me in the morning..." It's only a matter of time before we'll be hearing this from our doctors. The age of molecular medicine is truly upon us. This article is about just one specific "molecular" technology that has applications in medicine—DNA vaccines.

What are DNA vaccines? Well, let's start with what we mean by DNA. DNA stands for deoxyribonucleic acid. "Deoxy...what?!" You can see why we call DNA by its nickname rather than by its full name. But even without knowing its full name, you've no doubt met the term "DNA" before. As you may know, DNA is the genetic material of all living things on our planet. By genetic material, we mean that the biological traits that we can pass on to our offspring are encoded in our DNA. The biological information required so we can have curly hair, brown eyes, and brown skin are in our DNA. That is, unless you get a perm, wear contacts, and get a tan! But even if you get a perm, if your children inherit your DNA for straight hair, then they will still be born with straight hair, whether or not you had a perm when your child was conceived or while the mom was pregnant.

So now we know what DNA is, and we already know what vaccines are. DNA vaccines, then, are simply vaccines that have *only* DNA. Yet, while having less, DNA vaccines do more.

Most of the vaccines that we have now are composed of whole germs, or microbes, that cause the disease that we want to vaccinate ourselves against. For example, if we don't want to get the flu, caused by influenza viruses, then we'd get injected with the influenza virus. If we didn't want to get measles, we'd get injected with the measles virus. But why don't we get sick if we get injected with the actual germ or pathogen? Well, these germs are first weakened, or attenuated, sometimes even killed, before they're injected into us. In that way, our body's immune response will "see" the germ in its disarmed and harmless form and be able to recognize it so that the next time we "see" that germ in an actual naturally acquired infection, when the germ will be armed and harmful, our immune response already knows it's the enemy and will fight it, thus protecting us

Intellectual Property Office, Makati, Philippines.
celia.torresvillanueva@gmail.com

from the disease. It's like training an army (our immune response) with mock enemies in costume (the vaccine).

Not all the vaccines that we are getting are made of killed or weakened germs. Some are made up of only parts of germs. These are called subunit vaccines, and are made up of proteins of the germs we are vaccinating against. It's possible to use subunit vaccines rather than whole vaccines because, in fact, our immune response doesn't actually "see" germs in their entirety. Our immune response only sees parts of germs, generally just their proteins. It's like seeing or looking for just the insignia of the enemy or color of their uniform, without looking at the face of the enemy itself. So when we get surface antigen protein (subunit) of hepatitis B virus as vaccine, we're protected against the entire hepatitis B virus because our immune response, in attacking the protein which is attached to the virus, is pretty much attacking the virus itself.

The good thing about using killed or subunit vaccines is that you avoid complications arising from the supposedly harmless germ (in live attenuated vaccines) reverting back to its harmful self. The problem, though, with using just killed germs or just parts of germs, is that only one arm of our immune response gets activated.

You see, our immune response has two arms—the humoral response and the cellular response. You could think of the two as being analogous to the air force and the infantry of the army, respectively. The humoral response involves antibodies which act like missiles, attacking germs that they can attach to. The cellular response acts like the infantry in that cells of the immune response directly act on the germs, or the cells where the germs are hiding, such as in a viral infection.

Live attenuated viral vaccines can induce the cellular response because these live viruses infect cells and produce viral proteins, which the immune response can see. The immune response knows that these proteins don't really belong to the cell, but are foreign, i.e., enemy (to our immune response, everything foreign is an enemy). Thus, the infected cell gets "flagged," that it has been occupied by the enemy, and the cellular response can now go in and destroy that cell, thereby controlling the infection (pretty much like bombing your own ammunition factories when the enemy has gained control of them).

Now, when the immune response encounters only killed germs or just parts of germs in the vaccine, only the humoral response gets activated. For viral diseases, this is bad news because it's very important to activate the cellular response in order to hunt down viruses as they hide in infected cells, and to kill these infected cells before they produce more viruses.

DNA vaccines combine the best of both worlds. DNA vaccines act like subunit vaccines, in that they encode for only specific proteins without the entire germ, and are thus safe. However, unlike subunit vaccines, DNA vaccines act like live attenuated vaccines because the DNA gets into our cells, which then produce the proteins for the immune response to see. This way, the DNA vaccine is mimicking an actual viral infection, where the virus proteins get produced by the infected cell itself, thus marking the cell for destruction by our cellular immune response.

But that's not the end of the story. Aside from being more effective than killed or subunit vaccines and safer than live vaccines, DNA vaccines are also cheaper and more stable. To produce DNA vaccines, we only need a basic and standard molecular biology lab, where we can grow the DNA vaccine in bacterial cultures, purify the DNA from the bacteria, then inject the DNA directly into the patient. No need for complex synthesis, coupling with carriers, or culture of germs, all of which are very expensive. DNA vaccines are also more stable than their other counterparts. In a tropical country like ours, where conventional vaccines need to be transported to far-flung barrios in a "cold chain," i.e., constantly kept cold in expensive thermos containers on ice or refrigerated, transporting DNA vaccines in your pocket is much more practical and cheaper.

DNA vaccines are also greatly empowered by the technology of genetic engineering. There are no available vaccines for many, if not all of the tropical diseases that plague our poor country. Besides not being a good investment for multinational countries (which may be forced for humanitarian reasons to provide such vaccines free or at low prices), vaccines against malaria and dengue have not been developed due to the complexities of the germs that cause them. For reasons too complicated to get into here and now, genetic engineering and DNA vaccine technology can help address these problems.

Our lab at UP is currently doing research aimed at developing DNA vaccines for dengue and malaria. We're years away from producing commercially available vaccines, but we've taken the first steps and are moving forward. So when your children or grandchildren get DNA when they come in for their shots, you'll know what they're getting, and why they'll be better off than you and your old-fashioned types of vaccines!

April 14, 2005

Maria Corazon de Ungria

DNA Evidence in Paternity Cases

Deoxyribonucleic acid or DNA testing is the most accurate form of testing to prove paternity or exclude paternity when the identity of the biological father is under dispute. DNA-based paternity testing has been requested to support claims for child support, inheritance, immigration, and for peace in the family. More recently, DNA tests have been used to dispute false paternity claims that have already been decided in favor of the child's mother prior to the submission of DNA evidence in US courts.

Traditionally, paternity is determined using evidence derived from the woman's testimony, the defense of the alleged father where he claims sterility or that another man had had relations with the child's mother, a presumption of legitimacy if the child was born within a valid marriage and physical resemblance between the putative father and the supposed child. In addition, either party may present the results of serological tests, e.g., ABO and MN blood typing, to support their respective claims.

In more recent years, the use of selected DNA markers has become the procedure of choice over blood typing because of the increased level of polymorphism and the less susceptibility of DNA molecules to degradation compared to proteins. The molecular stability of DNA is particularly important when dealing with environmentally challenged samples, e.g., exhumed bones and degraded tissues. Of the different methods of DNA analysis, STR (Short Tandem Repeat) typing is currently the most widely used because this method allows unambiguous scoring of DNA profiles, rapid processing, and analysis.

DNA typing for paternity is done by first carefully extracting the DNA from the biological samples submitted by the alleged father, child with mother (paternity trio) or in the absence of the mother's sample (paternity duo). The DNA pattern from the child is analyzed given those of his mother (if available) and alleged father. The DNA type contributed by the child's real biological father should be observed in the alleged father. Then, the probability that the alleged father is the father of the child is calculated as a ratio between that of the alleged father and any random male in the population. Notably, testing without the mother's DNA profile (motherless case) was found to be less informative and five times more prone to paternity inclusions when testing seven STR markers than when the maternal DNA profile is made available. The current DNA Laboratory set-up at the University of the Philippines-Natural Sciences Research Institute

DNA Analysis Laboratory of the Natural Sciences Research Institute, University of the Philippines, Diliman, Philippines.
mcadu@uplink.com.ph

(UP-NSRI) Diliman campus uses some sixteen to twenty STR analysis which includes the FBI-defined Combined DNA Identification System (CODIS) markers for DNA typing (*www.dnaforensic.org*). At the UP-NSRI DNA Laboratory, the lack of information brought about the absence of the mother's DNA profile in motherless cases is minimized by increasing the number of DNA markers of the alleged father and child that are tested to twenty markers compared to the standard sixteen markers for paternity trio cases.

A mismatch suggests that the alleged father is excluded as the biological father of the child. In some cases, mutation results in a false mismatch between real fathers and their children, hence the standard accepted in most laboratories is to require a minimum of two mismatches prior to excluding a man from potentially fathering the child. On the other hand, a match between the DNA profile of the alleged father and the child does not necessarily establish paternity, but may be due to chance matches between totally unrelated individuals. To estimate the likelihood of paternity over non-paternity, a Probability of Paternity (W) is calculated based on the DNA profile of the father, mother, and child. The Supreme Court prescribed the minimum value of 99.9 percent for W.

Although the resolution of questioned paternity is normally a civil issue, it may also play an important role in criminal cases such as those involving rape when the victim also claims that the accused is the father of her child born out of the rape (criminal paternity). The first such case where DNA evidence was used was *People vs Paras* (1999) where blood typing and DNA profiling results conclusively excluded the accused from being the father of the victim's child. To the trial court, the date of the last incidence of rape stated by the victim is important since the child was born ten months after the said date. According to the trial court, "these facts would be in violation of the rule of nature."

With the rapid development of DNA-based paternity testing, it is inevitable that DNA evidence will be used more and more to support or argue against paternity in the courts of law. Initially, the strength of paternity tests lies primarily in its power to exclude the wrong man. However, the rapid development of STR typing technology has also increased its power to identify real fathers, thus providing objective evidence for a fair and swift resolution of civil and criminal cases.

August 31, 2006

Part 8

Philippine Biodiversity, Environmental Concerns and Issues

Island Magic: Understanding Philippine Biodiversity

Benjamin Vallejo Jr.

Two Englishmen in the mid-nineteenth century, Charles Darwin and Alfred Russel Wallace, established our modern ideas of biodiversity. Charles Darwin sailed around the world on the *HMS Beagle* and explored the Galapagos, South America, and Australia. Alfred Russel Wallace spent fifteen years in Indonesia collecting beetles, birds, butterflies, reptiles, etc. The two men observed the variety of living organisms in their explorations and independently concluded that the extraordinary biodiversity they had observed could only be explained by evolution and natural selection. This theory of evolution is central to our scientific understanding of biology and our history as a species.

Neither Wallace nor Darwin had the chance to visit the Philippines. The Philippines was then a Spanish colony and an English-led biological expedition would have invited suspicion from the Spanish authorities. But if Wallace and Darwin had visited, would they have reached their conclusion much sooner? Probably, they would have.

This is because the Philippines is a mega-diversity country. And like the Galapagos, each of our islands has its own set of unique plant and animal species found nowhere else in the world. We call these animals and plants "endemic." The islands are of differing ages and this provides the setting for an elegant natural experiment. Luzon, our oldest oceanic island, is about sixty million years old and has the most number of endemic species. Mindanao, the second oldest, is second in endemics. The other younger islands, Negros, Panay, Cebu, Leyte, and Samar, have their own set of endemics. Sulu and Tawi-Tawi are the youngest and smallest, thus they have the fewest endemics. Palawan is unique, being once connected to Asia before becoming separated or isolated. Its flora and fauna are similar to those of Malaysia but, in many ways, are also endemic. Each island may therefore be considered a unique biological treasure in itself. But how did the natural experiment come about?

Geology has some answers. In the ice age, sea levels were much lower than today's levels. The islands formed bigger islands and animals were able to move between them. Greater Panay and Negros included Masbate and Cebu. Greater Luzon included Marinduque, and Greater Mindanao included Leyte, Samar, Dinagat, and Bohol. However, some islands like Sibuyan and Camiguin were never connected at all. This explains why each island contains animals found nowhere else. This is why Bohol and Mindanao have the tarsier and *kagwang*, while Luzon has cloud rats. Our unique biota

Institute of Environmental Science and Meteorology, University of the Philippines, Diliman, Philippines.
benjamin_jr.vallejo@up.edu.ph

was largely derived from mainland Asia. Larry Heaney, the zoologist who described these wonders, estimated that the Asian colonizers successfully made it to our islands once every 20,000 years. That in itself is nearly miraculous.

Only evolution can explain in a scientific way why we have these living treasures. From these few original migrants, evolution waved its magic wand and over time, many new unique species came into being. Habitat complexity and long isolation promoted the diversification of our flora and fauna. Our biotic history had eons of time before man appeared. The first people probably set foot on our islands in the middle of the Pleistocene Age, about 500,000 years ago. This was the height of the ice age. The seas separating the islands were narrower as sea levels were much lower.

The first people of the islands modified their environment. But for a long time, we had very little impact on the environment until the time of Spanish colonization. One of the results of colonization is that Filipinos began to live in townships and practiced more permanent forms of agriculture. Western political, religious, and cultural structures were then imposed. This caused populations to increase.

Then the widespread clearing of the forests began. But the effects were negligible at first, except in the more populous Luzon. Negros had 80 percent or more forest cover at the end of Spanish colonization. It was only after independence in 1946 that forest clearing rapidly increased on all the islands. Today, primary forest cover is less than 10 percent and islands like Cebu have been completely deforested. What most of us believe to be virgin forests is, in fact, secondary growth.

The forests and seas were the sites of evolution. They provided the habitat complexity and isolation that promoted new species to arise. Forest destruction has resulted in extinction of island species. The Ticao island hornbill in Masbate is now gone since that island has lost its primary forests. Two bird species, the Cebu black shama and flowerpecker, are on the edge of extinction as they barely hang on in Cebu's last forest. In the seas, reef destruction has decreased the habitat complexity that supports marine biodiversity. Darwin and Wallace recognized that habitat diversity and isolation are the raw material on which evolution works its wonders on living organisms.

But it is not too late. Filipino scientists and their foreign counterparts have stepped up efforts to document our biodiversity. In the last two decades, our scientists from the National Museum have worked with zoologists from the Chicago Field Museum of Natural History in numerous expeditions to Bicol, Quezon, Mindanao, and Panay. They still discover new species even though the forests in these places are almost gone. In June of 2004, I was part of an expedition to document marine biodiversity. Scientists from many countries, including a sizable Filipino group, headed by Professor Philippe Bouchet from the National Museum of Natural History in Paris, documented the scale of marine biodiversity on Panglao Island, Bohol. Five thousand mollusk species were recorded. There were about 1,200 crab and shrimp species collected. In comparison, the whole of Japan only has 1,600 species. The Philippines probably has several thousand more

Panglaos, all of which are relatively unstudied. This is a sign of hope since Panglao and other islands have had a history of reef degradation in past decades. It is only recently that people have started to protect their reefs. It is now realized that the role of local communities is indispensable in biodiversity conservation.

We hope that this new information will contribute to a fuller understanding of the biotic history of our country. If we understand this history, then we can implement sound management and conservation strategies and protect whatever is left. And if we are successful, perhaps we can reverse the damage and have a more ecologically sound restoration strategy. We also hope that we are able to attract young people to consider careers in biodiversity. While this may mean being out of the city for months on end, this is a worthwhile endeavor and also an exciting adventure. A few students have begun working for PhD and MSc degrees in biodiversity here and abroad. What is encouraging is that some of them are from regional universities and government agencies. Hopefully, they will return and contribute to the unfinished tasks ahead. They will help save the magical Philippine islands that are as wondrous as Darwin's Galapagos.

December 9, 2004

Francisco Heralde III

Panglao Marine Biodiversity Project 2004—Science Unraveling Nature's Beauty

When Dr. Baldomero "Toto" Olivera, world-renowned conotoxin scientist, invited me to join the Panglao Expedition last June 6, 2004, mixed thoughts rushed through my head on what to expect, more so because it meant taking a three-week leave from a busy lab work schedule at the University of the Philippines National Institute of Molecular Biology and Biotechnology (UP-NIMBB). Nevertheless, my excitement about the new learning experience was pervasive; hence, on June 17, I was off to Panglao.

Panglao is a tourist paradise island on the southwestern tip of Bohol. Scientists from nineteen countries gathered in that special spot of Bohol through the initiative of the University of San Carlos (USC), Cebu and the French Museum National d'Histoire Naturelle (MNHN) to study the biodiversity of marine life in the municipalities of Panglao, Dauis, Baclayon and Cavite and Tagbilaran City from May 29 to July 9, 2004. The project, known as the "Panglao Marine Biodiversity Project 2004," operated through the funding of Total Foundation, Inc.

I arrived at the heart of the operations in Crystal Coast Resort in Barangay Tawala, Panglao where a laboratory was built for the expedition's use. Members of the media (GMA-7 and *Philippine Daily Inquirer*) who were my incidental travel mates began their coverage of the project as we were welcomed by the expedition leader, Dr. Philippe Bouchet of MNHN. Two weeks had passed since the start of the expedition and approximately 1,200 species of gastropods, 500 species of crustaceans and 800 species of nudibranchs were already documented and much more were awaited to be encountered as the days progressed, according to Bouchet.

I was assigned to work with Dr. Ellen Strong of the University of Minnesota and Dr. Yuri Kantor of the Russian Academy of Sciences to process specimens for barcoding and morphoanatomy. I was given a briefing and demo and no later than that same afternoon, I was on my way to processing specimens. The expedition organizers had high expectations of the project as it employed special techniques, equipment, and manpower to generate a massive turnout on biodiversity assessment. This was the product of their several expedition experiences with the New Caledonia site as their latest in 2002.

Department of Biochemistry and Molecular Biology, College of Medicine, University of the Philippines Manila, Philippines.
fheralde@yahoo.com

Indeed, as could be gleaned from the daily schedule of participants (i.e., work begins at 6 a.m. and ends at 10 p.m.) as well as the samples that came in, ten collection techniques were employed, namely brushing of rubble (B), dredge (D), *lumun-lumun* (L), intertidal (M), traps (N), tangle nets (P), scattered specimens taken during dive (R), suction sampler (S), trawl (T) and yabby pumping (Y). The capital letters following the collection type were used as sample identifiers by the collection group as soon as the specimens were laid out on a receiving table in front of the laboratory. A number was assigned by the database group to indicate the order of collection; this was cross-referenced to the GPS (geographical positioning system) data and the specimens processed for barcoding, morphoanatomy, photography, or drawing.

The system by which the specimens were documented and processed followed a factory-like layout. The laboratory was designed to be multipurpose in handling different marine specimens and allowed ample time for processing. Six groups were housed in the lab, namely: 1) the database group led by Mike Cusi and Ryard Paig, 2) the microgastropod group by Virginie Heros and Philippe Maestrati, 3) the nudibranch group by Marina Poddubetskaia and Terry Gosliner, 4) the crustacean group by Peter Ng and Joelle Lai, 5) the macrogastropod group by Serge Gofas and Marco Oliverio, and 6) the mollusk barcoding group by Yuri Kantor, Ellen Strong, and Toto Olivera.

When the specimens arrived, they were laid out on a receiving table and given an area number. The different groups inspected the specimens to see if these have already been processed and barcoded. If not, the specimens were taken to their respective tables inside the lab where these were processed according to each group's procedures. For the macrogastropod and mollusk barcoding group, their specimens were placed in plastic containers and immersed in cold magnesium chloride to relax the specimens. This was done to collect tissues from the specimen for DNA analysis without breaking the shell. Similarly, the relaxation process allowed the preservative (e.g., ethyl alcohol) to penetrate the core of the shell and ensure inner tissue preservation. The specimens were identified by taxonomists like Henk Dekker of the Zoologisch Museum, Amsterdam at the species level if possible, but most often at the family level, especially those suspected to be new species.

Crustacean specimens were relaxed by storing them in a low-temperature cabinet. As the specimen hibernated and gradually died, a tissue sample was obtained for DNA analysis and a photograph was taken. The specimen was subsequently preserved in formalin and packed in a sealed plastic bag. The nudibranch group took photographs of their specimens underwater while collected specimens were tissue-sampled for DNA work. The microgastropod group was intriguing as a lot of the new species were found among them. Sand obtained from trawling, brushing or suction was subjected to sequential sieving on a screen of varying mesh sizes. The sieved sand was sorted out microscopically, collecting the live mollusk and those with intact shells. The collected specimens were pooled; similar ones grouped together and given to the mollusk barcoding group for processing. The specimens were inspected microscopically by looking at the shape of the protoconch (snail apical tip) and the whorl (side lobe) architecture. Similar ones were grouped together with one or two specimens kept as voucher and the others cracked for DNA work.

Photographs were taken of both macro and micro specimens from time to time, especially if these were new finds or the specimens were beautiful. This was done by making the specimen glide or move through a glass support with or without seawater on a photography table where a mounted digital camera was used to take pictures. Aquaria were maintained to house nudibranch and other mollusk specimens for overnight observation. Noteworthy were the efforts to study eulimids, a group of parasitic mollusks. Specimens of these were processed together with their host (e.g., starfish and sea urchin). Anders Varen, a Swedish taxonomist, and Ludivina Labe, of National Fisheries Research and Development Institute (NFRDI) worked along this area.

Some of the interesting finds of the expedition included a tiny marine snail that is said to be a living fossil as claimed by Bouchet, and a living specimen of microscopic snail which, according to Varen, has not been found in the last twenty-eight years. Also interesting were the observations of a "vampire" snail in the aquarium setup of Olivera that feeds on the blood of parrotfish through its extendable proboscis, as well as a "naked" snail viewed under the microscope of Kantor that spontaneously went out of its shell. Surprises happened everyday, as treasures of the marine world were unraveled. Much of what I have talked about here is viewable at *www.panglao-hotspot.org*. This website was created and maintained by Ryard, a computer science graduate of Holy Name University in Tagbilaran, Bohol.

The Panglao expedition was the first systematic marine biodiversity assessment ever performed in the Philippines. Even as the expedition progressed, the turnout in terms of the magnitude of species richness and recovery of some rare finds demonstrated the optimistic outcome of the project. Most of the preserved specimens were taken to MNHN in France for further taxonomic studies, while the crustacean specimens were brought to the National University in Singapore. According to Marivene Manuel, a crustacean specialist, once the taxonomic identity of the specimens is ascertained, a holotype (i.e., representative sample) of such will be transferred to the National Museum of the Philippines. Two Filipino savants will assist the taxonomic work in France, namely Noel Saguil, a marine biologist who pioneered the groundwork activities of the expedition, and Dave Valles of the University of San Carlos. Similarly, Manuel of the National Museum will perform corresponding work on crustaceans in Singapore.

My three weeks of participation in the expedition resulted in a massive exposure to different marine organisms, their behavior, special characteristics, and names. It was also an opportunity for interaction with leading scientific experts in the field. My experience in the expedition was best captured by a poem which I happened to write one late afternoon while I was enjoying the magnificent view on a ridge overlooking the beach of Tawala, as follows:

Panglao Biodiversity

Rustling waves on a jagged cliff,
Lull the mind in hypnotic shift...
Forgetfulness instilled from a busy world's motion,
A moment of bliss in harmonious unison...

Alas! The soulful howling wind,
In pervading silence, a message send...
Bidding one to come and be,
In Mother Nature's bosom immerse and hear her plea...

"Thou my son, savor thy beauty,
Know their kind and awe on diversity...
Yet thy sole desire thou must remember,
Preserve their make to last forever..."

FMHIII 6/26/04
Panglao 6:15 p.m.

February 3, 2005

Wolfgang Reichardt

Coating Matters in the "Invisible Sea": Biofilms, Marine Snow, and Bioengineering

Fish Kill Reveals Invisible Forces

In February 2002, after many months of dry and sunny weather, people in the province of Pangasinan witnessed a massive fish kill of *bangus* (milkfish) being raised in pens and net cages in Bolinao. Masses of dead fish discharged into the open water were soon after washed ashore with incoming tides. With a second mass kill of wild fish and shell fish following on the next day, a considerable stretch of the shoreline was to remain for weeks without the living resources which nutrient-rich intertidal flats are usually expected to sustain.

More recently, coastal marine waters, the everlasting domain of fish "hunters," is increasingly shared with fish "farmers." In economic terms, this makes sense, at least in theory. Could Philippine fish farmers eventually enjoy a similarly sustainable resource base as their cousins in the Ifugao Rice Terraces? Ecocatastrophic events like the one in Bolinao may provoke skepticism. On the other hand, large-scale fish farming has just begun, and we have to ask first: What do we really know about energy flow, carrying capacities, and about Nature's pathways of waste disposal and recycling? How can one reliably assess environmental risks?

Growing interest in food supply for ocean fisheries had paved the way for what later evolved as marine science. One of the early milestones in this development was the design of plankton nets to quantify all drifting organisms (plankton) that could be perceived as a source of fish feed. Defined mesh sizes were used to assess and predict fish production in the ocean. Sunlight-driven (photosynthetic) production of phyto-plankton biomass is mainly fueled by inorganic nutrients. As these become rapidly exhausted, their pools must always be replenished. This is achieved in half-cycles of remineralization that are linked with the formation of living biomass. According to early models, most of the dying plankton has to sink to the seafloor first. Here inorganic nutrients will be regenerated and eventually resurface to sunlight-exposed water layers to fuel another round of fish feed production. When the first models for oceanic ecosystems were applied several decades

The Marine Science Institute, University of the Philippines, Diliman, Quezon City, 1101, Philippines.
wtreichardt@yahoo.com

ago, they paradoxically predicted fish stocks would never be sustained due to lack of food, even under the most optimistic (but completely unrealistic) scenario that fish were able to consume 100 percent of the primary (photosynthetic) production. Apparently, these models were too simple. They relied only on the most visible organisms and their most evident functions.

"The Invisible Sea"

At about the same time, scientists and the public alike became gradually aware of missing links in the prevailing knowledge. *Invisible sea* became the title of a film that addressed the role of the microscopically small organisms. Until then, public interest had shunned this hidden and hardly imaginative aspect of marine ecology. Three decades later, marine microbiology has now made huge leaps forward. Meanwhile, molecular genetics enable us to identify microorganisms directly in environmental samples, no tedious cultivation required anymore. The 99 percent majority of bacteria that used to evade most sophisticated isolation and cultivation attempts can at least be tracked down by their genomic signatures. But how has marine microbial ecology changed our perception of fish feed production and fisheries in the ocean? Which lessons have ecosystem modelers been taught? And, above all, how can prokaryotic microbes, these smallest forms of life and direct descendants of the oldest creatures on earth, be involved in the control of life-sustaining processes in the oceans? Could microbes eventually hold a key to sustainable food supply in the ocean?

Excess fish feed and feces undergoing enzymatic burning (respiration) with the help of dissolved molecular oxygen will rapidly deplete the dissolved oxygen in the water column, and even more so, at sediment surfaces on which sinking detritus particles are deposited. This does not mean, however, that recycling or remineralization of organic matter would have to stop, when dissolved oxygen pools have been exhausted. A sequence of alternative pathways replacing oxygen-dependent respiration will be at hand. In the sea environment, the most sustainable recycling pathway for organic matter under oxygen depletion depends on sulfate as an abundant component of seawater salts.

Organic matter recycling via this so-called "sulfate respiration" of marine microbes produces hydrogen sulfide, an easily recognized component with the smell of rotten eggs. Hydrogen sulfide (H_2S) is toxic to most animals living in the sea, unless they can manage to produce detoxifying compounds binding it. Most importantly, this toxicant also binds heavy metals. By doing so, it helps to deposit and to detoxify another group of environmental toxicants. In oxygen-rich waters, hydrogen sulfide will disappear rapidly. Therefore, clams, crabs, and other sediment-inhabiting animals can survive in hydrogen sulfide-rich sediments so long as they manage to irrigate their surroundings with water containing oxygen. Only permanently high rates of hydrogen sulfide production will eventually erase sediment-inhabiting communities.

Bioengineering at the Seafloor

Sediment-dwelling creatures are instrumental to the entire process of organic matter recycling; many of them work as very effective bioengineers by producing a multitude of tubes, burrows, and other structures with aerated (oxidized) interfaces. At these interfaces microbial biofilms are forming structured communities of microbes that are imbedded in a sticky matrix of (polysaccharide and protein) biopolymers. These prime sites of mineralization dispose of immobilized biocatalysts (ectoenzymes) that are needed to degrade large organic molecules. These biofilms constitute crucial "relais stations" for organic matter recycling. Where the seafloor has been purged of its sediment-dwelling fauna of bioengineers by excessive production of hydrogen sulfide, those biofilms will become defunct. This means microbial recycling of organic matter will be deprived of the mediating role of fauna-made biofilms.

This scenario can be observed on the seafloor below excessively fed fish cages in Bolinao Bay (Pangasinan). Where hydrogen sulfide seeps out of the sediment and mixes with traces of oxygen in the water, a niche is created that attracts the filaments of hydrogen sulfide-oxidizing bacteria. These giants in the world of microbes can create thick yellowish-white lawns covering the seafloor: the color stems from elemental sulfur, an oxidation product of hydrogen sulfide that is stored in the filaments. These sulfide-oxidizing microbial lawns initiate the restoration of the sediment as a residential area for bottom-living animals. As a "hot spot" of organic matter recycling, an ecologically sound seafloor is virtually perforated with the aerated burrows. Their biofilm-coated walls house the microbe-driven machinery of "biochemical relais stations" as required for organic matter recycling.

"Marine Snow"

Anyone familiar with biological sewage treatment plants can imagine that marine sediments are not the only sites of organic matter recycling in the sea. The more natural counterpart of activated sludge flocs in seawater is called "marine snow." The high salt content of seawater facilitates adhesion and attachment of microbial cells to interfaces. Abundant attachment sites on freely suspended particles lead to the formation of aggregates of partially decaying plankton forming marine snow. These marine snow aggregates (of living cells and decaying plankton debris) support the formation of planktonic biofilms containing a battery of immobilized enzymes with recycling functions. Representing a conspicuous biochemical factory in the water column, marine snow is responsible for short-circuited organic matter recycling. This kind of "on-site-recycling" stands in contrast to "outsourcing strategies" that are characterized by sedimentation of dying plankton cells to the sea floor. Predominance of the latter in traditional ecosystem models has now been challenged. The role of microbial biofilms in sustaining ecosystem functions is gradually emerging, as previously "invisible" or neglected pieces of Nature's recycling machinery receive more attention.

A walk over slippery intertidal rocks can always remind us of the real existence of biofilms, integral parts of Nature's "bioengineering" capacity that account for the self-purification of coastal marine waters. The abundance of biofilms in marine environments also complicates monitoring of environmental health. Anyone who tries to detect certain microbes in the sea must be aware that microbial biofilms serve as protected refuge and survival sites in microbial life cycles. Analyses of water samples would even become useless, if the target organisms to be monitored reside in biofilms at interfaces and suspended materials.

Can Biofilms Be Controlled?

In a multitude of man-made constructs other than waste water treatment devices, however, biofilm formation is viewed as undesirable. Biofilms in general cause adverse effects on a variety of liquid-exposed objects, from ocean oil drilling gear to medical catheters. Hence, many new insights into structures and functions of biofilms stem from researchers who are exploring ways and means to prevent them, including the medical profession.

A few years ago, it was discovered that certain seaweeds produce chemical compounds of the furanone family. As signal analogues, these molecules block the signal compounds required in gene expression to initiate the bacterial colonization of surfaces. This signaling system is based on measuring critical cell densities. This finding aroused interest not only among applied, but also among academic researchers. "Cell to cell communication" in microbial biofilms via chemical signals has stirred speculations about the scope of "supraorganismic" capacities among the smallest forms of life with by far the longest history of evolution. On the applied side, a non-antibiotic compound that merely prevents microbial cells from colonizing surfaces, might be turned into an environment-friendly weapon against biofouling in marine environments.

Most Marine Organisms Wear "Microbial Coats"

Despite increasing organic pollution along our coasts, the short supply of organic nutrients and energy in bulk offshore ocean water continues to limit the growth of freely suspended microorganisms. Hence, microbes that require organic matter for growth take advantage from attaching themselves to interfaces with adsorbed nutrients. Enhanced nutrient levels at metabolically active surfaces select for colonizers with particularly high nutrient demands. This feasting type of bacteria is typically associated with higher marine organisms and is relatively easy to isolate from seawater. So it is not surprising that most of the known (cultivated) marine bacteria are found among the feasting biofilm associates of marine animals and plants.

"Hot spots" of microbial growth and activity in the sea are associated with metabolism (as well as decay) of marine animals and plants. The specific role of most microbial biofilms in the life of higher organisms remains yet to be investigated. A few conspicuous cases of symbiosis highlight the role of rather specific biofilms; for example, the "light organs" of squids are run by symbiotic light-emitting bacteria.

The capacity to decompose chitin, the polysaccharide construction element of crab and copepod shells, is common among bacterial colonizers of these animals. While healthy chitin-armored animals are hardly affected, stressed and weakened specimens may invite chitin-degrading colonizers to intrude through lesions in their outer "skin." At this point, the primarily harmless colonizer has turned into an opportunist pathogen. Extraordinary stress in intensive shrimp mariculture can render cultivated shrimp vulnerable to attacks by such opportunistic pathogens. Probably all pathogenic bacteria of shellfish and fish are opportunistic scavengers and surface colonizers.

In principle, however, bacterial lawns covering marine animals need not pose a particular health risk for their host. On the contrary, several detritus-feeding animals inhabiting marine sediments seem to depend on their "bacterial gardens" as a source of protein-rich diet. Other sea floor dwellers owe their survival in toxic hydrogen sulfide-rich sediments to "fur coats" composed of a sulfide-oxidizing bacterium.

Thus, aquatic ecosystems and especially the "invisible sea" rely on microbial biofilms. These are instrumental for sustaining indispensable ecosystem functions, such as organic matter recycling and detoxification. Some of these natural designs have already been copied and exploited for man-made technologies, like in biological waste water treatment. On the other hand, numerous man-made constructs suffer from economically undesirable, material-deteriorating effects of biofilms. Support and development of biotechnologically desirable biofilms is counterbalanced by prevention of undesirable biofilm formation on man-made constructs, including those in medicine. Therefore, microbial biofilms constitute a unique case in which support of research can be expected from two opposing ends: From the conservationists and copiers of Nature's design to preventionist protectors of man-made constructions. While "biofilm" has become a popular buzz word in today's world of microbiology, it is already an "old acquaintance" for marine scientists. Nevertheless, to those concerned about environmental health and marine resources management, microbial strategies of biofilm formation will continue to teach a lesson.

Printed in two parts on September 2 and September 9, 2004

Rhodora Azanza

"Red Tides," "Brown Tides," "Green Tides," and Poisonings

When the "sea turns to red," the uninformed public expectedly becomes panicky and very stressed. This discoloration of the seawater is not necessarily linked to the rising or receding tide. It is a phenomenon which is not always red but can be brown, green, or any "discoloration," depending on the kind of organism which is in bloom. Microscopic organisms in the water called phytoplankton or microalgae can increase their density and reach a bloom stage, like being present at a particular time and area at about one million individuals per liter of the water.

The last decade of the last millennium ushered in the rise of "red tides" globally. The public, however, need not panic when the blue waters turn to red, green, brown, or orange. When there is microalgal bloom in an aquatic system, they serve as increased food supply to the next group of organisms in the food chain, i.e., the zooplankton, which then positively impacts the next group in the chain and therefore effects increased marine productivity. The danger, however, is in the type of organism or species that could sometimes bloom. There are about 5,000 marine microalgal species in the world and only about 200 of them could cause harm. There are less than a hundred causative organisms that could cause poisonings. The route of poisoning is usually through the consumption of shellfish, e.g., green mussels (*Perna viridis*), oysters (*Crassostrea spp.*) or finfish, collected or harvested in the area where the bloom occurred. The shellfish or finfish are vectors of the poison because they accumulate the microalgae and their toxins are passed to the consumers (like human beings and other mammals). In other countries, marine mammals like manatees have been poisoned by these toxins.

"Harmful algal blooms" (HABs) is a generic term which refers to all algal blooms that can cause harm. A wide range of organisms is involved, including dinoflagellates, other flagellates, cyanobacteria, diatoms, and other phytoplankton. The term HAB is not always synonymous to the term "red tide" because not all harmful algal bloom events can be considered as "red tides" which refer to the discoloration of water due to the significant accumulation of biomass by the microalgae that has proliferated. However, not all harmful algal events involve the development of a significant accumulation of biomass. Many HAB species are harmful at very low densities and therefore do not form discoloration or red tides but cause harm by producing toxins that can be transferred to the food chain.

The Marine Science Institute, University of the Philippines, Diliman, Quezon City, 1101, Philippines.
rhod@upmsi.ph or rhodaazanza@yahoo.com

There is cause for alarm when it comes to HABs. It has been surveyed that HABs have been increasing in occurrence and impacts globally. Aside from causing a growing number of seafood poisoning cases and deaths, the accompanying economic loss to people dependent on the coastal ecosystem as their only means of livelihood is a cause of concern. The cause of its spread has not been completely elucidated. One of the factors scientists are looking into is a part of the life cycle of the microalgae which is the "cyst." Cysts can serve as seeds for future blooms by germinating when conditions become favorable. These cysts can be transported via currents and ships' ballast water from one part of the world to another where they can invade local ecosystems, bloom, and eventually cause harm. It is no wonder therefore that a global convention has been held on the management of ship ballast water.

Poisonings from harmful algal blooms result in negative effects on the gastro-intestinal system such as diarrhea, nausea, and vomiting; and the nervous system, such as paralysis, and loss of memory. In extreme cases, they may even result in the death of the victim. These poisonings can be classified as Paralytic Shellfish Poisoning (PSP), Amnesic Shellfish Poisoning (ASP), Diarrhetic Shellfish Poisoning (DSP), Neurotoxic Shellfish Poisoning (NSP) and Ciguatera Fish Poisoning (CFP) and may only happen when the specific causative organisms bloom and the shellfish/finfish that accumulated them are consumed.

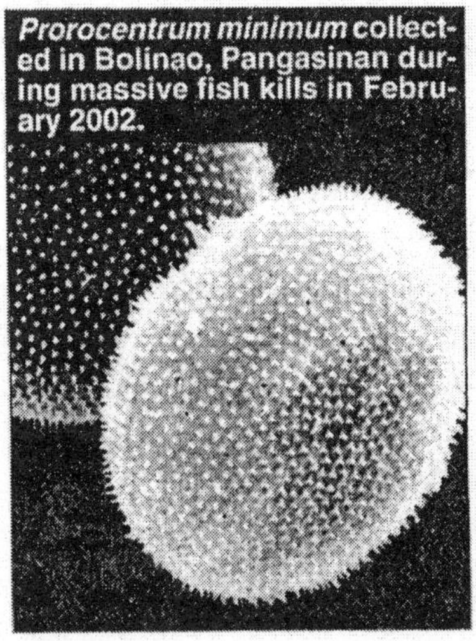

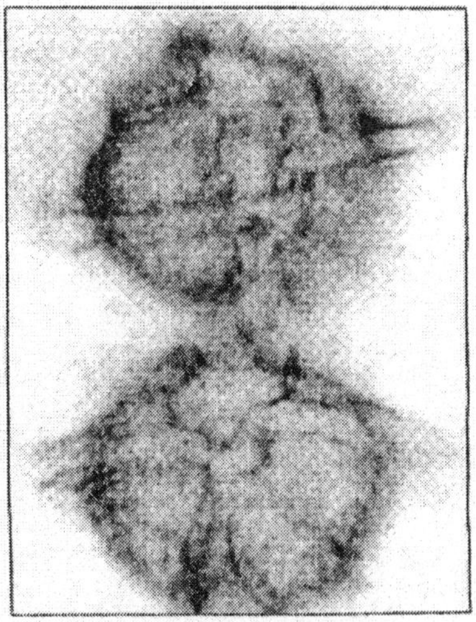

Pyrodinium bahamense var. compressum (right), filtered by molluscs such as *tahong* and subsequently transferred to consumers, causing Paralytic Shellfish Poisoning (PSP). Over 2,000 PSP cases and 200 deaths in the country have been associated with the organism since 1983.

Manila Bay and some other bodies of waters in the Philippines and the rest of the world have experienced both harmless and harmful red tides. As early as 1905, a priest reported the occurence of "red tide in Manila Bay" which was apparently harmless because there was no report of accompanying poisoning. In 1983, in Samar and Leyte, waters were hit by a harmful algal bloom event that caused Paralytic Shellfish Poisoning to hundreds and death to twenty-three people. From then on, to most Filipinos, the term "red tide" has been synonymous to toxicity and economic disaster. An uninformed public does not want to buy and consume marine food products, including those not affected by the phenomenon. This ripple effect can generate an estimated US$300,000 loss of income from fish/fishery produce. Public education is part of the management scheme that can alleviate or mitigate economic disasters and prevent poisoning due to harmful algal blooms. Research and development is ongoing at the regional, national, and international levels to enhance management of HABs.

Some water discolorations may be caused by the bloom of phytoplankton which are harmful only to fish, thereby causing "fish kills." *Chattonella marina* and *Prorocentrum minimum* were found associated with the fish kills in Bolinao in 2004 and 2002, respectively. In the recent water discoloration in Morong, Bataan where a "fish kill" was also reported, *Noctiluca scintillans* was in bloom.

For more information on harmful algal blooms, visit our website *http://portal.unesco.org/habsea*. The HABSEA portal is a gateway to HAB information in Asia initially emanating from the Southeast Asian countries. It offers information on the varied interpretation of harmful algal blooms, a scientific background on causative organisms, their life cycles, interaction with the environment and degree of occurrence, management strategies instituted, and varied source learning to enhance understanding of the phenomenon.

August 12, 2004

A Fighting Coral Fights for Life in Its Hometown, Puerto Galera

WILFREDO LICUANAN

PUERTO GALERA is synonymous with a lot of things (beaches, beauties, and bars are some words that come to mind). For some reef scientists, however, Puerto Galera is also known for the coral *Anacropora puertogalerae*, an exquisite and fragile animal, first discovered there and uncommon elsewhere in the country.

Yes, corals are animals, with each colony made up of tens to thousands of clones called polyps. In the movie *Finding Nemo*, Nemo and his dad lived in a giant polyp of an anemone, a close relative of corals. And like anemones, polyps sting to defend themselves or capture food. Nemo would have been stung if he didn't regularly brush against the anemones' tentacles to build up his immunity (remember a scene in the movie where Marlen was teaching his son to do this as part of his morning routine before leaving the anemone?—now you know why).

A close-up of a healthy colony of *Anacropora puertogalerae*

Biology Department, De La Salle University, Manila, Philippines.
licuananw@dlsu.edu.ph

A discarded carbouy lies among *Anacropora puertogalerae*

Anyway, like a lot of animals living at the bottom of the sea, corals need space. This is because space means more area to feed in (they gather food that drifts by with the currents), and gather light (for the tiny plants that they farm in their own bodies), among others. And so corals fight, sometimes to the death, to get or defend space. Most people don't notice this because often the fighting occurs at night. But at daytime, evidence of this fighting can be seen all over the reef as tissue damage among the combatants, especially in those who lost or are losing this slow-moving battle. To wage this battle, corals and their kin use a variety of weapons. Most use arms (actually tentacles) laden with special stinging cells than can inject venom. Others rely on chemical warfare, releasing toxic chemicals to the water to harm those that are downstream.

Anacropora uses a unique strategy. It surrounds and besieges its adversaries and starves them to death! Because of its twig-like appearance, large fragments of the coral do break off from a mother colony when it matures. Hence, the offspring is found next to its parents and, given enough time, it forms patches five to ten meters across. When patches of this coral grow too big, the ones in the middle start dying, probably because the ones that surround them have picked up or filtered off the food from the water before it gets to them. This effect is most obvious when *Anacropora* patches, which are usually taller, reach and surround those of other corals. You will see the polyps of the besieged coral die slowly, starting from the base (where the water is most stagnant and presumably stripped of food) upwards towards the branch tips.

I recently got a chance to go back to Puerto Galera to gather data to help the municipal government and the people, with assistance from a conservation NGO, and to set up a coastal conservation and resource management program. As the ferry entered the bay, I was heartened to see that human activities and resort developments appeared not to have changed the landscape much over the ten years since I had been to the area. As part of my dissertation research, I used to come here every two weeks for eighteen months, making three to four dives a day for five days each time, rain or shine. This is how I learned these things I am writing here.

Unfortunately during my return, the landscape masked a deeper (relatively and figuratively) and more troubling development. *Anacropora puertogalerae*, the coral species first discovered here by the late Prof. Francisco Nemenzo in 1964, is disappearing from its hometown ("type locality" in scientific parlance). Many of the large patches of corals on the First and Second "Plateaus" (actually reefs in Muelle Bay) at 10- to 15-meter depth are now just rubble.

What may have happened to them? Like many things in the environment, the story will involve a number of interrelated factors. The quality of the water may have changed. Most likely, it would involve the roads cut through to *Sabang* and the sediments carried from them by rain runoff. It would also involve the sewage from the town proper. Colleagues have already documented how, at certain times of the year, the sewage leads to blooms of an algae that smother the corals. Damage from anchors and fishing nets have been noted before, and actually incorporated in computer models that I have made. *Anacropora* is so fragile; you can break branch tips by just swimming over them and not touching them!

So what can we do about the loss of this coral? Again, like many things about the environment, the solution will require attacking the complex problems from several directions. We (Puerto Galera residents, resort owners, boat operators, government officials, visitors, and other stakeholders) have to agree on a clear and transparent course of action. At some point, a management plan will have to be agreed on (this is different from the plan of attack). Whatever we do will require concerted, well-focused, and directed efforts involving several sectors. Ideally, a native of Puerto Galera should lead the effort to protect one of his own.

November 3, 2005

Helen Yap

When the Coral Reefs Turn White

"El Niño" has now become a household word. It is associated with a dry, hot climate. Farmers fear it because it means drought and disaster for water-demanding crops such as rice. Even housewives know the El Niño. They blame it for water shortages and power interruptions. (Paradoxically, even while this is happening in the Philippines, some other parts of the world, like the United States, may be experiencing unusual torrential rains!) The El Niño is a climatic, nearly global phenomenon, that occurs regularly. Unfortunately for us, its frequency of occurrence has been increasing (why and how should be the subject of an article on climate change).

What very few people in this country know is that the effects of El Niño can also be profound and dramatic in the ocean, that is, beneath the surface of the water where we do not see them. I am one of a team of professors who teach an undergraduate course at UP Diliman called "Introduction to the Marine Sciences." What really astounded me is the fact that not a single one of my students had heard of the catastrophic El Niño of 1998, when many coral reefs in all oceans of the world bleached, meaning, they turned white. But then again, why would I expect them to have heard of it? I try hard to recall the media coverage of the event at that time. Now that I think hard about it, there seemed to be almost none at all. On the other hand, we marine scientists were running around frantically, sending teams out to the field, and communicating with our colleagues abroad because of the sheer magnitude and seriousness of the event. After all, the Philippines has very extensive and rich coral reefs (this, at least, should be public knowledge by now). Coral reefs are extremely important natural resources in the country, generating much of the fish we eat. In 1998, they were under severe threat. Perhaps it is partly our fault that we failed to properly inform the public about this through the media, maybe because the alarm bells were ringing so loudly in our ears that we could not think of anything else but the scientific tasks immediately on hand. But we were on top of the event, if you could call it that. Our oceanographers were continuously monitoring a mass of heated water that was slowly moving southward from the northwestern part of the Philippines, in the South China Sea. This was made possible because of close collaboration with scientists abroad who had real-time access to satellite images of the ocean. As the hot water mass progressed southward, the reefs along the way were bleaching almost in a systematic fashion.

But let me move back in time a bit, to the beginnings of the 1998 El Niño episode. Since the El Niño, as already mentioned, is a global climatic event, it takes time to evolve

The Marine Science Institute, University of the Philippines, Diliman, Quezon City, 1101, Philippines.
hty@upmsi.ph

and develop. With present day knowledge and technology, it is now possible for scientists to predict, with accuracy, whether an El Niño will occur the following year. This exercise is usually a form of cooperation between scientists studying the atmosphere and those involved with the ocean. A year before an El Niño event occurs, the central part of the Pacific Ocean starts warming up. This process is accompanied by unusually large swings in atmospheric pressure over parts of the central and southern Pacific Ocean. (This is why the more complete and correct term is "El Niño-Southern Oscillation" or ENSO). These features are the readily detectable and measurable ones, but are directly related to more distant phenomena such as the strength of the trade winds.

In the autumn of 1997, climate and ocean scientists noticed that the central Pacific was getting really warm. In other words, not only was there a coming El Niño, but this one was also going to be a whopper. Much later, after the event had passed, scientists reported that the positive temperature anomaly (to use scientific jargon) was an incredible five degrees Centigrade. For the surface of the ocean, in this case, the vast Pacific, to warm up that much must really take a huge, huge amount of heat. It is for this reason that the 1998 El Niño is known to be the biggest ever in living human memory and documented history.

It so happened that in December of 1997, I took my parents for a visit to the Bolinao Marine Laboratory, our field station. One of the highlights of that trip was an excursion to the ocean nursery of the giant clam project run by Professor Ed Gomez. My parents remarked how big and beautiful the clams were, so clearly visible in the shallow water. But more importantly, they said, how peaceful their world was. For my part, on the other hand, my thoughts started racing: clear water, shallow ... my God, if and when the El Niño does strike, these poor creatures would suffer incredibly.

Why? After a roundabout route, I come to the heart of the bleaching story. Giant clams, like the hard corals that build our reefs, have a special relationship with tiny plant-like creatures called dinoflagellates (this is what we refer to, in technical terms, as a symbiosis). Hard corals, like the clams, are animals, even if some of them look like they harbor flowers, forming fantastic, colorful gardens underwater. The dinoflagellates, which are one-celled algae, live inside the cells of clams and corals. They contribute to the vivid colors that many coral and clam species are known for. More importantly, they are an in-house source of food for the host animals. Through a process known as photosynthesis, they are able to convert raw materials such as water and carbon dioxide, with the help of energy from the sun, into simple sugars, fats, and proteins.

When conditions in the surrounding environment, particularly levels of temperature and light, rise above normal, the animals become stressed, and tend to expel their resident, symbiotic algae into the surrounding water! The obvious results are: the animals (the corals and clams) lose their color and become white, and they proceed to starve because they lose an important food source.

The first real alarm that sounded the fact that the El Niño had finally arrived in the Philippines was a report by a graduate student that dead coral reef fish were seen floating on the surface of waters of a reef a few kilometers from the marine station. This was around June of 1998. Measurements were immediately taken of the temperature of the water in the area, along with the dissolved oxygen (DO) content of the water. Sure enough, temperature was higher than normal, and DO was low to actually be stressful for animals that need to breathe oxygen. In the following days, more and more dead fish started to float up to the surface, even the ones that normally would be hiding among the rocks and crevices of coral reefs. They had died from either heat stress, or suffocation, or both.

But the most dramatic scenes were yet to come. With warning of a coming El Niño, the station staff had started an evacuation effort for the giant clams, to relocate as many of these wonderful creatures as possible to deeper waters. But each mature clam can weigh more than a hundred kilos! Hence, the process was slow. I swam out with a team once during this rescue effort, and beheld a sight I had never seen before. The remaining clams in shallow water had turned a stark *white*. A visiting scientist from the US called them "boiled clams," but they were still alive.

During that summer, I snorkeled and dived around the familiar coral reef flats and slopes in a daze of disbelief. Whole colonies, entire expanses of reef, had turned white. Corals on the reef flat measuring a meter or more in diameter looked like white-washed tombs, so that the whole area appeared like a cemetery. The deeper slopes covered with bleached coral resembled white, snow-covered Alpine slopes.

After a terrific environmental stress like that, marine creatures are hardly expected to recover. Sure enough, a lot of the clams and corals died afterwards. After a year, the sublime underwater landscapes that used to be sculptured by coral were covered by an amorphous mass of algae. The algae had used the dead coral as substrate on which to gain a foothold and grow. The colorful fish that used to shelter and feed around the coral were replaced by those that eat algae, the herbivores. This scene was repeated in many parts of the world, in the Pacific, Indian, and Atlantic oceans. Many years later, the recovery of affected reefs by means of recruitment of new coral is patchy.

One of the questions a layman would surely pose is: As a scientist, if you know that an El Niño is coming, what should you do? What a difficult question! How does one prepare for a natural event, the effects of which cover an entire country (and many parts of the world)? How does one minimize the negative impacts? These considerations are now part of ongoing research programs undertaken by the international scientific community. But one lesson is clear: reefs that are already abused because of other reasons, such as pollution or overfishing, are more vulnerable to an additional stress such as warming, than are healthy reefs.

Perhaps the best form of action the scientists can undertake, hand in hand with the public, is to protect our remaining reefs from environmental abuse, and keep them as healthy as possible.

August 19, 2004

Maria Lourdes San Diego-McGlone

Nutrients in Seawater—Good and Bad

My daughter has a problem with her aquarium. She tends to overfeed her fish which causes algae to grow, and if not cleaned, will cause the fish to die. It seems Nature likes a certain state of balance, and when this balance is disturbed, Nature will adjust to a new balance. From a human perception, sometimes these adjustments are for the better, or quite often as in the case of excessive food (nutrients) in my daughter's aquarium, for the worse.

Nutrients are defined as substances needed for growth. They are, of course, present in food and from supplementary sources such as vitamins for animals, including humans. They are also found in soil and fertilizers for plants. Further, they may be present in organic wastes (such as sewage) and inorganic wastes (such as fertilizer runoff) that result from human production and consumption activities. In the marine environment, nitrogen and phosphorus are the "traditional" nutrients of interest. Typically in the marine environment, the limiting nutrient is nitrogen. Nutrients find their way to the sea from land via river discharge and nonpoint source inputs such as runoff from agriculture and unsewered areas. They can also be introduced through atmospheric deposition, as in the case of nitrogen.

Nutrients are good. Being food for plants and animals, nutrients are essential for life. Nutrients feed the phytoplankton (tiny plants) of the sea, the phytoplankton feed the fish, and the fish feed man. Intuitively, more nutrients should be great for the food chain. But then again, too much of a good thing could be bad. Here's why.

Nutrients and human activities are linked. The continued increase in human population brings in attendant problems associated with human activities. With dwindling space for waste disposal in land, the sea has become a convenient repository. For example, domestic sewage, mostly untreated, often ends up in coastal waters such as Manila Bay. The organic matter in sewage decomposes and releases the inorganic constituents, among which are nitrogen and phosphorus, which now become available as food for the phytoplankton. Plenty of nutrients translate to more phytoplankton produced, often resulting in bloom conditions. The occurrence and die-off of blooms are actually part of the cycle of life in the sea. Increased frequency of blooms, however, quite often larger in magnitude, disrupts the balance and affects the cycle. Furthermore, the

The Marine Science Institute, University of the Philippines, Diliman, Quezon City, 1101, Philippines.
mcglonem@upmsi.ph

presence and abundance of toxin-forming phytoplankton, such as in red tide blooms, is a problem with a negative impact on the economy and health of the populace.

Nutrients and fish kills are linked. When the phytoplankton bloom dies, decomposition sets in, requiring oxygen for the process to proceed. A large amount of organic material to decompose would require substantial amounts of oxygen, thus compromising the supply of oxygen in the water column. Fish need oxygen to live, and with low levels or no oxygen, they die. Massive fish kills have been associated with depleted oxygen in the water column. A case in point is the fish kill of February 2002 in Bolinao, Pangasinan that caused major losses in bangus production amounting to approximately ₱500 million. This fish kill event had been preceded by the occurrence of a bloom of phytoplankton *Prorocentrum minimum*, toxic to shellfish, and possibly to fish and man. Low oxygen, and quite likely, the phytoplankton bloom, may be attributed principally to the proliferation of fishcage and fishpen structures far exceeding the carrying capacity of the area. Furthermore, some of these structures were sited in places not designated as appropriate, for instance, in poorly flushed locations.

More than half of the protein requirement of Filipinos comes from fish. The demand for fish has been increasing but the ability of the capture fishery to meet this demand has been pushed to the limit as catch per unit effort has shown drastic declines in many fishing grounds. Fish farming through aquaculture (fish ponds) and mariculture (fishpens and fishcages) will need to play an increasingly significant role in fish production. A recent trend has been toward expansion of mariculture, particularly fishcages, in our coastal waters. This fast development, however, has lacked parallel investigation into the environmental impact of the activity. Intensive mariculture releases large amounts of nutrients to the environment in the form of wasted feed and fecal material. These wastes have the potential to enrich the sea when they are located in semi-enclosed coastal basins that have restricted exchange with the more open coastal waters. Anticipated problems associated with the discharge of these wastes include algal blooms, oxygen depletion, and fish kills.

Isn't this a macro view of what happens in my daughter's aquarium?

August 26, 2004

Rhodora Azanza

Space Technology and Applications in the Philippines and Other Developing Countries

Definitely, even developing or Third World countries have been using space technology and its applications for many years now. In fact, the Philippines has been utilizing them to some extent in vital activities like weather forecasting, environmental planning and management, and geological and marine research and development. Data and information from remote sensing technology have been applied to enhance health, food, education, and other sectors of our society. The Philippines, however, has to fully utilize or try all the other possibilities in the use of space technology because we have not benefited much from the wealth of data and information that can be generated from the use of space systems.

A few weeks ago (September 13 to 16, 2005), I participated in the United Nations/Austria/European Space Agency Symposium on "Space Systems: Protecting and Restoring Water Resources" in Graz, Austria which was attended mostly by participants from developing countries. The conference served to recognize how space technology can provide tools necessary to collect various types of real-time data and information that planners and decision-makers in charge of management of water resources can use in understanding water systems and managing available and potential water resources effectively. With the use of remote sensing and telecommunications capacities, critical data and information on water resources can be accessed, analyzed, and disseminated in a timely manner. Pilot projects on the use of space technology in water resource management can be mostly operational and sustainable. Further, information can be customized to be made understandable to decision-makers and end-users involved in water resource management.

Some activities where space systems have been beneficial to developing countries like the Honduras, Thailand, Vietnam, India, and South African nations in protecting and restoring water resources are as follows:

- Protecting and restoring water resources in the Americas, i.e., areas affected by landslides by the Ministry of Agriculture, Honduras;
- Remote sensing and Geographic Information Systems (GIS) integration in the model for river basin environment management in northern Vietnam and also in Africa;

The Marine Science Institute, University of the Philippines, Diliman, Quezon City, 1101, Philippines.
rhod@upmsi.ph or rhodaazanza@yahoo.com

- Low-cost remote sensing data for water resource management in Thailand;
- Developing projects that use space technologies for provision of safe drinking water in India with focus on tsunami-affected areas; and
- Space capabilities for improving water sanitation in India.

Here in the Philippines, the Department of Science and Technology (DOST), through the Philippine Council for Advanced Science and Technology Research and Development (PCASTRD), is organizing the First National Congress on Space Technology Applications and Research (NC-STAR) to be held on November 15, 2005, at the Philippine Institute of Volcanology and Seismology. The national congress will draw up an action agenda and propose flagship projects that will address specific problems using space technology applications and research. Specific concerns to be addressed include natural resources accounting, environmental management, disaster monitoring, poverty alleviation, and sustainable development planning. Pre-congress workshops were held in Luzon, the Visayas, and Mindanao that covered topics on satellite communication applications, meteorological satellite applications, natural hazard monitoring, disaster management using satellite-based information and communication technologies, and special education development.

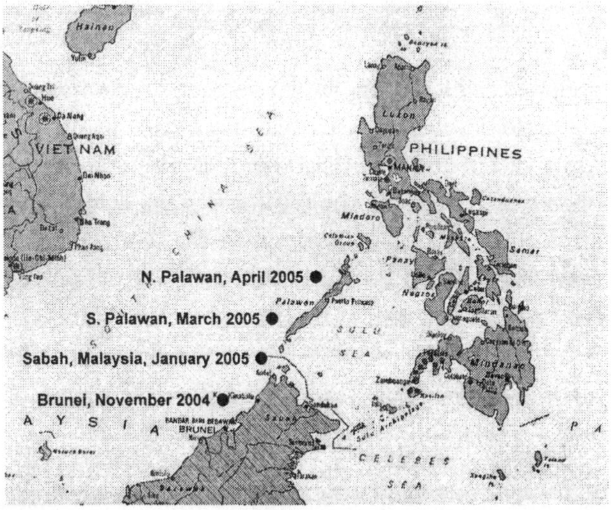

Map of the Philippines indicating red tide events along the east coast of South China Sea from November 2004 to April 2005 involving coastlines of three countries: Brunei, Malaysia, and the Philippines

As a marine scientist, I will show how space technology is being used for research and development in the coastal areas and oceans. Just like in the land habitat, it can be used for coastal utilization and management and also for disaster management.

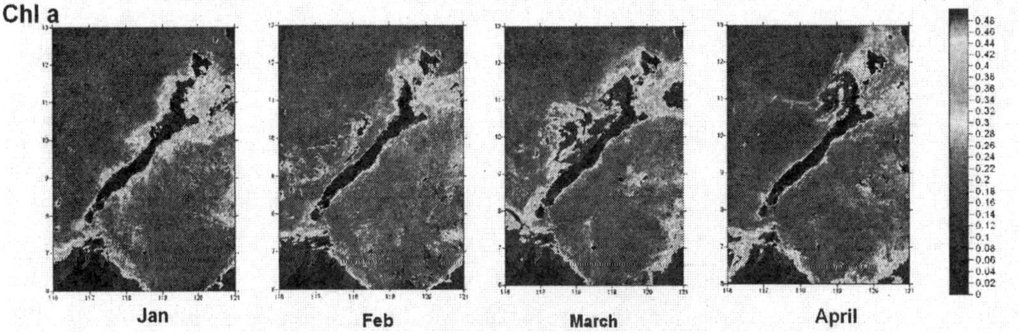

MODIS satellite images from January to April 2005 showing the chl a/algal bloom in the coastlines of three countries (Source: Azanza RV, David LT & Borja R, 2005).

Take, for example, the latest case of "red tide" along the west coast of Palawan. In April 2005, local fishermen were alarmed when they witnessed fish kills along the southwest coast of Palawan coinciding with streaks of red patches in their waters. Water analysis showed dense concentration of *Cochlodinium polykrikoides* (500 to 11,000 cells per liter), a naked dinoflagellate species that has been known to reach bloom levels and cause fish kills in Korea and Japan. The dense bloom subsequently moved from southwest to northwest of Palawan until May 2005. Satellite images from Moderate resolution Imaging Spectroradiometer (MODIS) Aqua Level 2 data (one-kilometer resolution) were obtained for Palawan and its surrounding waters from January to April 2005 and processed. Mosaics were constructed out of the available images per month to represent the monthly distribution of chlorophyll-a. The composite image for February shows patches of water with high chlorophyll content ($\geq$0.5 mg/m3) offshore of west and southwest Palawan. By March, a large portion of the western coast, from the northern tip of the mainland down to Eran Bay, showed high chlorophyll values. At its peak, the bloom was shown to extend 310 kms in length and 80 kms in width. In April, the bloom appeared to be less intense and these waters appeared to have migrated northward, occurring largely off the northern tip of mainland Palawan down to Imuruan Bay. There was a corresponding waxing and waning of chlorophyll values off the western coast of Sabah, Malaysia as well, evidence that this particular productivity bloom was not limited to the coast of Palawan. It could have originated from water masses coming from the rest of South China Sea following ocean circulation patterns. Warning systems that are in place to alert adjacent countries of this harmful event could be enhanced by using space technology.

November 10, 2005

Josefino Comiso

Arctic Sea Ice and Greenhouse Warming

In a *Philippine Star* article published last year, I discussed my research on global warming signals as detected by satellite sensors in the polar regions. I mentioned the rapid decline of an important component of the Arctic sea ice cover called the perennial ice which consists mainly of the thick multiyear ice floes (averaging about three meters in thickness) that determine the thickness distribution and physical characteristics of the ice cover. The multiyear ice floes are in fact the very entities that have kept ships from moving freely in the Arctic and made it impossible to have a Northwest or Northeast Passage that would have greatly shortened the routes between East and West. The perennial ice cover in the Arctic is now about 27 percent less than what it was when we started monitoring the region continuously with satellite data twenty-seven years ago. The shortcut from East to West (e.g., Japan to New York or Europe) is obviously becoming a reality and even during the jet-age era, it could still be an important development for big cargo ships. While the perennial ice is providing a strong warming signal, there has been a puzzling inconsistency in that the winter ice maximum extent had been relatively stable and declining at only a moderate rate of 1.5 to 2 percent per decade. Modelers have projected that warming effects due to increases in atmospheric greenhouse gases would be most pronounced in winter months since during these months, the Arctic is basically in darkness and long wave radiation (~11 microns) dominates. It is this long wave radiation that is emitted from the Earth's surface, absorbed by greenhouse gases in the atmosphere, and re-emitted back to Earth that causes what we now call greenhouse warming. The greenhouse gases serve as a blanket that keeps this long wave radiation from being lost to outer space.

The new Arctic study that appeared in the journal *Geophysical Research Letters* last month provides insights into what might be an answer to this puzzle. The results from this study suggest that the greenhouse phenomenon is actually becoming apparent in the Arctic during winter. It has been late, for reasons that we don't completely understand yet, but a winter warming signal is finally being observed. In 2005, the maximum winter ice extent was a record low and was about 6 percent less than the average value. In 2006 the winter ice cover was even lower in extent than that of 2005. During the twenty-seven-year period of satellite coverage, the occurrence of record low values for two successive years has not happened until the last two years.

NASA Goddard Space Flight Center, USA.
Josefino.C.Comiso@nasa.gov

The impacts of wind and surface temperature on the winter sea ice cover were studied to gain an understanding of processes behind this phenomenon. As expected, the ice cover was anomalously low in the seasonal regions during the winter period. The effect of wind forcing was apparent as the ice retreated in areas where strong winds carrying warm air temperatures from the south hit the ice edge. But wind direction and strength varied from one month to the next and distinctly abnormal wind patterns associated with 2005 and 2006 were not apparent. Studies of surface temperatures inferred from satellite infrared data, which represented a measurement of long wave radiation from the surface, indicated abnormally warm temperatures in the Arctic basin during the winter periods of 2005 and 2006. The most abnormal month was January 2006 when the average temperature in the entire Arctic basin was substantially warmer than usual. This coincided with considerably reduced ice cover in the eastern region making the perennial ice less protected and even more vulnerable.

A big concern is that there are many physical reasons that would lead us to believe that this winter phenomenon will continue. Among the key reasons is that the length of the melt season has been increasing by about fifteen days per decade which means that the length of the ice growth period in the seasonal ice regions is shrinking as well. In other words, ice freeze-up occurs at a later period and ice melt occurs earlier than usual. This implies lower total ice production and less time for the ice to advance to the south. There is also a positive trend of sea surface temperature (SST) in the northern hemisphere. The region greater than 60°N latitude has been warming at about 0.15°C (0.22 °F) per decade and this also inhibits the advance of sea ice during the winter period. Furthermore, there is the so-called ice-albedo feedback effect. What this means is that as the ice cover declines, the surface gets warmer because open water absorbs solar heat much more efficiently than ice since ice has a much higher reflectivity than water. This in turn means a warmer SST as the –2°C isotherm (i.e., freezing point for sea ice) goes further north and therefore causes more vulnerability for the sea ice cover. A warmer SST and a decreasing ice cover may also impact the global ocean circulation, the consequence of which would be profound since the ocean is such a big entity in the global heat-energy balance.

There are, of course, lots of complications, including the presence of oscillatory patterns in the atmosphere such as the so-called "Arctic Oscillation (AO)," "the North Atlantic Oscillation (NAO)," and the "El Niño Southern Oscillation (ENSO)." The AO indices have been shown to be correlated with the volume of the sea ice cover in the Arctic but in the last nine years they have gone neutral while the ice cover has declined almost in a linear fashion. While some scientists interpret this as a change in phase of the AO, others are skeptical and postulate other reasons such as the possible impact of greenhouse warming. I have done correlation analyses of the ice cover and temperature data with indices derived from these phenomena and I get correlation coefficients of only about 0.2 or less, which means that only a few percentages of the variances are associated with AO, NAO or ENSO. Thus, these oscillatory patterns are unlikely the prime factors affecting the observed decline in the sea ice cover.

Combined with the observed rapid decline in the perennial ice cover, the ice decline during the winter maxima constitutes a big warming signal in the Arctic region. A continued reduction of the Arctic winter ice cover would be a good confirmation of the warming effect of increasing greenhouse gases in the atmosphere as projected by modeling studies. It would at least validate our current understanding of the physics of the Arctic climate system incorporated in these models. These models are used to sort out quantitatively the magnitude of the temperature change associated with the increase of carbon dioxide (CO_2) concentration in the atmosphere during the last century. CO_2 is one of the greenhouse gases and is released to the atmosphere during the burning of fossil fuel. The latter is a commodity that is considered one of the driving forces behind the industrial revolution and restrictions on its use have been the subject of contentious debates among environmentalists, industrialists, and politicians. Some inconsistencies in the results from these models have been the target of many climate-change skeptics who question the impact of increases in CO_2. The physics of the climate system is admittedly complex and although the advent of high performance computers and satellite data have led to big advances in the development of these models, validation of these models is urgently needed so the public can have confidence in the accuracy of their predictions, most especially the impact of greenhouse gases on global temperatures and perhaps the future of planet Earth.

If the winter ice retreat continues, there is not enough space in this article to cite the many negative consequences, short-term and long-term, but I will cite one if only to catch your attention. The seasonal ice regions in the Arctic are among the most biologically productive regions in the world. Some of the richest fisheries are found in the region, partly because of colder water that contains more oxygen and more importantly, because of sea ice. Seasonal sea ice provides meltwater in spring that floats because of relatively low density and this layer is known to biologists as the ideal layer for phytoplankton growth because it is stable (does not sink) and there is plenty of sunlight for photosynthesis. Planktons are at the bottom of the food web and if their concentration goes down, animals at all trophic levels would be deprived of a basic source of food. The impacts on marine mammals, including seals, have already been observed and some unexpected negative effects on polar bears, which use sea ice as a platform in their search for food, are being reported. Seafood lovers among us (myself included) would be worried if the availability of planktons in the region gets so depleted that the population of fish and crustaceans goes down considerably.

October 5, 2006

Josefino Comiso

A Disappearing Polar Ice Cap

Only a century ago, explorers turned to the Arctic to find the means to make the Northeast or Northwest passage a reality. At the height of commercial shipping, it would have meant considerable savings in time and money to be able to navigate through the Arctic region instead of the Panama Canal. Many ended up bitterly disappointed and some, like Sir John Franklin and his team members, never made it back because of the ever presence of multiyear ice floes, the average thickness of which is about three meters. While not a big necessity for shipping anymore because of the success of commercial airlines and aircraft transport, recently observed declines in the Arctic sea ice cover are still intriguing not only for shipping enthusiasts but also for environmentalists, biologists, climatologists, businessmen, and politicians.

When I joined NASA in the late 1970s, there was already an awareness that the Earth may be warming as a result of human activities. Measurements in Mauna Loa, Hawaii have indicated that the carbon dioxide concentration in the atmosphere has been steadily going up, and together with other greenhouse gases, they would eventually cause a warming trend on the surface of the Earth. Compared to earlier years, there were already some indications of global warming at that time, but it is difficult to associate the latter with anthropogenic activities because the Earth's climate system is quite complex and it is not straightforward to separate greenhouse warming from natural climate variability. This was taken as a challenge by many institutions worldwide, but accurate and consistent global data are required, and so NASA decided to be involved in the study with a project called "the Mission to Planet Earth."

Our research group at National Aeronautics and Space Administration/Goddard Space Flight Center (NASA-GSFC) concentrated on studies in the polar regions because it was postulated that global warming would be enhanced in these regions because of feedbacks that are associated with the high reflectivity of snow and ice that blanket much of high latitude areas. To assess changes, one needs continuous and well-calibrated data sets from satellites. Although satellite data became available in the 1960s, it was not until the late 1970s that the sensors became mature and stable enough to produce accurate and consistent measurements that can be used for time-series studies. We have two systems that have produced long historical records, both of which are passive systems, namely, the infrared and the microwave systems. By "passive," we mean sensors that measure the intensity of radiation originally emitted from the surface of the Earth within the field-

NASA Goddard Space Flight Center, USA.
Josefino.C.Comiso@nasa.gov

of-view of the instrument. The thermal infrared and passive microwave data are now routinely used to assess spatial and temporal variations in surface temperatures and sea ice concentrations, respectively.

Having accumulated more than two decades of historical data, we could now do serious studies on how the Earth has been changing. With passive microwave data, the actual area of the sea ice cover in the entire Northern Hemisphere has been estimated by many research groups to be declining at the rate of 2 to 3 percent per decade. This is a moderate decline, hardly a big concern because of the large inter-annual variability of the sea ice cover and the unpredictable effects of winds and the oceans. But performing a more thorough examination of the data, I discovered that the area of the perennial ice cover has been declining at a much faster rate of 9 percent per decade. The perennial ice is the ice that survives the melt season during spring and summer and consists primarily of the aforementioned thick multiyear ice floes that resides in the Arctic for as long as seven years.

The rapid retreat of this ice type, which I reported in a scientific journal in 2002, caught the attention of scientists, environmentalists, and politicians around the world because the impact could be profound. The sea ice is part of the ice-albedo feedback that is expected to amplify warming and accelerate the demise of the former. The disappearance of sea ice during the height of solar insolation in the summer would dramatically alter the amount of heat absorbed by the ocean and the atmosphere. It would mean the loss of a habitat for many life forms from microorganisms to mammals and the transformation of the ecosystem of the region. It would mean a much higher biological productivity and an altered vertical distribution for the Arctic Ocean that could lead to a change in global ocean circulation and therefore, of the climate. The data used in the 2002 report was for the years 1978 to 2000. Presently, the perennial ice is still declining, with 2002 showing the least extensive perennial ice observed during the satellite era and with 2003 and 2004 also showing anomalously low values.

(By "solar insolation," I mean the amount of solar radiation incident on the surface of the Earth. With more open water in the Arctic in the summer, there is more solar radiation absorbed by the surface since open water has low reflectivity or albedo. The net effect is a warmer ocean that melts more ice and this, in turn, leads to even more open water. This is what we call the "ice-albedo feedback.")

Is this an indication of Arctic warming? The results of analysis of thermal infrared satellite data indicate that it is likely so, but the distribution in the warming trends is largely off center with respect to the North Pole and the positive trends are concentrated in the Western Arctic and in North America. In other regions, the warming is more moderate with some regions in Russia having even slight negative trends. Our analysis of the data indicates that the surface temperature in the entire Arctic Region has been increasing at the rate of 0.5 °C per decade, which is about eight times higher than that of the last century as measured by meteorological stations. This is partly because about sixty years ago, the Arctic was almost as warm as it is now. However, such previous warming was more localized, while the current one is more global in scope. In North America, the rate of warming is about 0.8 °C per decade. It should also be noted that these results are corroborated by the following signals from other parts of the Arctic: the glaciers are retreating, the ice sheets in Greenland are thinning, the snow cover is declining, the permafrost is thawing, and vegetation in the region is growing.

Are these signals associated with anthropogenic greenhouse warming? This is an issue that has spurred contentious debates among scientists as well as policymakers and environmentalists. The problem is that there is a natural climate variability associated with a changing sun and perturbations in the orbit and spin of the Earth and it is not so straightforward to separate such variability with that of warming from increasing greenhouse gases in the atmosphere. We hope to resolve this issue through modeling studies. In fact, recent studies suggest that the warming of sixty years ago appears to be a part of the natural climate variability, while the current one indicates association with anthropogenic greenhouse warming. The current warming has also been shown to be out of phase with the natural climate variability as inferred from long-term records from ice cores. There are currently more than thirteen global circulation models from various institutions that are used to study the effects of the increasing amount of greenhouse gases and all models predict global warming and an amplification of this warming in the polar regions. However, the results for the rate of warming differ substantially from one model to another and predictions for polar amplification range from two to six times the global warming. The lack of consistency suggests that current models are still not up to the task of making accurate predictions. However, it is encouraging that some of the modeling results are beginning to produce climate patterns similar to those of observations. A promising trend is the assimilation of satellite observations in models and with time, it will become increasingly possible to distinguish cyclical patterns from long-term trends and to separate anthropogenic from natural influences.

November 18, 2004

Laura David and
Cesar Villanoy

A Voyage into the Unknown: The Success and Demise of Ferdinand Magellan

"Don't believe in witnesses, believe instead in what the evidence tells you." This statement sounds like a line from *CSI*, a popular forensic TV series, but this, in fact, is the same philosophy behind every investigative science. Witnesses at best can be biased, whereas fact is fact. Take, for example, the story of Ferdinand Magellan.

The first circumnavigation around the globe was chronicled by Antonio Pigafetta, an Italian nobleman who made detailed observations of the people they encountered. Pigafetta, however, was definitely not a navigator. His log of the first recorded crossing of the Pacific was sketchy at best. From the point of view of navigation, the only really reliable data from his log were the dates of setting sail and anchoring. Magellan's own logbook was burned along with the boat *Concepcion* after his death at Mactan. It is, therefore, not surprising that what people recall of the whole adventure can often be surmised in a single sentence, "Magellan, a Portuguese, set sail under the Spanish flag to circumnavigate the world, but was killed in the Battle of Mactan by the local chieftain, Lapu-lapu." Many fail to realize that Magellan's journey across the Pacific is the greatest navigational feat of humankind, made possible by the natural state of the Pacific, and that forces of nature also played a part in his untimely demise.

Physical Oceanography (PO) is the study of the movement of water and one of the first lessons in PO is that wind, water, and the motion of the earth combine to form oceanic currents whose direction of flow is effectively unwavering. Even fuel-run navigational ships of today make use of this fact to plot their course across oceans to save time and fuel, and much like airplanes, they use jet streams.

Magellan's ships, like most ships of that era, had no choice but to either flow with currents or sail with the winds to their backs. They could not move in slack current or slack winds. It was necessary, therefore, to have favorable winds and current systems for Magellan's ships to have reached the Philippines. The Northern Pacific Basin is like a large washing machine with surface water rotating in a clockwise direction (called a gyre). A similar feature is also found in the South Pacific Basin but it rotates in the opposite direction. Near the equator, these two gyres meet and result in a dominant westward flow across the Pacific, the North, and South Equatorial Current. It is also in

The Marine Science Institute, University of the Philippines, Diliman, Quezon City 1101, Philippines.
ldavid@upmsi.ph; cesarv@upmsi.ph

these latitudes where the tradewinds are strongest. The most parsimonious route for the first documented Pacific crossing is, therefore, along this region. In fact, Samar, the first Philippine island seen by Magellan, lies directly downstream of the North Equatorial current system.

But Magellan would have needed to get to the equatorial region first from the southern tip of South America, the most southern landmass. An interpretation of Pigafetta's chronicles places Magellan's ships to have diagonally crossed the South Pacific Basin, sighting two isles in the South Pacific. Given what is known about the currents and winds over the Pacific at the present time, it is more likely that Magellan emerged from the Strait of Magellan and sailed off the western coast of South America to meet up with winds and currents that run northward—the Southeast Trades and the Peru Current.

That our history became entangled with Magellan's is, therefore, not a matter of chance but rather due to the favorable winds and currents with him all the way to the Philippines.

It is also interesting to note that for the wooden ships of that era, during times of storms or if they experienced wind blowing toward their bows, they were forced to strike their sails, to fold the sails, and sit still until the winds were again blowing in their favor. For Filipinos who experience tropical storms and typhoons almost year-round from the Pacific, we can easily imagine how hazardous and tedious the crossing must have been.

Here again Nature played a favorable hand. Magellan's crossing of the Pacific may not have been possible if ocean and weather conditions were different. Throughout their voyage across the Pacific, which took them around three and a half months, they did not encounter a single storm. Pigafetta notes in his journal that "*In three monethes and xx. dayes, they sayled foure thousande leaques in one goulfe by the sayde sea cauled Pacificum (that is) peaceable, whiche may well bee so cauled forasmuch as in all this tyme hauyng no syght of any lande, they had no misfortune of wynde or any other tempest... So that in fine, if god of his mercy had not gyuen them good wether, it was necessary that in this soo greate a sea they shuld all haue dyed for hunger*" (Fiske, John. *Discovery of America*. Reprint. Boston: Houghton, Mifflin Co., 1892.) In fact, the name Pacific Ocean was given by Magellan because of the calm conditions they encountered. These favorable conditions of the Pacific for sailing west exist mainly during the northeast monsoon in non-El Niño years; hence, had Magellan traversed the Strait of Magellan faster than the almost seven months it took them, storms in the western Pacific might have stopped them from reaching the Philippines.

But inasmuch as Nature played a huge role in Magellan's success, so it did, too, in his demise.

On that one fateful day three hours before daylight, Magellan and his men arrived at Mactan to do battle with Lapu-lapu instigated by a neighboring chieftain, Humabon. Ships with large pieces of artillery and armored men with crossbows and swords were ready to do battle at the break of dawn. Fortuitously, however, the ships could not come near shore for they encountered the extensive reef flat of Mactan, while the men had to abandon their leg armors for they had to wade through the shallow water. Picture full-grown men wearing upper body armors and brandishing swords and crossbows wading 50 to 100 meters to get to shore. It was a no-win situation, all on account of the conflict being instigated during a time when tides (the cyclical rising and lowering of ocean water) were unfavorable to boats rowing over the reef flat. Had the conflict arisen four days earlier, when bigger high tides could have at least allowed Magellan and his men to fight in full battle gear, history might have been different. For it was in his exposed leg that Magellan got shot with a poisoned arrow.

November 25, 2004

Raymond Tan and
Joel Tanchuco

On Decarbonizing Energy

There is a strong consensus in the international scientific community that global climate change is a reality. Russia's ratification of the Kyoto Protocol last year was met with celebration, particularly in the climate change-conscious countries of the European Union, as it meant that the treaty to reduce emissions of greenhouse gases responsible for climate change would take effect in 2005. This landmark event at least gives the world some hope of staving off the slow but almost inevitable effects of climate change. In Southeast Asia, and the Philippines in particular, public awareness of climate change remains wanting and policy directions have thus lagged behind those of the West.

Of all the greenhouse gases, it is carbon dioxide that accounts for the bulk of climate change effects. This gas is emitted in the combustion of fuels that contain carbon—which is to say, most of the major fuels currently in commercial use. Climate change concerns have served to stimulate research and development activities in low-carbon energy sources. Renewable sources such as wind or solar energy are the most well known. These technologies tap into naturally occurring flows and convert them into electricity. Bioenergy—energy derived from vegetable matter—is another low-carbon resource. Bioenergy systems can be designed and operated sustainably so that the carbon dioxide emissions generated from burning the vegetable matter or its derivatives are offset by the carbon dioxide absorbed through photosynthesis during plant growth. Thus, bioenergy is regarded as carbon dioxide (CO_2) neutral from a system-level standpoint.

Many biofuels can be derived from indigenous energy crops without the need for cutting-edge technology. Brazil, for example, has been using sugarcane-derived bioethanol as a blending agent for gasoline for several decades now. Their highly successful *ProAlcool* stands as an example of a robust alternative energy program that has stood the test of time. Critics of bioenergy raise the "food versus fuel" issue, arguing that agricultural land is best used for food production rather than as energy farms. The resolution to this debate is as much a social, political, and economic issue as it is a scientific and technological one. In any case, global trends have once again made alternative energy attractive enough to be seriously considered. An example is the recent effort by the Philippine government to develop bioethanol and biodiesel programs that utilize indigenous crops to produce substitutes for petroleum products. These measures

Raymond Tan: Chemical Engineering Department, De La Salle University-Manila, Philippines.
tanr_a@dlsu.edu.ph
Joel Tanchuco: Economics Department, De La Salle University-Manila, Philippines.
tanchucoj@dlsu.edu.ph

are particularly timely, with recent oil price trends suggesting that gasoline prices may reach ₱40/liter by the end of 2005.

Despite the wide array of low-carbon energy technologies available at commercial or near-commercial levels of development, the bulk of the world's energy is still derived from fossil fuels—natural gas, coal, and (especially for the transportation sector) oil. There is a certain pervasive technological inertia that makes it difficult to dislodge these dominant energy sources; economists refer to this effect as "network externalities." For example, most modern vehicles run on either gasoline or diesel oil. Many countries also have well-developed fuel supply chains to bring these fuels from refineries to depots and eventually to the end-users. The existence of established networks for producing and distributing these goods gives an implicit advantage to the incumbent technologies, since new competitors must penetrate the market without the benefit of such supporting infrastructure. In the case of alternative motor vehicle fuels, for example, the most viable candidates are those that are sufficiently similar to gasoline or diesel to be used interchangeably, or mixed, with these conventional fuels.

Elsewhere in the world, major energy companies are investing heavily in retrofitting existing fossil fuel-fired plants to be more climate-friendly. Some of the concepts being tested are so exotic that they seem to border on science fiction—and yet huge sums of money are being invested in their development. One such technology is carbon sequestration, or, more precisely, carbon capture and storage (CCS). CCS techniques involve removal (capture) of carbon dioxide gas from the exhaust of power plants, and then pumping the gas into some isolated reservoir (storage). Capture technologies include gas scrubbers that use chemical agents to isolate carbon dioxide from power plant stack gases. More innovative techniques include oxyfuel combustion—burning coal, oil or gas in pure oxygen to give an exhaust gas consisting mainly of carbon dioxide and water vapor—and precombustion capture through fuel gasification. Once the carbon dioxide gas is captured, it is compressed and prepared for storage. Candidate storage sites include impermeable underground geological formations—certainly seismic risk assessments are necessary to ensure long-term viability of the storage system. Deep-sea disposal is another option. At the extreme pressures found deep in the world's oceans, carbon dioxide liquefies into a dense, oily liquid that sinks to the ocean floor to form pools that are—in principle, at least—isolated from the atmosphere. As fanciful as these schemes might appear, carbon dioxide storage is now being used commercially, for example, in conjunction with enhanced oil recovery (EOR) in petroleum reservoirs. Some experts estimate that CCS power plants can be in commercial operation before the end of the next decade.

It remains to be seen exactly what will become the dominant energy resource of the twenty-first century. Thermodynamic arguments point to renewable energy as the only viable option in the long run, particularly when we think in terms of centuries rather than decades. On the other hand, fossil energy definitely has the incumbent's advantage of network externalities, and it will be necessary to make do with them for the foreseeable future. CCS may eventually turn out to be a viable and socially acceptable transition technology—or it may not. As for the futuristic concept of the "hydrogen economy," that will certainly bear some detailed discussion in another article.

November 17, 2005

Raymond Tan and
Alvin Culaba

Bioenergy, the Carbon Cycle, and the Photosynthetic Ceiling

In mid-2006, crude oil prices reached record levels of close to $80 per barrel, stimulating renewed interest in alternative fuel research. Although price levels have eased to about $60 lately, it is now widely recognized that the world oil market remains volatile, and that governments of net oil importing nations cannot afford to be complacent about the prospects of continuous supply of cheap energy. There is still considerable controversy as to exactly how much petroleum reserves remain in the world—with estimates ranging from a low of a few decades to a high of over one century. However, these reserves are in any case finite and nonrenewable, and the imminent threat of global climate change adds yet another driving force toward reducing the world's dependence on oil and fossil energy.

These threats have been recognized by the Philippine national government and much effort has been put into developing appropriate measures to reduce the impact of rising oil prices. The transport sector is particularly vulnerable, as it is in the peculiar position of being almost entirely dependent on petroleum products such as gasoline and diesel. In response to these threats, there is now pending legislation intended to introduce liquid fuels derived from plant matter to supplement conventional petroleum products. For gasoline, the fuel identified is ethanol—the same alcohol found in intoxicating beverages—to be blended at levels of up to 10 percent. Ethanol is produced through the fermentation of sugar, which can in turn be derived directly from sugar crops, or from the chemical conversion of starch or cellulose. As a result, a large number of potential indigenous raw materials can be used as potential feedstocks, including sugarcane, molasses, corn, and even fibrous material found in agricultural and municipal wastes. In the case of diesel, the most promising substitutes are chemical derivatives of vegetable oils (derived from traditional crops such as coconut, or dedicated energy crops such as *Jatropha curcas*) known as methyl esters. As a result of the chemical treatment, methyl esters closely resemble the properties of conventional diesel, which allows them to be used in existing engines. This last point clearly illustrates the phenomenon of technological inertia, wherein new fuels not enjoying the advantages of the incumbent energy sources

Alvin Culaba: Center for Engineering and Sustainable Development Research and Mechanical Engineering Department, De La Salle University-Manila, Philippines.
culabaa@dlsu.edu.ph
Raymond Tan: Chemical Engineering Department, De La Salle University-Manila, Philippines.
tanr_a@dlsu.edu.ph

must be made to dovetail neatly with the needs of existing vehicles in order to gain a foothold in the market.

There are some significant advantages to the use of biomass-derived liquid fuels. First of all, they can be derived from locally grown crops without requiring cutting-edge technology. Thus, for countries like the Philippines, they represent a workable means of developing significant indigenous energy sources. Brazil's three decades worth of experience with their *ProAlcool* fuel ethanol program show that properly managed biofuel systems can be sustained over long periods, and thus can provide a good model for the Philippines to study. In addition to enhancing energy security, biofuels are inherently climate-friendly. Carbon dioxide emissions from combustion of biofuels are offset significantly by photosynthetic carbon fixation during the growth of the crops used for fuel production. Thus, the replacement of fossil fuels with biofuels helps mitigate climate change effects of human activities, in effect by embedding the latter within the natural carbon cycle.

Biofuels such as ethanol and methyl esters are, in a very real sense, solar energy condensed into convenient liquid form. It is photosynthesis which allows diffused sunlight to be converted into concentrated chemical energy for use in motor vehicles. Hence, it is important to recognize the inherent limits imposed by natural processes, or else it is possible to be misled into believing that biofuels represent a simple, painless solution to the Philippines' energy concerns. In a paper published in *Renewable and Sustainable Energy Reviews* in 2005, Nonhebel estimated that based on global crop production statistics, a typical Third World country will require at least a tenfold increase in agricultural production, if its total energy demands are to be met using biofuel crops. Whether such an increase is met by improving productivity levels or by increasing the area of land cultivated, such a situation is likely to be untenable for a rapidly growing country such as the Philippines. Thus, while it is possible to envision liquid biofuels as being significant contributors to the national energy mix in the near future, and in particular playing the valuable role of providing much-needed energy diversity to the transport sector, it must be made clear that for thermodynamic considerations they cannot be expected to be a dominant energy in the long run, without also considering major changes in current energy consumption patterns. For the Philippines, a sustainable energy strategy will have to rely on an integrated approach consisting of both energy efficiency measures and energy resources diversification.

Note: This article is partly based on discussions during the Biofuels Symposium held jointly by the National Academy of Science and Technology, the National Research Council of the Philippines and the Department of Energy in Manila last September 15, 2006.

October 19, 2006

Jocelyn Ng

What to Do with *Basura*?

DRIVING THROUGH the Black Forest ("Schwarzwald") in Southern Germany, you realize why they call it that. The forests are so dense that they look, not green but, black. Germany is one country that puts a high priority on the environment. Natural forested land as well as farm and grazing land is so important that inhabited land—cities and towns—comprise less than 15 percent of their land area. It has become one of the most advanced countries when it comes to recycling waste. In every household or apartment building, there are large waste receptacles for paper, biological waste, and regular waste. All throughout cities and towns are waste dumpsters for the following: paper, aluminum, plastic, clear glass, brown (or amber) glass, green glass, and also old clothes and shoes (see *http://www.howtogermany.com/pages/recycling.html*). Manufacturers also have participated in the effort not only by repackaging their products with recyclable material but also by reducing the amount of packaging used per se. As a result, normal, everyday garbage or trash has been reduced by at least 50 percent.

Can we learn anything from this? *The Philippine Star*'s editorial dated August 27, 2004 points out that "indiscriminate garbage disposal" clogged up Manila's drainage system and caused the floods during typhoon "Marce." For such a densely populated city as Manila (at least 10 million), even if only for health concerns, waste management should be a key issue. Firstly, the mindset should be to minimize waste, i.e., the amount of garbage we produce, to reuse materials, and to recycle. Secondly, for the garbage that we do produce, we should have the appropriate technologies available to process or simply handle them: recycling, composting, incineration, and landfilling.

Many think a dumpsite and a landfill are one and the same. A dumpsite is an area where garbage is dumped without land preparation or concern for what happens to neighboring communities or the underground water table. A landfill involves several technologies working hand in hand. Before construction begins, the geology of the land is surveyed as well as where the water tables are, the flow and direction of any bodies of water in the vicinity, since the garbage has to be properly and safely contained and should eventually blend into the landscape. A landfill is broken up into units or cells. A key factor in its construction is the liner system: up to a meter of impermeable clay and/or high-density plastic is used to insulate the waste from the surroundings, followed by a layer of gravel or other suitable material as a protective layer so that the liner is not damaged when waste is put into these landfill units. The capacity of these units should be

CSIRO Molecular and Health Technologies, Parkville, Victoria, Australia.
Email: jngmelb@yahoo.com

precalculated and landscape restoration predetermined when permission to construct a landfill is issued. Waste is introduced into a unit in layers where each layer is compacted and covered up with soil. Units should be filled one at a time. Another unit is prepared only when a previous one is full. Landfill gas is 50-60 percent methane and 35-40 percent carbon dioxide and has a characteristic smell. They are safe to vent into the atmosphere when they are produced at very low levels. But since their production rate is not constant, a pipeline can be installed to collect these gases to be burnt off or to generate electricity. Rainfall on a landfill unit, when contaminated with pollutants, can be harmful and can leach out. "Leachate" levels should be monitored and excess quantities should be tankered for off-site treatment. Even when a unit is full and capped, decomposition continues and hence both leachate and biogas need continued collection and treatment. A properly engineered landscape will be expensive but worth it.

Recycling not only reduces the amount of garbage per se but also is a viable and economical solution, especially in industries where the raw materials require more energy and capital to process. Aluminum is the prime example. Recycling uses only 5 percent of the energy it takes to process aluminum ingots from bauxite ore where the resulting aluminum is only 4 percent by weight of the original ore. In simplified terms, instead of mining or importing, refining, chemical treatment, and electrolytic reduction, recycling involves cleaning, melting, and recasting. Recycled glass is broken up into smaller pieces called cullets, crushed, sorted, cleaned, and mixed with other raw materials and made into glass again. (Colored glass cannot be decolorized and therefore should be presorted to produce new glass in the same color.) Cullet melts at a lower temperature and hence requires less energy and produces 20 percent less air pollution and 50 percent less water pollution than the standard process. The situation with plastic is more complex. Some claim all types of plastic are recyclable; some think otherwise. The difference in opinion stems from more complex sorting and processing steps. There are different types of plastics and they come in many colors and are easily contaminated to a greater extent with dirt or different materials. Nevertheless, the technologies are out there and (foreign) businesses do recycle plastics simply based on economics. They save money by reducing tipping and hauling fees associated with disposal. In paper recycling, after pulping, cleaning, decolorizing and deinking, it goes back to producing paper, albeit of a lower grade unless new raw material is added. With every cycle, the fibers are shortened and hence paper has a finite recycling life. However, there are other uses for recycled paper. In fact, for most recyclables, only a percentage is truly recycled back into its original form and the rest end up in other uses.

So what to do with *basura* (garbage)? There is a lot that can and should be done. The information is literally at our fingertips. The Internet has a lot of information on trash collection, processing of particular types of waste, and the primary and secondary markets for recyclabes. It also describes recycling for other types of waste, e.g., wood. If the *diaryo-bote* kids did it, think of what economies of scale can do. Like any public utility, waste recycling or processing can be in the hands of the private sector. Implementing changes can start small. And we can do what best suits the Philippine situation. In the meantime, we can ask ourselves: what to do with *my* basura? It doesn't go on the ground or out the window. Think not only to minimize and reuse, but also not to litter. There should be recycling but there should also be picking up after ourselves.

June 16, 2005

Francis de los Reyes III

How Good Microorganisms Clean Our Wastes: What Happens When You Flush

Mention "bacteria" and most people think of disease-causing organisms that should be eradicated at all costs, as evidenced by our massive use of antibiotics in all forms. However, microorganisms, such as bacteria and protozoa, have been on Earth much longer than humans (bacteria appeared about 3.5 billion years ago), and make up most of the Earth's biomass. Microorganisms can be divided into the three "domains" of life: Bacteria, Archaea, and Eukarya. Members of the Bacterial and Archaeal domains are prokaryotic cell forms, and mostly unicellular, while Eukaryal organisms have true nuclei, and can be multicellular. All three domains are involved in "cleaning up" the wastes that we humans produce.

The domestic and industrial wastewater, solid wastes, hazardous wastes, and air contaminants that human activity produces are consumed or converted by microorganisms to innocuous forms. How do microorganisms do this? As an example, let's take the case of human ("domestic" or municipal) wastewater. After your toilet is flushed, the wastewater flows to a septic tank (very common in Metro Manila) and/or through a series of sewer lines to a centralized wastewater treatment plant (e.g., the Waste Water Treatment Plant along Commonwealth Avenue in Quezon City serves the University of the Philippines Diliman area). A common measure of "degree of pollution" or waste strength in wastewater is the amount of organic carbon present in the wastewater. Because it is very difficult and time-consuming to determine all the specific chemicals in human wastes as well as in many waste streams, we measure the waste strength in terms of the amount of oxygen needed to convert the organic carbon to carbon dioxide (CO_2). We call this the "Biochemical Oxygen Demand" or BOD. The typical BOD of domestic waste is around 150-250 mg/L. Industrial wastewaters can have much higher BOD values, sometimes in the tens of thousands. Before discharging to receiving streams and rivers, the BOD should be as low as possible, around 10-20 mg/L (this depends on the applicable regulations). When high BOD wastewater is discharged into the river, microorganisms present in the river use up all the dissolved oxygen (DO) in the water. The river turns "septic;" it becomes black and starts to smell, as anaerobic processes take over. The Pasig River is the classic example of a river with depleted oxygen (O_2).

Department of Civil, Construction, and Environmental Engineering, North Carolina State University, USA.
fldelosr@eos.ncsu.edu

At the wastewater treatment plant, a variety of bacterial species (and some protozoa) are actively "grown" in tanks. These organisms consume the organics and lower the BOD. To do so, they have to be supplied with enough oxygen so they can perform the conversion of organics to CO_2 and more bacterial cells. Some of the cells are retained in the system; some are excess, and have to be removed. The excess biomass, or "sludge" solids, should be treated (to reduce pathogens and degrade the volatile component), dewatered, and disposed of properly. Typically, the solids are land-applied to take advantage of the nutrients (nitrogen and phosphorus) present. Environmental engineers and WWTP operators design and operate wastewater treatment plants so that the different factors (e.g., biomass concentrations, reactor size, flow rates, oxygen levels, etc.) are in balance and optimized.

Aside from BOD, microorganisms transform nutrients such as nitrogen and phosphorus and decrease their concentrations in receiving water bodies. Nitrogen is an important nutrient, and high levels in lakes and rivers may lead to eutrophication, wherein high productivity causes algal blooms that then die and deplete the dissolved oxygen as organisms use up the organic carbon. Some cyanobacteria also produce algal toxins (for example, red tide poisoning is a manifestation of cyanobacterial growth). Nitrate in drinking water may lead to methemoglobinemia (blue-baby syndrome), where nitrate is converted in human saliva to nitrite, which then competes with oxygen in hemoglobin, thus reducing the O_2 in the bloodstream.

In wastewater, nitrogen is usually in the form of ammonia or organic nitrogen, which is quickly converted to ammonia. Ammonia is used by autotrophic bacteria (e.g., *Nitrosomonas* and *Nitrosospira*) as an electron donor (substrate), and is oxidized in the presence of O_2 to nitrite. This occurs at the wastewater treatment plant by giving these relatively slow-growing bacteria enough time to grow in the reactors. The nitrite produced is then subsequently oxidized by nitrite-oxidizing bacteria (e.g., *Nitrobacter* and *Nitrospira*) to nitrate. The two-step process is collectively called nitrification. At this point, the nitrate is still dissolved in the wastewater. To remove nitrate, oxygen is removed from the reactors by turning off the air supply. Those heterotrophic bacteria that can denitrify then use the nitrate (instead of O_2) as electron acceptor, and convert the nitrate to gaseous N_2. Nitrogen gas is simply released to the atmosphere, which is 78 percent N_2 anyway. There are other nitrogen transformation processes that occur in Nature and in wastewater treatment plants, but for the most part, nitrification-denitrification is the major route for nitrogen in wastewater treatment.

Phosphorus is another nutrient that limits the productivity (read: eutrophication) of lakes and other water bodies. This is why phosphorus in the form of phosphates (P) has been removed as an ingredient in most detergents. However, domestic wastewater still has some P, and industrial cleaners (e.g., phosphoric acid) contribute phosphorus to waste streams. Phosphorus can be biologically removed by phosphate accumulating organisms (PAOs). These organisms apparently store polyphosphates (P polymers) as they are cycled through anaerobic and aerobic conditions. Under no oxygen conditions, these organisms take up volatile fatty acids (VFAs) such as acetate and release phosphate by cleaving polyphosphate molecules. However, in subsequent aerobic conditions, they take up more phosphate than they released, resulting in a net P removal from the liquid stream. The stored polyphosphates are removed when the solids are removed. When treated properly, the sludge can be used as fertilizer, and the accumulated phosphates can serve as plant nutrients.

In your backyard septic tank, the bioprocesses are slightly different. The deeper portions of the septic tank are anaerobic, and here fermentative bacteria convert the organics to reduced forms, such as volatile fatty acids. These VFAs are, in turn, converted to methane by methanogenic organisms that are members of the Archaea. If sulfate is present, the sulfate is reduced to sulfide by sulfate-reducing bacteria. Hydrogen sulfide is the gas that gives off the characteristic "rotten egg" smell. Most of the liquid flows out of the septic tank, ideally into a leachfield (a volume of soil of adequate proportions). Here, the organic material in wastewater is converted by soil bacteria to CO_2. Over time, the solid byproducts of anaerobic digestion accumulate in the tank, and that is the time homeowners have to call in the folks who extract the solids for a fee. These extracted solids would also have to be disposed of properly.

In Metro Manila, septic tank overflows are typically connected to the sewer system infrastructure. Very old sewer lines are typically clogged with sediment, tree roots, and are most likely full of cracks and breaks, and leaking into the surrounding soil. Since groundwater wells are common, the possibility of contamination of water sources is high. In general, the deeper the well, the less chance there is for contamination from septic tanks. Still, most homeowners in Metro Manila relying on groundwater pumps should have their drinking water tested for coliform bacteria, the indicators of fecal contamination. There is a drastic need for upgrading the very old sewer system infrastructure in Metro Manila.

How about solid waste? In your backyard compost pile, Bacteria and Eukarya such as fungi break down the complex material to macromolecules and by-products such as humus. The processes in compost piles are typically aerobic, and are simply "accelerated" processes of what would occur naturally (e.g., on forest floors). Typically, municipal solid waste is collected and hopefully disposed of in an engineered landfill. The landfill is really a giant anaerobic solid waste reactor, and anaerobic processes occur here. Complex material is broken down. For example, paper and other cellulose-based material are degraded by cellulolytic bacteria that ultimately convert the cellulose to fatty acids. Fermenting bacteria convert other complex carbohydrates, proteins, and fats to long chain and short chain fatty acids (such as acetate), hydrogen, and CO_2. Methanogenic bacteria then use the acetate and CO_2 to form methane. In properly designed landfills, the methane can be captured for energy, or at least burned to convert the methane to

CO_2. If left unburned, the released methane can act as a greenhouse gas (methane is twenty-one times deadlier as a greenhouse gas than CO_2). Once filled to design capacity after many years of operation, landfills can be capped and converted to parks and golf courses. In non-engineered systems, such as open dumps, all you would get is a mountain of smoking (that's the slowly burning methane) garbage. In either case, microorganisms are doing what they've been doing for a long time—simply eating to survive, grow, and produce more biomass. What engineers and scientists are doing, in both wastewater and solid waste treatment and disposal, is to harness these microorganisms to clean up our wastes in engineered, well-designed and operated systems. These waste treatment systems have to be designed, built, and operated properly. Without them, we will continue to have black, smelly rivers and smoking, open garbage dumps.

August 10, 2006

Terence P. N. "Butch" Talorete

Environmental Hormones

Don't look now, but can you remember what you just had for lunch? Who knows, you probably had a dash of environmental hormones. Plastics and polystyrene (Styrofoam) are used worldwide to pack food, and nowhere is this more common than in developing countries where viands change hands in clear plastic bags. In the Philippines and elsewhere, instant noodles are likewise sold in polystyrene bowls that require only hot water plus a three-minute waiting time.

However, unknown to many, these plastic and polystyrene containers contain compounds known as environmental hormones or endocrine disruptors that often leach out into food. This leaching process is expedited at high temperatures, i.e., while the food is hot. Moreover, these compounds can be stored in the body's fatty tissues where they can readily accumulate.

The simplest definition of an environmental hormone is that it is a compound that has hormone-like activity; hence, it adversely affects the endocrine system and consequently alters the development and/or reproduction of organisms, including humans. For instance, studies have shown that fish in rivers contaminated with effluent from sewage treatment plants have both male and female sex organs.

These compounds can act in various ways; however, most share the following modes of action: (1) they can recognize the binding sites of natural hormones, thereby mimicking their effects; (2) they can block the interaction of natural hormones with their binding sites; (3) they can react directly or indirectly with hormones; (4) they can alter the natural pattern of synthesis of hormones; and, (5) they can alter hormone receptor levels. (The simplest definition of a hormone is that it is a substance that regulates growth and development.)

Among the more common examples of environmental hormones are the insecticides DDT, methoxychlor and endosulfan, the herbicide atrazine, the termite killer chlordane, and kepone, the bait used in ant and roach traps. A component of electrical insulation called polychlorinated biphenyl or PCB is also a known environmental hormone.

But apart from those involved in weed or pest control, the focus of concern should center on compounds from plastics such as bisphenol A, a breakdown product of polycarbonate that leaches out into hot fluids, and nonylphenol, a plastic softener that readily leaches out into fluids even at room temperature. Bisphenol A can be found in

---

University of Tsukuba, Japan.
tpbnt@yahoo.com

food cans, beverage containers, and in dental sealant; on the other hand, nonylphenol is released from most plastics where it is used as an antioxidant.

Among others, work in our laboratory at the University of Tsukuba, Japan, has shown that alkylphenolic compounds, particularly nonylphenol and octylphenol, do not induce necrosis and apoptosis (programmed cell death) of neuronal cells at concentrations as high as 500 micromolar (molarity is a unit of concentration expressed as moles of solute per liter of solution). However, at concentrations as low as 0.8 micromolar, these alkylphenols disrupt the acetylcholinesterase activity of neuronal cells (acetylcholinesterase is an enzyme involved in nerve impulse transmission). This shows that compounds found in plastic can be implicated in neurological and behavioral disturbances in both animals and humans.

Concerns regarding DEHP or di-ethylhexyl phthalate, a compound used to make vinyl plastics soft and flexible, have also been expressed. Laboratory testing of forty-six vinyl medical products, including intravenous (IV) bags, syringes, tubing and catheters, showed that DEHP content ranged from 12 percent to 80 percent by weight. Sick infants exposed to DEHP during transfusion, haemodialysis or artificial respirations are particularly in danger. Laboratory animal studies of DEHP have shown it to induce both fetal malformations and adverse effects on the reproductive system.

On the other hand, maternal consumption of contaminated fish or food products has also been known to cause adverse reproductive and developmental effects in the offspring. For example, prenatal exposure to PCBs can cause lower birth weight and shorter gestation and has been correlated with deficits in IQ and memory as well as delayed neuromuscular development. Laboratory animals prenatally exposed to bisphenol A can also manifest structural and functional abnormalities in the reproductive tract.

Although data on the immunologic effects of environmental hormones on humans are inadequate to support any definitive conclusions, there is clear evidence of immunosuppression of innate and acquired immune responses in seals fed with contaminated fish from the Baltic Sea.

And finally, can environmental hormones cause cancer? Scientists D.L. Davis and H.L. Bradlow, in a paper published in *Scientific American* (October 1995), spelled out reasons for suspecting hormone-mimicking chemicals in many unexplained cases of breast cancer. They believe that since too much estrogen, a natural hormone, can cause breast cancer, prolonged exposure to estrogen mimics could also be harmful and might account for some cases that had no obvious cause.

They believe that while any given environmental hormone may enter the body in small amounts, because of their ubiquity, persistence, and ability to accumulate in fatty tissue, they may reach threshold levels that can cause breast epithelial cells to multiply uncontrollably, leading to cancer. Work in our own and in many other laboratories worldwide has shown that nonylphenol as well as the phytoestrogens genistein and diadzin enhance the proliferation of breast cancer cells in vitro.

So now, if you can't remember your lunch, you may want to think about your dinner. Will your food be packed hot and steaming at the neighborhood store in a cheap plastic bag? Or will it be microwaved in a recycled plastic container that also softens in the oven? And your fruits and vegetables, were they sprayed with pesticides? If you answer yes to any of these questions, then what do you get?

Simple. You get out of the way!

December 30, 2004

Part 9
Structural Biology, Biophysics, Bioinformatics, and Systems Biology

Eduardo Padlan

Looking at Molecules Using X-ray Vision

Physicians order X-rays taken when they want to see what's wrong inside you. Dentists do the same to see what's wrong with your teeth or to see what's happening underneath your gums. From X-ray pictures, doctors can tell if a particular part of your body is normal, or if it is diseased. X-rays have become a routine diagnostic tool because they penetrate tissues and reveal, in clear contrast, your various organs. And, of course, X-rays are used in airports and high-security installations to reveal what is inside your luggage, briefcase, wallet, shopping bag, etc.

(And then there is that Superhero who has X-ray vision and can see through cars and buildings and other objects—except those made of lead.)

But X-rays have other uses. For example, we can actually look at atoms and molecules using X-rays. This is because X-rays have short wavelengths.

X-rays, like visible light, are electromagnetic waves. Like all waves, X-rays are regular undulations of crests and valleys. Like all waves, X-rays are characterized by their speed (X-rays and visible light travel at the same speed), wavelength (the distance between successive crests), and amplitude (how high the crests are). The wavelengths associated with X-rays are about one ten-billionth of a meter—very short indeed. Now, the separation between the atoms in a molecule are about that distance, so that X-rays have just the right wavelength to enable us to look at molecules in atomic detail (the shorter the wavelength, the more detail is seen).

So X-rays have become a routine tool to study the structure of molecules, to compare the structures of normal molecules and those which malfunction, to study how molecules work, and so on.

Being electromagnetic waves, X-rays interact with electrical charges (e.g., electrons). Since the different elements have different numbers of electrons, we can distinguish the various atoms in a molecule from each other—if the X-ray picture has enough precision. The picture becomes more precise if the molecules are arranged in a regular array—as in a crystal.

Hundreds of thousands of molecules have been studied by X-ray crystallography, from simple salts (the very first structure studied was that of NaCl) to viruses (with

Formerly with the US National Institutes of Health and currently with The Marine Science Institute, University of the Philippines, Diliman, Quezon City, 1101, Philippines.
edpadlan@yahoo.com

molecular weights in the millions). And, of course, it was from X-ray photographs that Watson and Crick deduced the basic structure of DNA.

One could argue that the molecules in a crystal are far from being in their normal biological or physiological state. True. Yet, it has been shown in many instances that the molecules in a crystal can bind their specific ligand (in the case of receptors), or catalyze their specific reactions (in the case of enzymes), etc. In other words, the molecules in a crystal could display normal function, suggesting that their structure in the crystal is the same as, or is very similar to, their structure under physiological conditions. The first person to demonstrate this was a Filipino, Florante Quiocho, in the 1960s when he was still a graduate student at Yale.

The many structures that have become available from X-ray crystallography are being put to good use. For example, one drug that has been found to be effective in the treatment of AIDS is a molecule that inhibits the action of a protease that the AIDS virus needs to replicate itself; that inhibitor was designed after closely examining the structure of the protease. For another, the modifications done on the structure of antibodies in order to make them more effective in therapy are made possible by the detailed structures available for those molecules. Indeed, the engineering of proteins—a whole new field in itself—cannot be possible without the protein structures obtained using X-rays or other means.

Protein structures are usually deposited in the Protein Data Bank. For many years, when it was housed at the Brookhaven National Laboratories, the Protein Data Bank was managed by, again, a Filipino, Enrique Abola.

There is an ongoing effort to determine the structure of at least a third of all the proteins that human beings produce. It is felt—and rightfully so—that we need to know the structure if we are to understand the function, as well as the interactions, of the many molecules in our body. Most of the structure analyses will be done using X-rays. Enrique Abola is involved in this effort.

Incidentally, aside from Florante Quiocho and Enrique Abola (and his twin, Jaime), there are other Filipinos, in the Philippines and abroad, who look at molecules using X-rays.

So, step aside Superman. You are not the only one with X-ray vision.

February 16, 2006

Angel de Dios

Nuclear Magnetic Resonance (NMR) Spectroscopy: Understanding Antimalarial Drugs

Malaria is among the top causes of death in the developing world. The disease is responsible for about one to three million deaths per year. Almost half of the entire population of the world is exposed to various species of *Plasmodium*, the parasite responsible for this disease. This exposure translates into several hundreds of millions of people carrying the parasite. The disease is therefore a major concern in developing countries. Affordable medicine against the disease is imperative and its nonlucrative nature, unfortunately, does not draw interest among top pharmaceutical companies. For this reason, efforts in designing drugs against this deadly disease have been confined within academic research institutions in the United States, as an example. The widely used drug against malaria is chloroquine. Unfortunately, there are now strains in all three continents—Africa, Asia, and South America—that are resistant to chloroquine. With the emergence of drug-resistant strains, it has become timely to understand how antimalarial drugs work in the hope of intelligently designing new drugs against malaria.

The most common drugs used against malaria, including chloroquine, are based on quinoline, a heteroaromatic compound (similar to naphthalene, except with one carbon replaced by nitrogen). Its target is believed to be the heme fragment, a by-product of the metabolism of hemoglobin. The malaria parasite feeds on this human blood protein for its amino acid source. Unlike humans, the parasite cannot metabolize heme and this turns out to be toxic to the parasite. The heme, however, precipitates out and crystallizes into hemozoin, making it innocuous to the parasite. Crystals of hemozoin, commonly referred to as malaria pigment, are found in red blood cells infected by the malaria parasite. Antimalarial drugs such as quinine, chloroquine and amodiaquine are thought to inhibit heme crystallization. By understanding the underlying mechanisms and chemical properties behind the interaction between the drug and its target heme, it is hoped that new drugs can be designed in an effective and speedy manner. This desired increase in understanding could only come with a detailed knowledge of these molecules. The drugs and their target are very small, about a million times smaller than the parasite, preventing their direct observation by even the most powerful microscope. Spectroscopic

Department of Chemistry, Georgetown University, Washington DC, USA.
dediosa@georgetown.edu

techniques are required to probe systems at the atomic resolution level. To understand spectroscopy a little bit, a simple analogy can be made. These techniques are similar to listening carefully to what tunes atoms and molecules are singing. Hearing a love song brings us certain emotions. Similarly, the frequencies atoms and molecules display can reveal information regarding their environment and structure, allowing our naked eyes to probe into the ultramicroscopic world of atoms and molecules. Extraction of chemical information from these techniques, however, is not straightforward. Analysis is required and nowadays, computers are necessary to model the systems and to evaluate the interactions between molecules from first principles.

Our research group at Georgetown University employs nuclear magnetic resonance (NMR) spectroscopy and molecular simulations in the study of antimalarial drugs. NMR spectroscopy is a widely used technique in chemistry. This type of spectroscopy employs a magnet (similar to magnetic resonance imaging, MRI) in which samples containing the molecules to be studied are placed. Inside the magnet, the nuclei (the core of atoms that contain protons and neutrons) begin to spin at their characteristic frequencies, their unique tunes. Sophomores in college chemistry are usually introduced to this method as a tool for characterizing and identifying small organic compounds. In NMR spectroscopy each nucleus in a unique chemical setting will report a particular frequency, a "signature" NMR chemical shift. NMR spectroscopy, similar to other techniques in chemistry, provides us with a glimpse into the world of atoms our naked eyes would normally not see. Since its discovery in the 1950s, the NMR chemical shift, demonstrating a great sensitivity to the electronic environment around a nucleus, has likewise become of value in elucidating the secondary structures of proteins in solution. Proteins can be regarded as large molecules and the arrangement of their atoms, what chemists would refer to as structure, leads to the properties and functions these proteins exhibit in human health and disease. Changes in NMR chemical shifts can be observed by simply altering the solvent, the medium in which we place atoms or molecules. Interactions between molecules manifest in this ubiquitous observable. One example that demonstrates the sensitivity of the NMR chemical shift is the fact that nuclei in the vicinity of molecules that contain aromatic rings exhibit changes in the chemical shift, which can then provide clues regarding their position and distance from these aromatic centers. In chemistry, an aromatic molecule is one in which electrons are free to cycle around circular arrangements of atoms. Aromaticity is relevant to the study of malaria since most antimalarial drugs and their target, heme, are aromatic. The aromatic rings, due to the electrons freely circulating above and below a molecular plane, are able to perturb the NMR spectrum of a nearby molecule. Aromatic compounds make nuclei sing a different tune.

There are other parameters in NMR spectroscopy in addition to the chemical shift. Coupling constants, which arise from either a direct interaction between nuclear spins or from an indirect interaction mediated by electrons, are also excellent reporters of the environment in the vicinity of the nucleus. There are duets, trios, and even choral singing. Equally important are relaxation times, as these provide a wealth of information regarding the structure and dynamics of chemical systems. Relaxation times refer to the rate at which a nucleus returns to its equilibrium spin state after we have forced it to sing its tune. Nuclei are like singers that need some break before they could sing their tune again. Unlike other spectroscopies, relaxation in NMR is not spontaneous, as it requires agents or mechanisms, such as unpaired electrons, other nuclear spins, and molecular motion. And in the presence of unpaired electrons, nuclei relax a lot faster. And the closer they are to the source of unpaired electrons, the less time they take to relax.

At the center of heme is a ferric ion, which is paramagnetic (having unpaired electrons). One can clearly see this by simply seeing that the number of electrons in a ferric ion is not even, thus, no matter how these electrons are distributed, complete pairing is not possible. In molecules with unpaired electrons, how fast a nucleus can return to its equilibrium spin state can serve as a clue to how the various nuclei are situated with respect to the unpaired electron. The most widely studied nucleus in NMR is the proton. Biologically relevant compounds, like other organic compounds, contain plenty of hydrogen atoms. Thus, through an NMR spectrometer, one can view what is going on through the eyes of a hydrogen nucleus. And a compound usually will have several unique hydrogens, with each one offering its own vantage point. In the presence of heme, the NMR spectra of the antimalarial drugs are significantly perturbed. The paramagnetic heme affords an efficient relaxation mechanism for the quinoline nuclear spins so that their lines in the NMR spectrum become much broader when the quinoline molecule becomes closer to a heme fragment. In the presence of a relaxing agent, they can sing very often, but their tunes are not long-lasting. The enhancement in the relaxation rates can be carefully measured. Each hydrogen nucleus in the quinoline drug will have a relaxation time, which can be analyzed to extract its distance from the paramagnetic heme. With all the hydrogens in the antimalarial drug reporting a distance from the ferric center of heme, one can perform some sort of a triangulation procedure to draw the position of the drug molecule with respect to its target, the heme. Thus, it has been demonstrated that via NMR spectroscopy, one can determine solution structures, the three-dimensional arrangement of atoms, of heme-antimalarial drug complexes.

The NMR chemical shifts are likewise perturbed when aromatic systems grow close to each other. The interaction between heme and an antimalarial drug is suggested to be noncovalent (no chemical bonding is involved). The interaction is believed to arise from the aromatic rings of the porphyrin group in heme and the drug molecule. These aromatic rings are characterized by electrons lying primarily above and below a molecular plane. These electrons extend a bit further compared to other electrons found in systems that have only single bonds. These electrons can influence the NMR chemical shift of nearby nuclei. By monitoring the NMR chemical shift, one can also gain information regarding how far an aromatic ring is from any given nucleus or vantage point since these interactions are very much distance-dependent. Hence, NMR chemical shifts have likewise been utilized in deciphering how antimalarial drugs approach a heme fragment. Furthermore, NMR chemical shifts can be used to study drug-drug interactions as these antimalarial drugs all contain the aromatic quinoline ring. NMR spectroscopy has been utilized, for example, in the study of self-aggregation of various antimalarial drugs. Solution structures for chloroquine dimers have been proposed. Mixed aggregates are also important in cases where one drug interacts strongly with another, suggesting that a mixed cocktail of antimalarial drugs may not be a wise choice at all times.

The ultimate goal in these studies is to understand how antimalarial drugs work at the molecular level. With this understanding, it is hoped that new drugs can be designed in a systematic fashion. Work on antimalarial drugs involves computational work. Since the NMR chemical shift is electronic in origin, one can calculate this NMR parameter from first principles, those we learned from Newton, Schrödinger, Einstein, Dirac, Ramsey, and others. The NMR chemical shift reported at each site originates from the electronic or bonding makeup near that atom. Particularly interesting for the antimalarial drugs are the chemical shifts of the aromatic nuclei. Preliminary calculations indicate that there may be a heterogeneity or nonuniformity in the NMR chemical shifts of aromatic nuclei, an indication perhaps of the nonuniform distribution of electrons in the region above and below the aromatic plane. This nonuniform distribution of charges can lead us to the more familiar case of electrostatic interactions, the simple principle that states unlike charges attract. Substitutions on a quinoline ring as exemplified by the well-known drugs chloroquine, quinine, and amodiaquine, enhance this nonuniform distribution. These subtle differences may partly explain the interactions found between aromatic rings, the interaction believed to be at the heart of the drug-target complex in malaria. Therefore, NMR spectroscopy may not only provide detailed pictures of drug-target complexes at atomic resolution, but may likewise provide insights regarding the origin of the properties of these drug molecules. With each tune played by a nucleus, we learn more about what atoms see.

February 9, 2006

Leopold Ilag

Proteomics: Weighing the Evidence

Sequencing of the human genome was completed in 2001, a few years ahead of schedule. One of the implicit aims of this project was to identify disease markers. However, to date, we still grapple with the elusive causes of many diseases and the hunt for biomarkers is still on. This is rooted in the fundamental problem that the genome is a mere blueprint for a vastly complex network of interactions orchestrated principally by proteins. The protein complement of the genome is what we refer to as the proteome and the variety of disciplines aiming to survey and understand the proteome falls under the umbrella of proteomics.

The complexity of studying the proteome arises from the fact that proteins encoded by genes can undergo a number of processing events that are regulated in space and time. Although a particular organism is endowed with a single genome, there are several corresponding proteomes for each cell population at a certain time and under specific environmental influences. At the molecular level, a single gene encoding a protein, yields a product that is then processed (e.g., glycosylation, phosphorylation) amplifying the variants. In biological fluids like plasma, it is not uncommon to find extraneous proteins due to tissue leakage in cases of disease. Considering mature forms, degradation products and modified species, numbers could easily escalate to over a million protein products!

The techniques involved in proteomics include those that can be used for protein separation, identification, and quantification. More advanced applications allow for structural analysis, mapping modifications, defining complex interactions, and even tracking protein regulation in the cellular context. The most common strategies include 1D and 2D gel electrophoresis, various types of chromatography, and mass spectrometry (MS). The latter is powerful as a stand-alone technique in proteomics research, but maximal effectiveness is achieved when it is coupled with the above-mentioned separation techniques.

Mass spectrometry, as its name implies, basically measures the "weight" of molecules. It is amazing how much information one can get by merely weighing things. However, when one realizes that MS offers resolution enough to distinguish a mass difference of a single proton (1 Da), and in some cases even as much as resolving a mass

Department of Analytical Chemistry, Stockholm University and Stockholm University Proteomics Facility, Sweden.
leopold.ilag@anchem.su.se

difference in the milli-dalton range, then it is not surprising why this technique is so powerful.

Most of the variations we try to track as indicators of abnormal physiology can be related to a change in the mass profile of a sample. One can, for example, easily detect the presence of a "foreign" protein or disappearance of a resident enzyme. When two molecules interact or are transformed, there is a corresponding change in mass (except rearrangement reactions). When peptides or proteins are modified chemically, there would also be a change in mass seen as an increase attributed to the moiety added. Because MS allows directed fragmentation, even isomers can be distinguished based on the different fragments generated as a function of structure; hence, products of different masses are obtained and give insight into chemical structure.

Analytes or samples to be studied by MS can be solid, liquid, or gas as starting materials as long as charged analytes can be generated from them, because MS depends on gas-phase mass separation and detection of ions (charged molecules). Soft-ionization methods, namely electrospray ionization (ESI) and matrix-assisted laser desorption ionization (MALDI), allow routine analysis of biomolecules. These developments in MS were so important that they earned the Nobel Prize in Chemistry in 2002.

MALDI coupled with the time-of-flight analyzer has been a workhorse for the strategy called peptide mass fingerprinting. This is based on the fact that proteins having unique sequences also often have distinct patterns of cleavage sites for a given enzyme. The cleaved protein therefore yields peptides that reflect this unique distribution (distinct sequence), allowing the matching of the experimental digest with a theoretical digest stored in databases, thus aiding the identification of a protein. However, this is not a very high-resolution method and it often fails when you are looking at related proteins or isoforms. Therefore, it is still important to get actual sequence information and this can be obtained through *de novo* sequencing.

De novo sequencing can be achieved using tandem MS. This refers to the possibility of using a mass filter allowing the isolation of ions of a desired mass and subsequently subjecting these to collisions with inert gas molecules to cause fragmentation. In this way, structural information via MS could be obtained from peptides (or small molecules). The weakest bonds happen to be along the peptide bond, making the reading of the sequence from the mass differences of the various fragments rather straightforward. Fragment sizes differ by a discernable mass corresponding to known amino acids from which the sequence can be pieced together. An additional feature of this strategy is that it allows detection of modifications on proteins, for example, when one detects mass differences not corresponding to any of the essential amino acids. Typical modifications may include a mass increase of 80Da for phosphorylation or addition of 16Da for oxidation.

Mass spectrometry indeed is a powerful technique not limited to mere mass measurements for small molecules but could be powerfully employed for the identification, characterization, and structural analysis of any biomolecule, particularly proteins. By weighing molecules and their fragments, we could weigh the evidence for potential biomarkers related to disease and for understanding molecular and cellular networks.

February 8, 2007

Richelle Duque

Love Actually

Shakespeare once asked, "What t'is to love?"

I have always been a strong advocate of love. Whenever people have something to do in the name of love, I support them a hundred and ten percent! I've witnessed many a love team made and broken—most of the time from the beginning, some still ongoing, others until the end. For someone who claims having never fallen in love yet, I'm a confessed hopeless romantic.

Throughout history, it was philosophy, religion, and art that explored the phenomenon of romantic love. Now, you may be wondering why an article on love is being featured in *The Philippine Star*'s Star Science column because one does not hear the words love and science together often. But in the last century, science has taken part in discussing the nature and function of love. Only recently have scientists tried to define the once ineffable, "another dimension beyond reason" love—a mammalian drive as primal as hunger and thirst, rather than an emotion such as happiness.

Contrary to popular belief that it is the heart that loves, it is actually the brain that does. So which areas of the brain are active when people are in love? Andreas Bartels and Semir Zeki of University College, London, located the areas activated by romantic love by measuring the blood level in the brain, thereby knowing its activity or lack thereof. Functional magnetic resonance imaging (fMRI) scans of students who claimed they were madly in love showed that people in love have activity in the caudate nucleus, putamen, insula, ventral tegmental areas, and nucleus accumbens—brain parts associated with the reward system, making it primarily a motivational state to pursue a chosen partner. What is more surprising is that the amygdala, which is primarily linked with emotions such as depression, anxiety and fear, has little or no activity at all. Equally interesting is the decreased activity in parts of the brain involved in moral judgment. There is some truth to the saying "Love is blind" after all.

Studies of Helen Fisher at Rutgers University have shown that three emotion-motivation systems evolved in our brain as we courted, mated, and reared our young: lust, attraction, and attachment. Each stage of love corresponds to a brain system associated with different sets of neural networks and primary neurochemicals. These three systems are independent of one another, meaning each can operate simultaneously. It is possible then that you can be at two or even three stages at the same time, and it may be with the same partner or a different one.

College of Medicine, University of the Philippines Manila, Philippines.
richelle314@gmail.com

Lust—the craving for sexual gratification from anybody—is primarily associated with testosterone both in men and women. Yes, women, too, secrete testosterone although in minute amounts compared to men. It is produced in the thecal cells of the ovaries and the Leydig cells of the testis. Testosterone, a steroid hormone from the androgen group, can either be converted to 5a-dihydrotestosterone by the enzyme 5a-reductase or to estradiol by the enzyme aromatase. It then binds to its receptor, activating downstream pathways, thereby having either a virilizing or anabolic effect. The former includes enhanced libido, hence increasing the urge for sexual gratification.

Attraction—the elation, heightened energy, obsessive thinking, focused attention, lack of appetite—is the popular romantic love we know of. This is associated with increased levels of dopamine and norepinephrine and decreased levels of serotonin. Dopamine functions both as a precursor to norepinephrine and as a neurotransmitter. It activates the reward system in the brain, providing feelings of enjoyment and happiness and reinforcing motivation to proactively perform certain activities. It is the same system that is linked with addiction to drugs. So, when we say we are addicted to him/her, we really are! Moreover, schizophrenia is a disorder associated with abnormally high levels of dopamine, with symptoms including disordered thought, hallucinations, and social withdrawal—very much the same symptoms as falling in love. In addition, norepinephrine plays a large role in attention and focus, explaining why we have focused and undivided attention on our loved one. So much so that even though we are doing other things, we still have our focus on him/her. Lastly, serotonin regulates body temperature, sleep, mood, and appetite. A decreased level of serotonin then disrupts our normal sleeping and eating habits. They say we cannot fall asleep when we are in love because finally, reality is better than our dreams. It may be true; but mostly it is because of the serotonin. Serotonin is also responsible for the Obsessive-Compulsive Disorder (OCD), explaining our temporary insanity and obsession with our loved one. "Crazy for You" then is not just a song; there is a scientific basis to it. Some scientists have hypothesized that those who actually take antidepressants, which increase the levels of serotonin, may have less chances of falling in love.

Attachment—the long-term calm and emotional union with a partner to rear a young—is mainly associated with vasopressin (VP) and oxytocin (OT). These two hormones are similar in structure, differing only in the amino acids isoleucine (OT) or phenylalanine (VP) at position 3; and leucine (OT) or arginine (VP) at position 8. Although both contribute to the monogamous bonding of a couple, the protective attitude toward our partner is attributed to the former, while monogamy is associated with the latter. Vasopressin makes us aggressive with potential competitors (i.e., close friend of your girlfriend, best friend of your boyfriend). Next time you wonder why you are so jealous, it's the vasopressin. On the other hand, in an experiment involving monogamous prairie and polygamous montane voles (short-tailed stocky rodents), the oxytocin receptors in the reward-associated brain region of the prairie voles were shown to be responsible for monogamy; lack of oxytocin receptors in the montane voles then caused them to be polygamous as they had no memory of the other voles they had met. So, when you catch your boyfriend/girlfriend two-timing you, maybe he/she doesn't have enough oxytocin receptors to be with you alone. Both OT and VP are also related to territorial behavior, as you are with your loved one. In an experiment done by our very own Filipino scientists from the University of the Philippines, Lourdes J. Cruz, Victoria de Santos, Glenn C. Zafaralla, Cecilia A. Ramilo, and Baldomero M. Olivera, in collaboration with Regina Zeikusll and William R. Gray, oxytocin and vasopressin homologs were extracted from Philippine cone snails, namely Conus geographus and Conus striatus, and were intracerebrally injected into mice. A scratching response in the mice, which is a feature of territorial behavior, was observed.

Science isn't only interested in the upside of love but in its downside as well—rejection. When you love someone, there is always the risk of being rejected. When one is dumped, he/she undergoes two stages: protest and resignation/despair.

Protest involves obsession with winning your partner back. It is a typical mammalian response to a broken tie, resulting in increased dopamine and norepinephrine that stimulate your asking for help and calling for your loved one. It is indeed paradoxical—very typical of love—that as your darling slips away, the hormones associated with attraction actually increase. It is also possible that delaying the reward system actually prolongs its activities, explaining why you still feel you're in love. But this delay also has its threshold. When it finally sinks in that there is no chance that your loved one will come back to you, it is when resignation and despair set in. Dopamine levels drop, resulting in lethargy and moping. Depression also sinks in. Some may even be angry, thinking that anger switches off love. But this is not the case since anger and love are independent systems. You can be angry but still be very much in love.

Rejection devastates us in different degrees but time will come when we'll recover from it and be ready to fall in love again. Bouncing back from rejection is not easy but definitely possible. We make mistakes in life but it is never a mistake to have loved. Never lose faith in love for it is better to have loved and lost than not to have loved at all.

References

Aron A, Fisher H, Mashek DJ, Strong G, Li HF, and Brown LL. 2005. Reward, Motivation and Emotion Systems Associated with Early-Stage Intense Romantic Love: An fMRI study.

Journal of Neurophysiology 94:327-37. Bartels A and Zeki S. 2000. The Neural Basis of Romantic Love. NeuroReport 11 (17): 3829-3834.

Bartels A and Zeki S. 2004. The Neural Correlates of Maternal and Romantic Love. *NeuroImage* 21: 1155-1166.

Caldwell HK & Young WS. 2006. Oxytocin and Vasopressin: Genetics and Behavioral Implications. Springer-Verlag, Heidelberg, Berlin.

Cruz, LJ, de Santos, V, Zafaralla GC, Ramilo, CA, Zeikusll R, Gray WR , and Olivera BM. 1987. Invertebrate Oxytocin/Vasopressin Homologs. *The Journal of Biological Chemistry* 262 (33): 15821-15824.

Fisher, H. 1994. The Nature of Romantic Love. *The Journal of NIH Research* 6 (4):59-64.

Fisher, H. 1998. Lust, Attraction, and Attachment in Mammalian Reproduction. *Human Nature* 9 (1): 23-52.

Fisher, H. 2005. Dumped! The Nature of Romantic Rejection. *New Scientist* 181 (2434):41.

Fisher H, Aron A, and Brown LL. 2005. Romantic Love: An fMRI study of a neural mechanism for mate choice. *Journal of Comparative Neurology*, 493:58-62.

Fisher H, Aron A, Mashek D, Li H, Brown LL. 2002. Defining the Brain Systems of Lust, Romantic Attraction, and Attachment. *Sexual Behavior* 31(5): 413-19.

Fisher, H, Aron A, Mashek D, Strong G, Li H and Brown LL. 2002. The Neural Mechanisms of Mate Choice: A Hypothesis. *Neuroendocrinology Letters* 23: 92-97.

Fisher, H, Aron A, Mashek D, Strong G, Li H and Brown LL. 2002. Defining the Brain Systems of Lust, Romantic Attraction and Attachment. *Archives of Sexual Behavior* 5: 413-19.

www.harunyahya.com/miracle_of_hormones_02.php

www.kband.com/bluprnt/000538.html

www.sfn.org/skins/main/pdf/BrainBriefings/BrainBriefings_Dec2005.pdf

www.voxmagazine.com/story.php?ID=18260

http://counsellingresource.com/distress/autistic/autism-nimh-2.html

www.monkeytime.com/sciencemaster/galleries/brain/02.php

www.a1b2c3.com/drugs/pictures/brn002.htm

www.drugdevelopment-technology.com/projects/bifeprunox/bifeprunox3.html

www.a1b2c3.com/drugs/pictures/brn001.htm

www.thebrain.mcgill.ca/flash/i/i_03/i_03_m/i_03_m_que/i_03_m_que.html

www.squ.edu.om/mj/Oct2001/dopamine/index.html

http://missinglink.ucsf.edu/lm/IDS_104_dopamine_ILM/Dopamine/DAPathways.htm#

http://sprojects.mmi.mcgill.ca/gait/parkinson/biochemistry.asp

http://ardb.bjmu.edu.cn/main/ARLigand.htm

www.sineh.com/dopamine.htm

www.pitt.edu/SUPER1/lecture/lec2221/002.htm

www.wikipedia.com

December 28, 2006

Cherry Mae Ignacio

Biological Clocks

The Siberian hamster is a peculiar little creature. It has figured out a unique way to hide from its predators in the harsh Arctic landscape—it changes its fur color from brown in summer to white in winter to make it easier for it to hide from all the animals that want to eat it. How does it know when to change? It uses something we are all probably familiar with—biological clocks.

But have you ever pondered whether there really is a ticking clock in your body? True enough, the body does keep track of its own time—but probably not in the way most people think. Mankind has long suspected the presence of some sort of biological clock system. Yet only by looking at the biological clock through the eyes of science, more particularly, through molecular biology and genetic engineering, have we seen the ingenious way Nature has designed body clocks to anticipate daily changes of day and night.

It's a pretty well-known fact that the Earth revolves around the sun in more or less a predictable manner. Every day, our planet rotates on its axis, causing us inhabitants to experience day and night. As a result, we end up patterning our lives to the 24-hour day and to the subsequent changes in season also caused by the Earth's revolution around the sun. Because of this, we sleep at roughly the same time each night, bears hibernate in winter and bats hunt at night. All these actions are controlled by biological clocks present in all these animals.

Biological clocks are more commonly known as circadian clocks or circadian rhythms. The word circadian comes from the Latin words *circa*, meaning "around," and *diem*, meaning "day." Evolution favored the existence of these circadian rhythms because with it, an animal would know when it is most advantageous for it to look for grub, or when to hide from its predators; plants know when it can use light to make its own food. Those equipped with circadian systems are better able to cope with their environment, thus gaining higher chances of surviving and reproducing to pass on its traits to the next generation.

The circadian rhythm is an active cycle. Every single day, it takes in some sort of input from the animal's surroundings to synchronize itself, to make sure that its timekeeping is as accurate as possible. Scientists call this input a *zeitgeber*, German for "time giver." It refers to any sort of environmental cue that tells your body what time it is. The most important

National Institute of Molecular Biology and Biotechnology, University of the Philippines, Diliman, Quezon City, 1101, Philippines.
cherry_me5@yahoo.com

of these zeitgebers is light; another is temperature. The presence of light is detected by the eye, more specifically, special proteins in the retina called melanopsin and rhodopsin, and this message is passed on to the master timekeeper in the brain.

Apart from the input system used to register these zeitgebers, other major components of the animal circadian system include a master timekeeper and a way for the timekeeper to communicate the message of time to the rest of the body.

The center of the circadian system in humans is a pair of small bunches of brain cells located near the hypothalamus called the suprachiasmatic nuclei or SCN. It is what is called the circadian pacemaker—the master clock in your body. Containing roughly between 16,000 and 20,000 brain cells (called neurons) each, it tells the rest of your body what time it is.

The idea of the SCN as the circadian pacemaker was first verified in studies using laboratory mice in the early '90s. Anybody who has ever owned a pet mouse knows that they like to run in their exercise wheels more at night than during the day. Applying this information, scientists attached sensors to exercise wheels in mouse cages to monitor when they run. Monitoring this activity, 24 hours a day, seven days a week, had led them to establish a normal mouse activity pattern. This pattern of activity repeats itself every 24 hours.

They then destroyed the SCNs of these mice and measured their activity patterns again. These mice then exhibited arrhythmia, the complete loss of their original 24-hour activity rhythm. Moreover, transplanting healthy SCN to these mice caused them to reacquire their 24-hour activity rhythm.

Since the SCN is the master timekeeper, it is of utmost importance that it keeps the correct time. The SCN does something no different from what the average person, who is always unsure of the correct time, would do—it carries several clocks. In the case of the SCN, around 20,000 clocks—each one of the neurons that compose the SCN keeps its own time through a cellular clock. These neurons then send messages to each other to make sure that the SCN generates a single, consolidated output.

The third major component of the circadian system is also the least studied. Output pathways from your circadian pacemaker to the rest of the body are assumed to be quite complex. Scientists think that the SCN uses various ways to send the message of time to the rest of the body. One of these uses small molecules called neurotransmitters to send messages to different glands that secrete hormones into the blood. Eventually, all of these output pathways get the message of time to the different organs that need it.

Though these general components are present in all circadian rhythms, it is important to note that they have different effects on different organisms. To exemplify just how the circadian rhythm has varying effects on different organisms, let's look at something that we can all relate to—sleep. Many animals sleep, and the average total sleep time varies with each organism. At one end of the spectrum, bats sleep almost twenty hours per day; on the opposite end, giraffes sleep for just about two hours. In one way or another, these sleep patterns are affected by the organism's own circadian rhythms. (An interesting tidbit on sleep you may not know is that the ways animals sleep vary as well. Bats sleep upside down, rats sleep intermittently, dolphins sleep with only half of their brains at a time, and cattle can sleep standing up—often with their eyes wide open!)

Even with organisms of the same species, like humans, the fact that each of us has exactly the same type of circadian rhythm does not stop us from manifesting different time-dependent activities. Some of us classify ourselves as morning persons; others, evening persons. The circadian rhythm does not limit what kind of activities we do throughout the day. Rather, it is used more as a guide to help the body anticipate what we're likely to do next.

A common misconception about biological clocks is that there is only one clock inside the body. Actually, clock mechanisms are present in many different cells in the body. Cellular clocks, similar to those in the neurons of the SCN, are present in many major organs in the body like the liver, heart, and lungs. Each of these organs keeps its own clock for its own purpose. For example, the liver knows when to secrete bile in time to digest your dinner.

My sister and I went to California to visit our folks this last summer break. The trip led me to experience firsthand one of the more modern effects on human circadian rhythms—jet lag. Only in the past century has man been able to travel through fifteen time zones in half a day. This ability was not one that evolution anticipated. Because of this, it takes a long time for us to reset our biological clocks. Apart from the obvious disruption in regular sleep-wake patterns, other effects of jet lag can be changes in your eating patterns, indigestion, headaches, and a general feeling of weakness. All these symptoms are probably because of the desynchrony between your SCN's clock and the clocks present in the various organs of the body.

Say you went from San Francisco, California to Manila. By the time you get here, your SCN automatically knows that it is on Manila time. On the other hand, the rest of your body still thinks it's on San Francisco time. Though your SCN knows its lunchtime in Manila and you're supposed to be eating, as far as your liver knows, it's already 9 p.m. and you're about to go to bed. It takes several days for the rest of the body to get accustomed to the actual time. Studies in "jet-lagged" mice have shown that liver cells take as much as fourteen days to adjust themselves to the correct time. Cells in your major organs take such a long time to change because they have to receive signals from the SCN before they can change. With this, it's no wonder drinking espresso shots to keep myself awake in class caused my tummy to feel fuzzy! I eventually restored my regular body rhythm after several days.

At the heart of any clock, whether it be biological or mechanical, is a mechanism which can produce a reliable oscillation—a repetitive motion or occurrence of some sort that can be used to measure time accurately. For mechanical clocks such as watches, oscillation is achieved by the hands of the clock. The second hand moves around the clock once every minute, the minute hand every hour, and the hour hand completes a full cycle every 12 hours.

Cellular clocks, on the other hand, cycle proteins. Studies on these core clock components have been more extensively studied in the humble fruit fly *Drosophila melanogaster* than any other animal. In their cells, two proteins called "Period" and "Timeless" accumulate in the cell through the course of late afternoon and night. Both of these proteins are only stable inside the cell if they are paired together. Otherwise, either gets destroyed. At some point, these pairs accumulate enough numbers to stop the production of individual Period and Timeless components, thereby preventing more pairs from piling up inside the cell.

When dawn comes, Timeless proteins get destroyed by another protein, the light-sensitive Chryptochrome. With the absence of Timeless, the Period protein gets destroyed as well. Because Period-Timeless pairs are no longer present in the cell, it can now re-accumulate these pairs when nighttime comes. This whole cycle takes roughly around 24 hours and is what scientists call a delayed negative feedback loop. The cell takes note of the time of day by checking the amount of these proteins in the cell.

Going back to the cute Siberian hamster, how does it know when to change color? Apparently, it knows what season it is because of the changes in the amount of daylight. In the Arctic, the difference in the amount of daylight in summer and winter is much more pronounced than it is here. Daylight in the summer can last as long as six months. When cells in its body notice that daylight has been present for quite a long time, it knows that it is summer and thus, its coat color becomes brown. Conversely, a long period of darkness indicates winter and triggers the hamster to change its coat color to white.

If anything, the study of circadian rhythms has led us to understanding that time matters. Even the simplest actions we take for granted may be affected by our body clocks. This is why in scientific laboratories around the world, *when* one makes observations is as important as the observations themselves. Scientists studying all sorts of topics always consider time-dependent factors such as age and feeding time when designing their experiments. As much as possible, many of these factors are kept constant for the entire duration of the study.

The Laboratory of Molecular and Cell Biology, one of several laboratories in the National Institute of Molecular Biology and Biotechnology in UP Diliman, maintains a room filled with mice used for their various experiments on developmental biology. To make sure that all their mice have roughly the same circadian rhythms, they keep a strict light-dark cycle in their mouse room. Lights are turned on every morning at 8 a.m. and turned off at 8 p.m. at night. Lab members always feed the mice at 4 p.m., and all observations regarding mouse behavior are done during the same time every day.

The discovery of clock genes has led scientists to ask—what can we use this new information for? For instance, can this information be used to treat insomniacs? Not quite yet, but initial studies done in humans have determined that some sleep problems may be due to broken cellular clock components. Familial advanced sleep phase syndrome (or FASPS) is an inherited disease characterized by individuals who cannot help but sleep and wake up very early. This disease has been linked to mutations in a clock component called "human Period 2"—one of several human versions of the Period gene in the fruit fly. Maybe in the future we can somehow fix this protein to restore normal sleep patterns to these people.

One of the most promising applications of this information may be in cancer therapy. Certain clock components have been linked to the regulation of cell division. Hopefully, one day, we can use this information to stop cancer cells from proliferating—by fooling them into thinking that it is not time to divide yet. Nowadays, studies have also shown that taking anti-cancer medication at specific times of day makes them more effective in treating cancer.

The possible applications of information derived from studies with circadian rhythms are wide and varied. Who knows what other activities are controlled by the interesting and complex nature of biological timekeeping.

References

Harmer SL, Panda S, Kay SA. 2001. Molecular Bases of Circadian Rhythms. *Annual Review of Cell and Developmental Biology*. 17:215–53.

King DP, Takahashi JS. 2000. Molecular Genetics of Circadian Rhythms in Mammals. *Annual Review of Neuroscience* 23:713–42.

Konopka RJ, Benzer S. 1971. Clock Mutants of *Drosophila Melanogaster*. *PNAS USA* 68:2112-16.

Lowery PL, Takahashi JS. 2004. Mammalian Circadian Biology: Elucidating Genome-Wide Levels of Temporal Organization. *Annual Review of Genomics and Human Genetics*. 5:407-41.

Panda S, Hogenesch JB, Kay SA. 2002. Circadian rhythms from flies to human. *Nature* 417:329-35.

Ralph MR, Foster RG, Davis FC, Menaker M. 1990. Transplanted suprachiasmatic nucleus determines circadian period. *Science* 247:975–78.

Reghunandanan V, Reghunandanan R. 2006. Neurotransmitters of the suprachiasmatic nuclei. *Journal of Circadian Rhythms* 4:2.

Rosbash M, Takahashi JS. 2000. Clockwork Genes: Discoveries in Biological Time. Howard Hughes Medical Institute Holiday Lectures on Science.

Siegel LJ. 2006. The Time of Our Lives. *http://learn.genetics.utah.edu/features/clockgenes/*

Takahashi JS. 2004. Finding New Clock Components: Past and Future. *Journal of Biological Rhythms* 19:339–47.

Time Matters: Biological Clocks. 2006. HHMI Biointeractive Exhibit. *www.hhmi.org/biointeractive/ museum/exhibit00/00.html.*

Yu Q, Jacquier AC, Citri Y, Hamblen M, Hall JC, Rosbash M. 1987. Molecular mapping of point mutations that stop or speed up biological clocks. *Drosophila melanogaster. PNAS* 84:784–88.

(Special thanks to Dr. Cynthia Saloma for information regarding mouse room maintenance.)

December 7, 2006

Eduardo Mendoza

Mathematics and Cells

One of the leading mathematicians of the late nineteenth century, Henri Poincaré, once wrote "Mathematics is the art of giving the same name to different things." This process of abstraction is an essential part of mathematical activity and leads to concepts applicable to many different fields. One such concept is that of a "network"— simply a set of nodes or components connected by lines or arrows, which represent a relationship between the nodes. The mathematical theory of networks (often called "graph theory") has many applications—fields of study include social networks, information networks (e.g., the World Wide Web), technological networks (electric power grid, the Internet), etc. In recent years, complex networks—those with a large number of nodes—have become a focus of interest. Such networks are also almost always dynamic, meaning that they change over time.

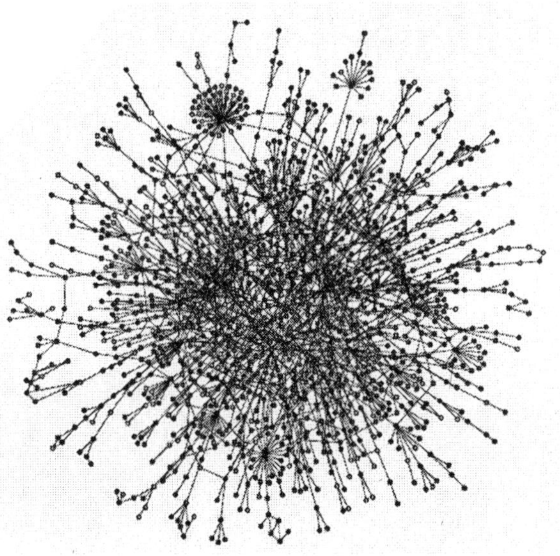

Fig. 1 Yeast protein interaction map (Jeong, et al. 2001)

The new field of "systems biology" uses the concept of networks in its attempt to understand the structure and dynamics of biological systems. In particular, a cell, the basic unit of life, is modeled as a complex, dynamic network—with millions of molecules as nodes and various interactions between these as the "arrows" (or connections). This

Physics Department and Center for NanoScience of the Ludwig-Maximilians-University in Munich, Germany.
Eduardo.Mendoza@physik.lmu.de or mendoza@lmu.de

model has proven to be useful, allowing insight into biological processes and the potential to address diseases, most of which are intimately related with malfunctioning of parts of this complex network. Systems biology is in a sense an extension of molecular biology; the latter has successfully studied in the last fifty years (particularly after the discovery of the double-helix structure of DNA in 1953 by James Watson and Francis Crick) the various "components" or biomolecules. While seemingly daunting in their complexity (see Fig. 1), cellular networks exhibit several essential kinds of structure and dynamics, allowing the use of more mathematical concepts and methods in their study.

Biological networks have been analyzed in terms of their *statistical* properties and show a number of surprising characteristics. The two important ones are:

- "Small worlds," i.e., there are short paths between any two nodes despite the large number of nodes. A subnetwork of protein interactions in yeast, for example, has 2,115 nodes but an average path length of only 6.8 (Newman 2003).
- In many cases, "scale-free networks," i.e., the probability of a node having n connections is simply n to a constant power. This, in particular, implies that there are only a few highly interconnected nodes and many poorly interconnected nodes, making them robust against random errors but vulnerable to targeted attacks.

One of the really surprising discoveries in the past five years was that most complex networks share these properties, regardless of whether they are social networks, technological ones like the Internet, or informational ones like a citation network. The study of complex dynamic networks is a very active field with researchers from many disciplines.

Another approach to coping with the complexity of these networks is the familiar "*divide et impera*": divide into parts and study the resulting subnetworks separately. One way of dividing is inspired by viewing cells as information processing systems: the genome (the set of all genes) handles the information storage; the proteome (set of all proteins) is responsible for information processing; and the metabolome (collection of all metabolites) finally for the execution phase. The resulting important subnetworks are the gene regulatory networks (interactions between genes and proteins regulating gene expression), the protein interaction networks (including signal transduction networks), and the metabolic networks. New experimental techniques from the corresponding "omics" (genomics, proteomics, metabolomics) are delivering large data sets which will contribute to detailing structure and dynamics of these networks. The second approach utilizes what Albert Barabasi (a physicist) and Zoltan Oltvai (a pathologist) have aptly termed "Life's Complexity Pyramid" where a cell network has a hierarchical structure of (at least) four levels: the system level (the whole cell network), the functional module level (subnetworks with a clear biological function, e.g., cell movement, cell division), the level of building blocks or recurring patterns (called "network motifs"), and then the well-known components or nodes. While the bacterium *E. coli* has an average of 2,500 active proteins, the module involving movement toward a chemical gradient (chemotaxis) involves only six to seven proteins. Research on motifs (which are small networks—usually up to four nodes—and fulfill an information processing function, not a biological one) has been particularly active.

Depending on the data available in publications and increasingly in well-structured, publicly accessible databases, several classes of mathematical models can be constructed. When sufficient dynamic or time-course data are available, nonlinear kinetic models may be pursued, particularly in the form of *differential equations*. However, computational requirements are still an issue with this approach, so that today only small-scale (up to ten to fifteen nodes) models (for functional modules) are available. If only a finite number of values are needed, a discrete, qualitative model can be built (for only two values, this is the familiar *Boolean* or "on/off" *model*). One should note, however, that fairly complex phenomena like the body segmentation of fruit flies have been successfully described by Boolean models. Finally, if only the underlying stoichiometry of the biochemical reactions (essentially the "bare" network) is known, one can apply an approach based on finding a feasible set of steady-state solutions based on physico-chemical and other constraints. This approach uses only mathematical methods from *linear algebra* and *convex analysis*. Such efforts have already resulted in an "*in silico*" model for *E. coli*, which incorporates about 25 percent of its 4,393 genes. However, this is just a very modest step toward the vision of a "Virtual Cell."

The complexity of biological networks mandates a very close cooperation between experimenters and modelers—the rigor and passion with which this is done is characteristic of the new field of systems biology. The formation of joint-experimenter-modeler (JEM) research teams is particularly effective in this regard. The broad multi-disciplinarity needed to understand biological networks has also led leading institutions like Massachusetts Institute of Technology (MIT) and the Swiss Federal Institute of Technology in Zürich (ETZ Zürich) to experiment with new forms of research and development. These initiatives—CSBi (Computational and Systems Biology Initiative) at MIT and BEST (BioEngineering, bioSystems, bioTechnology) at ETZ Zürich—are based on the concept of "community of practice," which has already found wide acceptance in many industrial organizations. At UP Diliman, researchers at several departments and institutes at the College of Science and the College of Engineering started the "Mathematical Life Sciences Initiative" (MLSI) last year with a series of joint-experimenter-modeler projects and a similar vision to establish a vibrant community of practice.

Systems Biology—the experimental and computational study of biological networks—is emerging as the basis for future advances in biomedicine and drug development. Research institutions and industrial companies worldwide recognize this potential and have begun to invest in the field.

For me, personally, it is gratifying to see large parts of what was called "pure mathematics" when I was doing my doctoral work in the seventies at Bonn University now finding application in the study of biological networks. Examples come as parts of algebraic geometry and topology. It is particularly amusing for me to note that in my thesis, I studied mathematical objects called "*cell complexes*" and now over twenty-five years later, I look at understanding *complex cells* with related methods.

References

H. Jeong et al. Lethality and centrality in protein networks, *Nature* 411 (May 2001).

M.E.J. Newman. The structure and function of complex networks, arXiv:cond-mat 0303516 V1, March 25, 2003.

April 7, 2005

BALTAZAR AGUDA

Excitement and Opportunities in Systems Biology

NOT SO many years ago, in February 2001, two groups in the US published the first drafts of the human genome sequence in the journals *Nature* and *Science*. The sequence is that of approximately three billion "letters" that run through strands of DNA, the primary component of our chromosomes where our genes reside. It has been known for a long time that a DNA letter can only be one of four—namely, A, T, C, or G—and it never ceases to amaze me that the linear sequence of these letters contains virtually all the information required to create a complex organism like us (see note 1 at the end of the article for meaning of acronyms). Recent estimates on human genes put their number at about 30,000, which corresponds to how many proteins our cells can make. The translation of the sequence of DNA letters to proteins requires the so-called "genetic code" that instructs cells what amino acids (the components of proteins) should be assembled for a given sequence of letters. The elucidation of the genetic code and understanding the molecular machinery for expressing genes to proteins are major achievements of molecular biology. The proteins are the workhorses of cells; they carry out various enzymatic functions such as catalyzing metabolic reactions, structural functions such as those of protein filaments and other cellular fibers, and they orchestrate physiological processes such as cell division and differentiation. So this means that the human genome sequence information tells us everything we need to know about human biology, right? Wrong.

To be fair, it is perfectly understandable why we should be excited by the availability of a human genome sequence. If we know where the genes and their sequences are, we can compare a normal gene from a mutated one (having at least one letter changed in the sequence), analyze the changes in the expressed protein, and then predict the consequences of this mutation on human health. A classic example is a mutation in the hemoglobin gene—namely, a specific GAG sequence in the gene is changed to GTG—which leads to the disease called sickle-cell anemia. So it is easy to see why there is a lot of expectation regarding the utility of the genome sequence toward developing cures for genetic diseases.

The discipline called *Bioinformatics* is concerned with the analysis of genome sequences, including discovering genes in the three-billion letter sequence of human DNA—work that is far from complete. But even when we get to know all of our genes,

Mathematical Biosciences Institute, Ohio State University, USA.
bdaguda@gmail.com

a far more daunting task will still be ahead of us. Genes can be turned "on" or "off"—meaning that they may or may not be producing proteins. It is the regulation of gene expression that is turning out to be horrendously complex. The making of proteins from DNA actually occurs via another class of molecules called RNA (see note 1). A gene sequence on the DNA is "transcribed" to the corresponding RNA molecule, and this RNA is "translated" to the protein. The complexity stems from the fact that information flow is not linear. Many proteins can bind to DNA and affect RNA transcription. Many proteins can also bind to RNA and influence the protein translation process. RNA molecules can also affect both transcriptional and translational processes. And as if the picture is not complex enough, proteins interact with other proteins, with metabolites or other molecules when they carry out their cellular functions. How are we to cope with such complexity?

First, we try to identify and classify all the players in the system. We are in the era of *Omics* revolutions brought about by high-throughput data-acquisition technologies. *Genomics* is churning out genome sequences of hundreds of organisms; *transcriptomics* is developing microarray technologies that monitor RNA levels (the transcriptome); and *proteomics* aims to identify and characterize the set of cellular proteins (the proteome). There is also *metabolomics* which aims to measure all the metabolites produced in a cell, and *interactomics* which assays protein-protein and other molecular interactions (yes, scientists do get carried away!). Essentially these *Omics* will provide us with a comprehensive parts list of our cells. Secondly, it will be the integration of these parts and the analysis of the dynamics of their interactions that ultimately will bring about a comprehensive understanding of biological systems—this is where *Systems Biology* enters the picture.

The basic premise of Systems Biology is that the whole is greater than the sum of its parts. In other words, there are properties attributed to the system as a whole which cannot be predicted from studying the parts in isolation because these properties emerge from the interactions of the parts. An unmistakable attribute of Systems Biology is that it is interdisciplinary. The complexity of biological systems demands collaboration among biologists, chemists, physicists, mathematicians, computer scientists, etc.; this explains the recent proliferation of institutes of systems biology in North America, Europe, and Japan.

I was trained as a physical chemist in graduate school and taught university chemistry courses for many years. After getting my tenure in the late 1990s, and with all the buzz about genomics and transcriptomics at that time, I saw the opportunity of refocusing my research to join in the fun. My re-education in molecular cell biology was wonderfully quick and painless—thanks to the excellent Internet infrastructure in my university. I cannot overemphasize this enough: the Internet is a great global equalizer in scientific research. With a good connection, anybody in the world can access a universe of information, instantly. There are many free or open-access online scientific journals (see note 2) and downloadable lecture notes and videos from eminent academic institutions such as the Massachusetts Institute of Technology in the US (for the open courseware, see note 3). There are even popular biology textbooks that one can access online (see note 4). To tell you the truth, I hardly go to our library now because *Google* searching (see note 5) is often sufficient! However, it is always a good practice to double check information obtained from the Web.

My pet research project is in an area called *cancer systems biology*. Cancer is a complex disease involving the interplay among many cellular processes, including the cell division cycle, apoptosis (programmed cell death), angiogenesis (formation of blood vessels required to feed tumors), senescence (aging of cells), and others. Years of research in the cancer community, and more recently the earlier mentioned *Omics* technologies, are providing unprecedented details of relevant molecular pathways. For those interested in cellular pathways in general, visit *pathguide* (see note 6) to access over 200 databases of biological pathways. I think the analysis of these pathways is where a lot of research and business opportunities lie, not just in cancer research but also in understanding other human diseases. It is no surprise that big pharmaceutical companies are keenly interested in reliable pathways databases and are actively developing computer software for analyzing these pathways. Sprouting from academia in the US are biotech companies that are established very cheaply by a few smart students with modest computing power (visit, for example, the website given in note 7). I believe that smart students in the Philippines can do the same. At the moment, a key problem in using online pathways databases is how to integrate them, cope with the uncertainties in the data, and extract network models for diseases of interest. In a recent international bioinformatics conference held in Brazil (see note 8), I presented my ideas on this issue in a set of tutorial lectures which can be downloaded from the website (see note 9). There is an urgent need for a network analysis software that could identify drug targets. I think multidrug therapeutic strategies will be the answer for cancer, and for many other diseases.

In some of those regular Friday night gatherings (of UPLB Chemical Society members) when I was still an undergraduate at the University of the Philippines at Los Baños (to play bidding bridge, the Society's unofficial card game at that time), I recall asking the question, "How can inanimate molecules organize and interact to create life?" Although I may not have been conscious about it, the longing for an answer steered the education and research I have pursued for almost three decades. My naivete was exposed as the years went by and I became more skeptical about whether I would ever hear an answer in my lifetime—until recently, when my optimism began to build up into excitement. Now I feel that the systems biology approach may ultimately provide us the opportunity to formulate a good answer.

Notes

1. DNA—deoxyribonucleic acid; "letters" A, T, C, G correspond to the first letter of the following chemical bases: adenine, thymine, cytosine, guanine. RNA—ribonucleic acid.
2. *www.plos.org* and *www.biomedcentral.com*
3. *ocw.mit.edu*
4. *www.ncbi.nlm.nih.gov/entrez/query.fcgi?db=Books*
5. *www.google.com*
6. *www.pathguide.org*
7. *www.gnsbiotech.com*
8. *www.ismb2006.cbi.cnptia.embrapa.br/tutorials.html*
9. *www.mbi.ohio-state.edu/publications/pub2006.html*

November 2, 2006

Marla Endriga

Bioinformatics for Everyone

In 2003, after a thirteen-year worldwide effort called the Human Genome Project, scientists were able to determine the complete sequence of the human DNA. It is composed of some three billion units called bases, represented by the letters G, A, T, and C. The sequence of these letters spells out the instructions that make us *us*, from the curve of our lip, to the shape of our eyes, even some of the diseases we are prone to. However, deducing the sequence of the human genome is tantamount to having a book written in a foreign language, this one with a four-letter alphabet. The task now is to unlock the meaning of the sequence that has been discovered.

Enter the field of bioinformatics, or the computational branch of biology. It is also referred to as a hybrid of computer science and biology, which may sound strange, since there is nothing biological about computers. Or is there?

Computers are very commonly used by bankers, students, and by those in the call center industry. They are also very useful for the scientist, and not only for typing up scientific articles. In addition to carrying out experiments within a living organism (in vivo) or in an artificial environment (in vitro), researchers can now also perform experiments on a computer (*in silico*) to answer biological questions. Thus, the union of the fields of computer science and biology.

Aside from human DNA sequence information, an offshoot of the Human Genome Project was the development of advanced technologies that vastly increased the speed of sequencing experiments. Thus, the DNA sequence of other organisms such as the chicken, cow, pufferfish, and many others have also been determined (see *http://www.ensembl.org/index.html*). Managing and organizing strings and strings of letters, or sequences numbering in the hundreds of thousands, such that they can be accessed and used even by those not familiar with programming and database management is no menial task, and this is one of the major areas of bioinformatics. The data that were generated from the Human Genome Project, for example, are stored in databases such as the Genbank *(http://www.ncbi.nlm.nih.gov/)*, the DNA Databank of Japan *(http://www.ddbj.nig.ac.jp/)* and the European Molecular Biology Laboratory *(http://www.ebi.ac.uk/embl/)*. These three databases contain the same information (scientists like to keep more than one copy of their data) and are scanned every day for errors. The information in these repositories of vast amounts of data is freely available to the public via the Internet, and there is no restriction on its use or redistribution. Scientists from all over the world who are able

Institute of Biology, University of the Philippines Diliman, Quezon City, 1101, Philippines
marla.endriga@up.edu.ph

to elucidate new sequences in their respective labs can likewise submit new information for inclusion in the databases, thus increasing the data available for use by the scientific community.

But of what use are these data?

One of the most common uses of sequence information is in the determination of evolutionary relationships among organisms. An undergraduate student, for example, may be curious to find out if he is more genetically similar to a frog, a mouse, or a fruit fly. Or, one may be curious if marine snails that possess similar shell patterns are indeed more genetically similar to one another than to those that have different shell designs.

There are also databases that contain not DNA sequence information, but protein sequence and other related information [for example, SwissProt *(http://www.expasy.org/sprot/)* and the Protein Data Bank *(http://www.rcsb.org/pdb/)*]. Proteins are those other molecules of interest in molecular biology. If DNA is composed of chains of bases, proteins are composed of twenty types of units called amino acids, strung together and folded in a particular way. Examples of protein molecules are the collagen in skin, insulin that breaks down sugar in the body, and amylase that aids in digestion of food in the mouth. Proteins range in size from a few (about twenty) amino acids to many thousands.

Each amino acid in a protein molecule is composed of atoms (say, nitrogen and carbon). Science is at a state where the position of each single atom of each amino acid is known, for tens of thousands of proteins. Thus it is now possible, using such information, to visualize how protein molecules appear in three-dimensional space. Software such as Rasmol and Protein Explorer *(http://www.umass.edu/microbio/rasmol/)* may be used for such a task. Using these freeware, it is even possible to turn the molecule around with the click of a mouse button, and see the protein from different angles.

Armed with protein structure and sequence information, molecular visualization tools, and a firm grasp of chemistry and physics, researchers are now able to design and construct their own protein molecules for use as therapeutic drugs.

Designer drugs are an exciting application of bioinformatics techniques. When making tailor-made molecules against diseases, scientists can control the specificity of the molecule. The idea is that the drug, once administered in the body, will react only with the specific molecule it was designed to interact with, thus decreasing side effects and increasing efficacy of the drug. An example of a work in progress are proposed structures of anti-inflammatory agents called COX-2 inhibitors, which are more potent versions of aspirin, designed at the Department of Physical Sciences and Mathematics at UP Manila. The use of bioinformatics tools in such projects decreases the need for the trial-and-error method traditionally used in screening would-be drug candidates.

One of the logical outcomes of studies at the cellular and molecular level is the eventual simulation of processes that occur simultaneously inside a living cell—the reactions that take place inside it, how it produces energy to perform its functions, how it communicates with other cells around it or some distance from it, what happens when it encounters an intruder, and processes like aging and cell death. This type of work requires not only knowledge of biology and chemistry, but of higher physics and a lot of mathematics as well. Sometime in the future, scientists may well come up with a fully simulated cell, thus increasing understanding of the unit of life. That would certainly lead to more advances in the biological and medical sciences.

As with any scientific field, bioinformatics continues to evolve. Though the initial focus has been on molecular applications, bioinformatics approaches are now also used in organismic and ecosystem-level studies. Aside from UP Manila, bioinformatics techniques are used in other local institutions like St. Luke's Medical Center, UP Diliman, Ateneo de Manila University, the Advanced Science and Technology Institute of the Department of Science and Technology as well as other universities and research institutions, in many research projects in human health, agriculture, and other areas of interest. A number of graduate and undergraduate theses that use bioinformatics techniques have also been completed. Bioinformatics not only involves the biologist and the computer scientist, but the chemist, physicist, mathematician, and statistician as well. Getting into bioinformatics is not difficult at all, as even undergraduate students can participate. All that's needed to start is a computer with a good Internet connection, an inquisitive mind, and a healthy dose of imagination.

May 11, 2006

Maria Pamela David

Going Loco over Bioinfo

Bioinformatics? How (un?)interesting does it sound to you? How interesting could looking at and analyzing letters upon letters of DNA and protein sequences and giving a meaning to it be, or finding patterns or consensus among thousands of sequences and deducing its significance, or defining the proximity or distance between organisms, or finding the significance of a peptide from a sea of proteins—perhaps tracing the lineage of a creature? These are just some of the diverse tasks that the field of bioinformatics would aim to perform—quickly. With the influx of millions of sequences from experimental efforts, it is no wonder that bioinformatics—a marriage of the computational and biological sciences—is emerging as one of the most important fields today. Basic knowledge of it—at the very least, knowledge of sequence alignment and related applications—is fast becoming a necessity in the biological field. Without it, it would be impossible for all the data from much hard, experimental work to be managed, explored, and used to its full potential.

Despite this knowledge, my involvement in the bioinformatics effort of our lab was initially so-so. I was partly, but not wholly, interested in it. It just seemed too dry—literally and figuratively speaking—possibly a mere academic exercise whose practical applications seemed far, far away. Modeling. Mathematics. Maybe Phylogenetics. These were simply not my stuff, in street parlance. Apparently, most of those who perform lab experiments in biology or molecular biology, including myself, were more interested in wet lab than dry lab. Even as a student, I pretty much shrugged off bioinformatics as an accessory, rather than considering it as an emerging necessity.

Everything changed when a challenge came. Our group was tasked to analyze a bunch of protein sequences for mutations and attempt to find general rules for these, or at least, mutational preferences in these. It was not a matter of similarity or homology searching, or even structure prediction, at which most in our group were adept. It was more a matter of looking at particular mutations at specific positions and tallying them in accordance with certain criteria. Does this mutation occur at a buried or an exposed residue? How many were there of a particular mutation? How many of the other? What were the implications of each? What are the mutations that occur for a given position? These were just some of the questions that we were supposed to answer, and with the hundreds of sequences, we hardly had a clue. There was no beating around the bush: no preexisting software, legal or illegal, licensed or cracked, could do the task. You could

University of the Philippines, Diliman, Philippines.
pamela_barbieri@yahoo.it

do the analysis manually (with a little help from preexisting spreadsheet applications), get very cross-eyed, develop carpal-tunnel syndrome, eventually discover mistakes that would make you wonder if you didn't make mistakes farther back, and be so frustrated that you want to give the pile of sequences a vicious kick; considering that every error matters, this is a difficult and even dangerous option. Or you could make a program. The right decision appears obvious and easy—only that we—I, in particular, didn't know how to use the only free compiler, PERL, available. It's one thing to learn a language—it's another to make a multitasking program in a week's time. Initially, we had to perform manual analysis while learning the language, since both the initial results of the analysis and the program were supposed to be finished or nearly finished within a week. In our case, we were lucky: we managed to make the program, and it yielded the same results as our manual analysis. I had a former classmate who learned the language before I did, and he helped me considerably by giving examples. It was easier to learn that way, than strictly by the book. Through sample programs—at least in PERL—you would be able to see what works in a particular manner, and what works in another way. It helped that the syntax of PERL was relatively easy to understand: a line, for instance, that says, "while ($protein = ~ /A/ig) {$A++}" basically would check each letter, which, in turn, represents an amino acid in the protein sequence stored in $protein; while it is an 'A' (for alanine), it adds a value to the variable $A. As such, it would perform that task of counting all alanine residues in a given sequence.

Right now, we have improved the program by giving it more tasks. It could now analyze particular segments of a set of sequences and give results for each—these, with only a few additions to the code. And we are getting interesting results. It's the beauty of programming—that you could have so many tasks done by invoking just a few commands.

Probably, the key to appreciating bioinformatics and programming in general lies in organization, logic, and an imaginative outlook. It was, at least for me, a matter of envisioning the kind of data that you want to get, and outlining the ways by which you could get it. It's best to come up with realistic and defined targets, and to attempt to find a variety of simple solutions for these. Coming up with a variety of approaches to a problem would give on back-ups, and would make things move more quickly and easily in case of failure. Once you know the syntax of a programming language, much in the same way that you have mastered the rudiments of Filipino or English, it's somewhat easy to translate your logic into the language. Of course, one would initially encounter lots of syntax errors that would require debugging, but making a program is not an impossible task. Just don't let it eat you. Usually, the simpler, the better: there would be less room for error, and debugging would be considerably easier. It may be equally frustrating as manual analysis, but once you get over the hurdles, the results are often rewarding. Even if you don't get anything really useful from the program from which you had high expectations, at least you could be proud of the fact that your program worked. And at least, you know that your data set could no longer be expected to yield the information that you wanted, in which case you could more easily move to another data set. Sadly, the frustrations of having to repeat a series of experiments to generate that other data set is another matter.

Having sat for a considerable time in front of a computer doing bioinformatics work, I have found a degree of unexpected fulfillment in it that equals the fulfillment that I have experienced in performing or participating in a well-done experiment. It's probably a matter of time for more people—especially biologists—to appreciate bioinformatics too, since it's slowly becoming an indispensable tool and, as such, it is highly probable that it would be inescapable as well. Most scientists are learning to write their own programs. Better to put a degree of fun and garner interest in it while you can, than be forced to learn it by circumstance. Learning bioinformatics and allied programming applications at your own pace is much easier than being forced to learn it, though a little pressure may help. Do not be intimidated by PERL, the bioinformatics language of choice, since it is quite a logical language—as a high school student some years ago, I nearly flunked the first quarter of Pascal, yet I could now work with PERL. If programming is really something that gets on your nerves, at least, try to familiarize yourself with pre-existing applications that could be applied to your field of science. For a start, there's DNAsis for DNA and protein sequence analysis, QSAR for binding kinetics and drug design, NCBI-BLAST for sequence alignment and the Swiss-PDB viewer and PSIPRED server for protein structural analysis, structure prediction and modeling. For most standard applications, there would usually be pre-existing programs that one could play around with. Some may be bought for a song, and usually, if one really, really needs the application, an urge (and a necessity) to learn programming is created. A few words of advice: an academic e-mail address is usually, but not always, a passport to the free use of some web-based bioinformatics tools, and cracking a licensed program would still not allow you to use it for a publication.

Before totally shutting out bioinformatics, try to find a niche that excites you. It has a lot to offer. Who knows? Maybe what you unravel will earn you a PhD!

October 14, 2004

Alfonso Albano

Comets, Chaos, Brains, and the Human Heart

In 1986, right on schedule, Halley's comet once again came close enough to Earth to be visible to the naked eye. Its name honors Edmund Halley who calculated its trajectory in 1705 using laws of motion that his friend, Isaac Newton, had formulated some fifteen years earlier. Later historical research suggests that it was the same comet recorded by Chinese observers in 240 BC, described in Babylonian clay tablets in 164 BC, and shown in the Bayeux tapestry that depicts the Norman conquest of England in 1066. Like clockwork, it has been visiting the Earth's neighborhood more or less every 76 years for a long, long time. It is due to visit us again in 2061.

We have come to take for granted the precision with which some phenomena are predicted and controlled using Newton's physics. Eclipses of the sun and of the moon occur and comets reappear just as they had been predicted years ago. Spacecraft are guided toward the moon, Jupiter, or Mars.

Yet, there are phenomena which cannot be predicted with the same kind of long-term precision as that possible with comets and planets, even if all these are subject to the same physical laws. An everyday example is the weather. Ultimately, the motions of the atmosphere are subject to the same laws that govern the motions of comets, yet weather forecasting, even with today's computers, is a rather imprecise art. Short-term weather forecasts are usually more or less accurate, but detailed forecasts that go weeks into the future are so unreliable, they are usually considered not worth making.

We may claim that the comparison is not quite fair, as the weather involves the complicated interactions of many elements, while the most significant interactions of a comet involve only a few objects—mainly the sun and the larger planets. It turns out, however, that this inability to make precise predictions can occur even when there are only a few elements engaged in relatively simple interactions. One famous example is the gravitational interaction of three bodies, famously tackled but not solved by the mathematician Henri Poincaré at the turn of the last century.

One of Poincaré's results is that if a small object were moving in the neighborhood of two large ones, then it would not necessarily follow the orderly, repetitive trajectories of planets and comets moving around a solitary sun. It could instead follow an irregular orbit that never repeats itself. Having been at one location moving with some velocity, it will never find itself there again, moving as it did the last time around. If the object

Department of Physics, Bryn Mawr College, Pennsylvania, USA.
aalbano@brynmawr.edu or alfonso.albano@gmail.com

were a planet moving near a binary star, the length of its year would change seemingly capriciously, without pattern or order.

This seemingly random motion which is nevertheless governed by well-defined rules was called *chaos* by the mathematician James Yorke some three quarters of a century after Poincaré. But it is not the helter-skelter, formless chaos of old. There are rules. And because there are rules, well-defined initial conditions lead to predictable futures—at least in principle. What makes it complicated in practice is that in this newly defined chaotic regime, slightly different starting conditions can lead to vastly different outcomes.

Small changes in the speed of Halley's comet caused, say, by spacecraft sent to study its composition the last time it was near Earth will cause only small changes in the time it takes to come back for its next visit. Cataclysmic volcanic eruptions or the detonations of nuclear bombs on Earth lead to no measurable changes in the duration of our year. But slight local changes in temperature or wind patterns can have large effects on earthly weather. Meteorologist Edward Lorenz once asked, "Does the flap of a butterfly's wings in Brazil set off a tornado in Texas?" The disturbance caused by the butterfly in Brazil is not likely to be large enough to rise above the random fluctuations of wind speeds in its neighborhood. It cannot plausibly be blamed for the tornado, but the sensitivity of chaotic motions to small disturbances makes it legitimate to ask the question. Small changes can lead to dramatically different results; one additional straw can break the camel's back.

The other side of this coin is that there may be order lurking behind some of those seemingly random sequences of numbers measured from physical and biological systems. This realization has led to the development of computational tools for the analysis of just such complex behaviors. These tools do not depend on the precise nature of any underlying rules, so they are applicable to a large variety of phenomena. They have been used to study such disparate things as insect populations, stock market prices, fluid flow, epidemics, the human heart, nervous systems, chemical reactions, highly excited atoms and, of course, the weather and astronomical phenomena. The rubric that encompasses all these goes by the rather unprepossessing name of *nonlinear dynamics* and has become a truly interdisciplinary field. Its practitioners include mathematicians, physicists, chemists, engineers, clinicians and medical researchers, biologists, economists, psychologists, and ecologists. The list keeps getting longer. Although these scientists are concerned about very different systems, they are now beginning to share a common technical language and a common set of computational tools. This has led to an almost unprecedented degree of conversations and collaborations across disciplinary borders.

One of the more interesting cross-disciplinary collaborations has involved the study of electrical signals generated by the human body. Many of these collaborations have involved studies of the signals generated by the brain and the heart.

An electroencephalogram (EEG) is a recording of the time course of the electrical activity of the brain. This is truly the stuff of life—of human life, that is. Clinical death is now defined as the absence of brain electrical activity rather than the cessation of respiration or heartbeat. An EEG recording is a series of wiggles, its peaks and valleys occurring in various sizes and separated by varying intervals. Sometimes these come in bursts of large oscillations separated by more quiescent periods, sometimes there are long stretches of small-amplitude irregular oscillations. But they are never still, except in death.

Many features of the EEG change with levels of mental activity, or with stages of sleep, or with the state of health. Just by looking at brain recordings, expert encephalographers can tell when the subject has her eyes open or closed, or if she is drowsy, or asleep.

Questions naturally arise: Can nonlinear analysis of the EEG tease out even more useful information? Can it enhance its utility as a medical tool?

- It would be a great boon to epileptics if they can anticipate an impending seizure. There are some who perceive an "aura," warning them that a seizure is about to occur, but this is not true of most epileptics. Could there be some information hiding in the EEG moments before a seizure that can be used to produce an early warning signal—a "computerized aura" of sorts?
- The brain activity of a person telling a lie must surely differ from that of a person telling the truth. Is the difference, if it exists, reflected by computable differences in the EEG? Could this be more accurate than current lie detector systems?
- How does an anesthetist know when enough anesthesia has been administered to a surgical patient? Can the EEG improve how the depth of anesthesia is determined?
- Can the brain's electrical response to visual cues be used for communication by someone who can neither move nor talk?

There are partial answers and even more questions. Sooner or later, somebody is bound to ask, what about consciousness, or memory, or love? These, too, must have correlates in the electrical activity of the brain. These are deep questions, and are very difficult to answer. The brain is such a complicated thing.

The heart, though, is not quite as complicated as the brain. It is a pump after all, and pump is pretty much all it does. The electrocardiogram (ECG/EKG), a record of its electrical activity, is so much simpler and displays much more regularity than the EEG. Physicians are much more confident in its use as a diagnostic tool. They can make precise diagnoses by just studying its deviations from what is considered normal behavior.

The heart's behavior is not quite as regular as that of a comet, however. That, we all know. It beats regularly when one is at peace, but it sometimes beats wildly at the thought of a loved one, or in anticipation of danger. However, we do associate regular heartbeat under relaxed conditions with a healthy, normal heart. It, therefore, came as quite a surprise when it was discovered that an excessively regular heartbeat may signal impending trouble. Studies by cardiologist Ary Goldberger and his colleagues showed that the beating of a normal healthy heart has many of the characteristics associated with chaos. They also showed that an increase in the regularity of the heartbeat is often a precursor of sudden cardiac death. Could it be that chaos is healthy?

In *The Dunciad*, Alexander Pope addresses chaos:

Lo! thy dread Empire CHAOS! is restor'd;
Light dies before thy uncreating word;
Thy hand, great Anarch! lets the curtain fall;
And Universal Darkness buries All.

The new, improved version of chaos is not quite to be dreaded as much as the chaos described by Pope. Pope's *Chaos*! threatens to bury all in darkness. If handled properly, this new chaos could well shed new light on some scientific problems that remain unsolved.

Printed in two parts on November 11 and December 2, 2004

Alfonso Albano

Do Leeches Have Free Will?

If a leech (*limatek,* aka *Hirudo medicinalis*) is poked on its side, sometimes it swims, sometimes it crawls, sometimes it does nothing at all. For a supposedly simple animal, that is quite a remarkable repertoire of responses to the same stimulus. How does it decide what to do? Does it actually decide? Indeed, is "decision" an appropriate term to use in connection with a leech?

When this behavior is shown by a leech in the wild, we may wonder if the poke is the only stimulus provoking its reaction. Does its response depend also on the temperature of the water in which it finds itself? Does it depend on the motion, or lack of motion of the water? Maybe it depends on the way in which the poke was delivered? After all, just "poking" is hardly a controlled experimental procedure.

Mindful of the need to study the leech's behavior under controlled circumstances, biologists have devised a way of studying the animal's nervous system in isolation. By performing a delicate surgery, they detach the leech's nervous system from the rest of its body, they throw away the latter, and study the isolated network of nerve cells, giving it an "out of body experience," as it were.

The state of the isolated nervous system is indicated by the electrical signals that the various nerve cells generate and transmit to each other. Although muscles are no longer there to perform swimming or crawling motions, motor neurons still generate signals that biologists can identify uniquely as commands for the no longer existent muscles to swim, or to crawl. "Fictive swimming" and "fictive crawling," they call these signals.

One reason for the interest in the leech is that its nervous system is relatively simple. Almost every nerve cell and its interconnections have been identified. On the other hand, it is complicated enough that it can display the variety of behaviors that made us wonder earlier if it is, in fact, capable of making decisions.

Once the leech's nervous system has been isolated and mounted on a dish, instead of poking it on its side, an experimenter can now deliver a precisely controlled amount of electrical charge to a neuron in one of the leech's brains (it has two brains, by the way, one at its front end, another at the rear). When stimulated in this manner, under tightly controlled conditions, it is still so that sometimes it displays fictive swimming, sometimes fictive crawling, sometimes nothing at all…

I first heard of this phenomenon from a colleague, Peter Brodfuehrer, who teaches biology at Bryn Mawr College. My immediate reaction was only half in jest—"Surely, this shows that the leech has free will!"

Department of Physics, Bryn Mawr College, Pennsylvania, USA.
aalbano@brynmawr.edu or alfonso.albano@gmail.com

"Shut up," Peter said. "You're a physicist. Go calculate something."

Thus began a collaboration between Peter and his students and me and my students. The biologists do the electrophysiological experiments, the physicists do the numerical analysis of the biologists' results. It has worked quite well. The biologists enjoy manipulating the leech's nervous system, and the physicists, some of whom do not wish to get any closer to leeches than the CDs on which leech data have been recorded, happily deal with the numbers.

So, what has this motley crew found? So far, maybe just the beginnings of what could well be a very long and complex story. Before the stimulus is delivered, the level of electrical activity in the ventral cord of those that will eventually swim (let's call them the "swimmers") significantly exceeds the activity in those that will eventually crawl (the "crawlers") which, in turn, is slightly greater than in those who will do nothing (the "do-nothings"). It is as if there is a threshold level of activity for swimming or crawling to occur and if, before stimulation, the activity is already high enough, it would become easier for the stimulus to bring it up to the threshold. But it gets more complicated than that.

Once the stimulus is delivered, the nervous system activity in all cases suddenly jumps. The swimmer and crawler signals rise to the same maximum level, then they both decline following pretty much indistinguishable trajectories. After about a second, however, the swimmer signal rises again, diverging from the crawler signal, which now has leveled off. Some three seconds on average after the divergence of the signals, with the swimmer signal still rising, swimming begins. Crawling also occurs at about this time.

The signal from the "do-nothings," on the other hand, does not rise to the same maximum value as those of the swimmers and the crawlers. It starts declining at the same time as the other two, and levels off when the crawler signal does, but it does so at a significantly lower value.

The "decision," if we can call it that, to do something rather than nothing seems to be made at stimulus onset and may well be dictated by the pre-stimulus level of the ventral cord electrical activity. The response of the do-nothings to the stimulus is consistently lower than that of the swimmers or the crawlers.

The swim/crawl decision comes later, when the swimmer signal diverges from the crawler signal. It is not clear if this decision could also have been dictated by the pre-stimulus level of the ventral cord activity. Between stimulus onset and the swim/crawl decision, the indistinguishability of the swimmer and crawler signals suggests that these signals do not carry any memory of their pre-stimulus levels. Or do they?

So, what determines the swim/crawl decision?

The study of nonlinear dynamical systems has shown that for systems capable of what is now called chaotic motion, very small differences in initial conditions can lead to widely divergent futures. Edward Lorenz, one of the pioneers in the study of chaos, is famously quoted as asking, "Can a butterfly flapping its wings in Brazil cause a hurricane in Texas?"

Could it be that this is what's happening here? Could it be that whether the leech swims or crawls ultimately depends on a cause as seemingly insignificant in global meteorological terms as the flapping of a butterfly's wings in Brazil?

This is by no means an original speculation. In the mid-1980s, when nonlinear dynamics was becoming a fashionable scientific field, a group of young pioneers—James Crutchfield, Doyne Farmer, Norman Packard, and Robert Shaw—wondered if perhaps "... chaos provides a mechanism that allows for free will within a world governed by deterministic laws."

Crutchfield and his friends were, of course, referring to free will in people... but what about leeches?

October 12, 2006

Part 10

Mathematics, Physics, and the Engineering Sciences

Eusebio L. Koh

Math and Aftermath

Many years ago while taking an electrical engineering class at the University of the Philippines in Diliman, I was thoroughly impressed by my professor, Bartolome Blanco, who solved a differential equation with initial conditions by some sort of a "trick." He called his trick the Laplace transform.

With thirtysome bright-eyed classmates agog at learning new ideas, I was wondering if he was pulling our legs. Ever quick on the uptake, Prof. Blanco said, "If any of you wish to impugn what I taught here, you can check out my solution by plugging it into the equation or you can run to the Math Department to see what was going on."

Wow! Some of us had brushed elbows with the simplest differential equations, never heard of transforms, except in reference to converting high voltage to low voltage in AC circuits. And "impugn"? I had read that word somewhere in a long forgotten book but that was the first time I heard it used. Now you understand why I was impressed with my "Prof" and had visions of following in his footsteps.

So much so that I jumped on the opportunity to teach engineering at UP upon my graduation, when Dean C. Ortigas offered me an instructorship. At the same time, I took more math classes on the side: a class on differential equations here, a class on advanced engineering math there. It certainly gave me a step up when UP later sent me abroad for graduate studies.

Now fast forward to what I ended up doing. My doctoral dissertation at the State University of New York at Stony Brook was on the Hankel transformation of generalized functions with applications to sorts of differential equations. This is really a mouthful and I am going too far ahead of myself. With your indulgence, let me start from scratch.

1. Mathematics is the study of relations and their properties. A relation is a correspondence between two sets of objects. These objects can be anything: numbers, figures, sets, or even relations, too. When the correspondence assigns a unique object from the second set to each element of the first set, we call that relation a function. A function is usually restricted to a relation between sets of numbers. If the relation is between sets of functions, we call it an operator. If the relation is from a set of functions to a set of numbers, we call it a functional. Here are some examples.

Functions: linear, quadratic, rational, exponential, trigonometric, special.
Operators: differential operator - d/dx, integral operator- $\int \ldots dx$.
Functionals: definite integrals, Dirac delta.

Mathematics Department at the University of Regina, USA.
dony@accesscomm.ca

2. Calculus is a branch of math dealing with change. The rate of change of a function f is its derivative f' and the process of finding the derivative is called differentiation. Because our world is always in a state of flux, processes, natural or otherwise, can be described by equations involving derivatives. These are called differential equations, which are creatures more intractable than algebraic equations. Their solutions are functions while the latter has numbers for solutions. For example:

Algebraic equation: $x^2+4x+3=0$ is solved by $x = -1$ or $x = -3$.
Differential equation: $y''+4y'+3y=0$ is solved by $y=a[\exp(-x)]+b[\exp(-3x)]$,
a & b constants.

The last equation describes a simple damped spring mass system or a simple electric circuit with a battery, a capacitor and a resistance and the solution is a combination of two decaying exponential functions. The solution is found by the observation that an exponential function replicates itself on differentiation. Thus the task is reduced to solving an algebraic equation. More complicated equations are solved by such techniques as infinite series, integral transformation, Fourier series, special functions and numerical methods and computer codes.

3. An integral transformation is an operator that takes a set of functions and maps it into a set of transforms via integration. The most famous is the Laplace transform given by $F(s) = L[f(t)] = \int \exp(-st)f(t)dt$ with the important property $L[f'(t)] = sL[f(t)]-f(0)$. Thus when $f(0) = 0$, differentiation corresponds to multiplication by s in the transform domain. The Laplace transform has been used to justify Heaviside's calculus used by engineers. The term exp(-st) is called the kernel of the transform. By changing the kernel as well as the range of integration, other integral transforms were developed and used for solving other equations. The Hankel transform replaces exp(-st) with $\sqrt{st}$ Ju(st) where Ju(st) is Bessel function of order u. This transform is suitable for differential equations with variable coefficients that arise in physical problems with axial symmetry. Other transforms go by such names as Fourier, Weierstrass, Stieltjes, Meijer, Hilbert, Jacobi, convolution, etc., each with its suitability for certain differential operators. The definitions, properties, and applications are enough to fill several books.

4. In 1926, Dirac introduced his d-function as $d(x)=0$ when $x \neq 0$, $\int f(x)\, d(a-x)dx=f(a)$ and used it in quantum mechanics. This is not a function in the classical sense. To be sure there were other improper functions such as Hadamard's pseudofunction pf x, Cauchy's pv of a divergent integral, derivatives of Dirac - δ that have appeared in some physicists' works. In the 1950s, Laurent Schwartz came up with his Theory of Distributions to explain all these as continuous linear functionals on test functions with compact supports. In the '60s, the Russians Gelfand and Shilov came up with their four-volume treatise on Generalized Functions not drastically different from Schwartz.

5. In the '60s and '70s, Armen Zemanian and others developed various integral transformations of generalized functions (gfs). Zemanian and I developed the generalized Hankel transform under some restrictions. Later, I went farther than our work going into representations of Hankel transforms, transforms of negative orders, transforms in higher dimensions. I later developed the generalized Meijer transform (with my student M. Ali), Hankel Transforms of Banach-space-valued gfs (with my student C. K. Li).

6. There are two methods of extending the integral transformation to generalized functions. The Adjoint Method uses a Parseval-type relation to define the transform of a gf F as the application of F on the transform of a suitable test function. Schwartz used this method for the Fourier transform, Gelfand and Shilov for the Hilbert transform, Zemanian for the Hankel and other transforms. What's involved is the construction of two test function spaces such that one is the transform of the other. Details can be found in their books.

The other method defines the transform as the application of the generalized function F on the kernel as an element of a test function space. This necessitates the construction of a suitable complete semi-normed space that contains the kernel. Zemanian used this method for the Laplace transform and I did it for the Hankel transform. Details may be found in Zemanian's book and Brychkov and Prudnikov (1998).

The University of the Philippines is a great source of inspiration for our youth despite its lack of material resources. It is, however, rich in a faculty that is a fountain of wisdom and knowledge. I credit Prof. Blanco in igniting a spark to my career. But there were many excellent professors I had. I would be remiss not to acknowledge Professor Josefina Constantino who instilled in me a love of the English language. From the time I became Professor Emeritus in 1999, I have published short stories, poems, and essays online and in print. In these writings my math background seems to creep in.

American poet Adelaide Crapsey invented a verse form in the 1920s called cinquain, which is a poem of five iambic lines of two, four, six, eight, and two syllables. There are no restrictions on rhyme but there is a central theme in the cinquain. In 1927, Angela M. Gloria published three cinquains in the *Philippine Herald*. Here are some unpublished cinquains of mine on scientists:

NEWTON

The man
Felt the apple
Dropped on his head with such
Gravity, he couldn't help yell,
"Aha!"

Einstein

He did
Tinker with his
Relatives; speed boggles
The mind as energy and mass
Equate.

Archimedes

He ran
Out in the buff
Hollering "Eureka!"
For he had found that buoyant force
Pushed up.

Leibniz

Symbols
He used for thoughts,
Logic was the richer,
And calculus was brought to the
Limit.

August 24, 2006

Carlene Pilar-Arceo

The Zero Mystique—Why Not?

A symbol for what is not there, an emptiness that increases any number it's added to, an inexhaustible and indispensable paradox...—Robert Kaplan, The Nothing That Is: A Natural History of Zero (October 2000)

At home one lunchbreak to nurse my then three-month-old daughter, I found myself watching Kris Aquino's *Pilipinas, Game Ka Na Ba*? Always an interesting show, it became extra interesting when a math question came up, something about "the number with no Roman numeral representation." Almost mechanically, I said to myself "Zero," a second before a contestant got it, too. In my mind I congratulated the contestant for knowing her numbers. Then I wondered, "Does she know why zero has no Roman numeral representation?" Then I wondered more, "How many of those watching or listening know the reason?" And let me wonder now—how many readers of this column, or of this newspaper even know the reason? How many would even bother to find out? Believe me, those who would bother will find the road to discovery very rewarding.

If I were handling a general education math class this semester, that noon break would have given me a very stimulating homework question. In such a class I handled a few semesters ago, I made my students track the history of precisely that number, zero. Zero being zero, it is usually not given much notice, not until it gives students headaches in calculus class when they take up, among others, indeterminate forms and limits of functions.

Surprisingly, such was already the case from the heyday of bigwigs like Pythagoras and Archimedes and way into and past the glory days of Rome. For the longest time, the brilliant minds of Greek and other early Western civilizations actually refused to acknowledge the existence of zero! Take the Paradox of Zero. The story relates how Achilles will never meet up with a tortoise plodding ahead of him, if he gives that tortoise a head start. It argues that whenever Achilles is about to catch up with the tortoise, the tortoise would have added a new distance between them, no matter how small, and so there will always be some distance, no matter how small, between Achilles and the tortoise! That he would "zero in" on the tortoise, though thoroughly logical, was inconceivable and unacceptable because it was simply unexplainable ... to Western minds, that is. Had they welcomed zero way back then, the Greeks might have been able to discover calculus themselves, and not Newton and Leibniz centuries later.

To borrow from Kris Aquino again, "Why not?" Why did the early Westerners not, nay, refuse to, study zero? Mathematics and philosophy have very close ties. Most

Mathematics Department, University of the Philippines Diliman, Quezon City, 1101, Philippines.
cayenp_arceo@yahoo.com.ph

early mathematicians were philosophers and vice versa. For them to consider zero, and consequently infinity, at that time was to go directly against Aristotelian teaching, which was the Bible of that era. Ergo, to consider zero at that time was to invite ridicule, ostracism, excommunication, things they could not afford.

There was no such fear for the Eastern philosophers. Unlike their Western counterparts, they embraced zero and infinity. Did they not believe in reincarnation and the never-ending circle of life? And so the Egyptians, the Babylonians, the Hindus, etc. had zero well ensconced in their number systems. Enter the Greeks, who under Alexander swept through Egypt and the Babylonian, Assyrian, Persian, Hindu civilizations. Among the precious loot they carried back home was the concept of zero, which finally gained acceptance and earned its proper important place in the West. Its oblate shape is taken from the Hindu system, hence helping form what we now know as the Hindu-Arabic numeration system. Note that the oblate shape connotes the absence of a beginning and an end, as does the symbol for infinity.

Imagine life without zero. Imagine writing thousands and millions using Roman numerals. The number 3,888 is already hideous when written in Roman: MMMDCCCLXXXVIII. What more 3,888,000!

Furthermore, imagine doing arithmetic using Roman numerals. Standing alone zero may mean "nothing," but it makes possible a wide variety of calculations. To begin with, it made arithmetic much easier. Computers actually function through a language using only zeros and ones.

How about financial matters? Do we not measure the significance of one's salary by the number of zeros in it, and thus one's financial advancement by declaring "*Nadagdagan ng zero ang kita ko!*" In which case zero definitely does not mean "nothing."

Indeed, zero also makes for picturesque speech. The urgency of the moment is emphasized by the phrase "zero hour;" the closeness to a target by the phrase "zero in;" the flattened site of a bomb explosion by the phrase "ground zero."

Zero. I could go on and on about something which ironically is taken to mean "nothing." A paradox? A mystery? So the story of zero goes: it is nothing but it is everything, it is neither positive nor negative, it is empty yet it makes possible the infinite. This is the zero mystique.

May 18, 2006

Carlene Pilar-Arceo

The License Plate Game or Something to Think About While Caught in Traffic

One activity I entertained myself with as I was growing up was looking at vehicular license plates. Sitting through a traffic jam, I would look at a license plate's last three digits and, oblivious to my parents' complaints about the traffic, happily spin numbers in my head. It made me an ideal passenger, always quiet and content.

At that time, plate numbers were still of the "letter-digit, digit-digit-digit" combination. Sometime in my teen years the license plate combinations changed to "letter-letter-letter, digit-digit-digit." Not bothering then with the reasons for the change, I expanded my number game to trying to count the total number of three-letter combinations.

As my study of mathematics deepened (meaning, as I learned how to count properly mathematically), I realized that the change from the five-character combination to the six-character one implied a huge increase in possible vehicle registration, and that traffic jams that would ensue would be just as huge! Now I no longer wonder why traffic jams are inevitable, especially in Metro Manila. This would give my parents, and all commuters, more to complain about.

Let us dissect the present license plate configurations, and concentrate on those from Metro Manila, a.k.a. the National Capital Region (NCR). First, let us look at the first three figures: the three letters. To make computation less complicated, I will fix the first letter, say, N. NCR license plates originally began with N, with the exception of course of government vehicles or "red" plates which start with S, and other special registration plates. (Region I license plates begin with A, Region II license plates with B, Region III license plates with C, and so on.) Next, we note that the letters I, O and Q are never used in license plates. That means that the second and third letters may be any of the 23 remaining letters of the alphabet (letter repetition is allowed). Rules of counting dictate that the total number of three-letter combinations with the above restrictions is 1x23x23 or 529.

Next we look at the last three figures: a three-digit number. This automatically bars 0 as the first digit. However, repetition is allowed so the total number of three-digit numbers is 9x10x10 or 900. (We may avoid the hard math for this by just figuring out how many numbers there are from 100 to 999.)

Mathematics Department, University of the Philippines Diliman, Quezon City, 1101, Philippines.
cayenp_arceo@yahoo.com.ph

Again, counting rules dictate that if we want the total number of six-figure combinations with all the above restrictions, that is, three letters where the first one is N and a three-digit number, then the total number of such license plates is 529 x 900 or 476,100.

I am sure you are aware that license plates in the region have gone through several different first letters already, namely, N, P, T, U, and W. When I began writing this article new license plates still started with X. Now, I am seeing plates starting with Z! This means that vehicle registration in NCR has used up 6 x 476,100 or 2,856,600 license plates! This translates to 4,560 vehicles per square kilometer of Metro Manila's 636 square kilometers. As I noted earlier, this still excludes government "red" plates, customized/personalized plates that start with letters other than N, P, S, T, U, W, and X, and possibly diplomatic "blue" plates that follow an entirely different configuration.

Fortunately for our health and travel concerns, theory and actual fact differ significantly. Not all of these almost three million vehicles are in use. I expect that quite a number have found their way to different endings—some sitting and rotting away in streets, garages and junkshops, some impounded, some "chop-chopped." Then there is the Department of Transportation and Communication (DOTC) directive implemented last year wherein jeepneys and buses aged fifteen years and older can no longer have their franchises renewed. Hooray! I hope the implementation has been successful. In any case, the DOTC-Land Transportation Franchise Regulatory Board (LTFRB) website says that the actual number of registered motor vehicles in Metro Manila, as of September 30, 2004, was just 852,045 (32 percent of total motor vehicle registration for the entire country). I say "just 852,045" relative to my initial estimate of almost three million vehicles. Then again, this is on the assumption that all vehicles have been properly registered. Still and all, 852,045 is as big as it looks and sounds.

Imagining 852,045 vehicles moving around in a space that is roughly only 636 square kilometers overwhelms me. Metro Manila covers only 0.2 percent of the Philippines' total land area, yet it is home to almost one-third of all motor vehicles in the country. Sadly, this is not imagination. This is for real. The DOTC-LTFRB figures translate to 1,339 vehicles per square kilometer! Now should we wonder why we have perennial traffic jams and worsening pollution? Why the Metro Manila Development Authority (MMDA) is continually cooking up schemes such as color-coding? The color-coding scheme aims to reduce vehicular traffic by 20 percent. This 20 percent already means 170,409 vehicles; but a whopping 681,636 remain! Moreover, even with the number of flyovers and flyunders being constructed, traffic still does not "fly" through these thoroughfares. Though they are welcome attempts at easing traffic flow, most usually turn out to be "crawl-overs" and "crawl-unders." The biggest culprit is sheer vehicle volume. Perhaps the government should also look for schemes of regulating vehicle registration instead of easing traffic flow, which is near impossible if registration is unlimited within a very limited land area.

I have observed that Region I has not used up its A-plates, Region II its B-plates, Region III its C-plates, and so on. Meanwhile in NCR, we are now onto our Z-plates (remember, we started with N-plates and proceeded to P, T, U, W, X)! By the time the regions use up their plates, and we in NCR use up our Z-plates, we can already imagine the volume of vehicle registration we have surpassed. Meanwhile, what we cannot yet imagine is the next license plate scheme. Will it have more characters than three letters and a three-digit number? This is something worth waiting and seeing ... and counting!

May 25, 2006

Carlene Pilar-Arceo

Pure Luck

Millions of us line up on Saturday nights, hoping and praying that six carefully chosen numbers from 1 to 42 will bring us riches galore. The truth is that no matter how much care we take in choosing those six numbers, the chance that all of them (nay, that even just one of them!) will be drawn is very, very small. First of all, I wish to make clear that this article does not intend to discourage you from betting, lest I incur the ire of the Philippine Charity Sweepstakes Office (PCSO). It merely wants to inform you of the total number of ways of getting a combination of six distinct objects, in any order, from a set of 42 distinct objects. In plain English, we want to answer the question, "In how many ways can we get six numbers from 1 to 42?" More to the point, "What are the chances of my bet(s) being drawn?"

It's not as daunting as it sounds. Think of the *tambiolo* (I don't know what they call the machine that spews out the balls with numbers written on them so kindly allow me to use the well-known term, tambiolo.) Think of what happens to the 42 different balls as one by one, six of them are drawn, and not returned to the tambiolo.

Unang bola. Of course, at the very start, there are obviously *42* different choices inside the tambiolo. After one ball is drawn from/spewed out by the tambiolo, obviously, 41 balls remain.

Ikalawang bola. Now, there are *41* different choices. After another ball is drawn from/spewed out by the tambiolo, obviously, 40 balls remain.

Ikatlong bola. Now, there are *40* different choices. After another ball is drawn from/spewed out by the tambiolo, obviously, 39 balls remain.

This goes on until...

Ikaanim na bola. Now, there are *37* different choices. After another ball is drawn from/spewed out by the tambiolo, obviously, 36 balls remain.

I hope you're still with me, because the math just starts here. The rules of counting tell us that to get the number of ways six such balls can be drawn, we must first multiply the numbers 42, 41, 40, down to 37. Thus, we get $42x41x40x39x38x37 = 3{,}776{,}965{,}920$. Whew, almost four billion possibilities! That calls for a brief pause in the computing. (Meaning, this is not the answer yet.)

To help us catch our breath, let's go back to the betting game. Although we may not follow any particular order in writing down our six chosen numbers, most of us encircle them (on the betting form), in ascending order (by row, left to right, top to bottom). For

Mathematics Department, University of the Philippines Diliman, Quezon City, 1101, Philippines.
cayenp_arceo@yahoo.com.ph

example, I lay a bet on the combination 36, 37, 1, 4, 6, 5 and I encircle the numbers in the order 1, 4, 5, 6, 36, 37. Indeed, on the stub returned to me the numbers are written in that order—1, 4, 5, 6, 36, 37. Does this alter my bet in any way? Not at all! In fact, if these six numbers do come out, maybe they won't even come out in those two earlier arrangements. Let us say they come out in the order 4, 5, 37, 6, 1, 36. Does this mean I didn't win? On the contrary, I still did win! As long as the six numbers show up, the order in which they come out does not matter. In mathematics we call this a combination, a selection, or grouping wherein order is not important.

Since we have mentioned mathematics, let us resume computation. We have just established that the arrangement 36, 37, 1, 4, 5, 6, in this discussion, is no different from 1, 4, 5, 6, 36, 37, from 4, 5, 37, 6, 1, 36, and any other arrangement of the six given numbers. However, in our initial computation (3,776,965,920, remember?) they are all accounted for. Meaning, if there are 100 possible arrangements of 1, 4, 5, 6, 36, 37 (in mathematics these are called permutations), all 100 possibilities should be counted only once; also meaning that among the 3,776,965,920, they have been double-counted 99 times! To do away with this double-count, rules of counting tell us that we should divide 3,776,965,920 by the total number of possible arrangements of the six numbers. By a procedure similar to the tambiolo draw, we get *6x5x4x3x2x1* possible ways for the six numbers to be arranged, a.k.a. six-factorial, or 720 ways. This means there were 719 double-counts! Proceeding with the required division, *3,776,965,920 / 720*, we get 5,245,786. This is the answer we set out to find.

So, without further ado...

"In how many ways can we get six numbers, in any order, from 1 to 42?" In *5,245,786* ways.

"What are the chances of my bet(s) being drawn?" If you bet on one combination, one out of 5,245,786, or 0.00002 percent. If you bet on two combinations, two out of 5,245,786 or 0.00004 percent. If you bet on three combinations... If you bet on 2,622,893 combinations, 2,622,893 out of 5,245,786, or 50 percent...If you bet on all 5,245,786 possible combinations, 5,245,786 out of 5,245,786, or 100 percent!

It is inevitably true that the more you bet, the more chances you have of winning. Even if you fail, the PCSO guarantees your hard-earned money will be put to good use so more bets translate to more goodwill. (In that sense you still win.)

Some points to ponder:

Let us say you bet on all 5,245,786 possibilities, are you sure you will get back all the money you bet—more than ₱50 million—and the money of all the other bettors? Think again and don't jump for joy yet. What if someone else also bet on the winning combination? *Ouch!*

An article on the Internet says that your chances of being struck by lighting in a given year may vary between one in 400,000 and one in 240,000, or between 0.00025 percent and 0.00042 percent. (You can boost those figures by going golfing or swimming during a thunderstorm.) That's up to 21 times higher than your chances of winning this lottery!

Finally, no matter how small those percent figures are, no matter how close to zero they are, it is still valid to say they are not equal to zero. So can you get struck by lightning? Sure, it happens. Can you win the lottery? Sure. With pure luck.

September 28, 2006

Amador Muriel

Turbulence—The Last Unsolved Problem in Classical Physics

In November 2003, the Scientific Committee of the International Center for Scientific Culture-World Laboratory, including Nobel Prize winner Tsung-Dao Lee, met at the CERN premises in Geneva to approve a modest project called the "Molecular Theory of Turbulence." CERN stands for Centre Europeen de Recherche Nucleare, which means what it sounds. The name CERN was used as a model by novelist Dan Brown in his thriller *Angels and Demons*. The decision by the Committee is considered routine, as the World Laboratory has made many similar decisions, some of which have contributed to détente between the US and the former USSR. What makes the decision unusual is the title of the project, "Molecular Theory of Turbulence."

Turbulence, since the time of Reynolds in the eighteenth century, has not been considered a problem in molecular physics. Yet, molecules are old hat to chemists. The most interesting chemical and biological phenomena are definitely molecular, but engineers who work with hydrodynamics and turbulence could do without it. It is enough to know that density and viscosity, or *lapot*, should be sufficient to characterize turbulence, until I expressed my heretical opinion in 1995, that turbulence is not a classical physics problem, but in fact a reflection of the quantum nature of matter. In the beginning, the idea was ridiculed because turbulence was viewed for a hundred years as a classical problem, the last unexplained classical phenomenon. There even exists a million-dollar prize posted by the Clay Institute of Boston to anyone who can show that it arises from the Navier-Stokes equation, a classical continuum equation. The best mathematical physicists of the world quietly attack it with the prize in mind. So far no one has succeeded.

In fact, my own preposterous view is that it will not be possible to derive turbulence from the Navier-Stokes equation because the problem is incorrectly posed as a mathematical problem. I view it as a physical problem arising from the following observation: in every experiment that studies the onset of turbulence, each molecule acquires enough energy to excite the vibrational, rotational and, in general, quantum states of the molecules, collectively called internal degrees of freedom, rendering the Navier-Stokes equation inapplicable, since it is only a crude approximation. "What took you so long to realize this?" chemists ask me. The explanation is that the overwhelming difficulty of solving the Navier-Stokes equation to show turbulence has overshadowed the physical picture that chemists already know. Not every physicist or engineer lives

Formerly with the European Organization for Nuclear Research (CERN),
Amador.Muriel@cern.ch or amador_muriel@msn.com

with chemistry, one would be surprised to find out. Anyway, to string together the words "molecular theory of turbulence" is an appellation questioned by engineers and the typical turbulence researcher. Turbulence, it has been stated, is a continuum problem, and molecules are unnecessary. I have found this terribly questionable, although my doubts about the continuum description of turbulence have been met with disbelief, incredulity, and ridicule. This is not to say that the classical view is useless; we have airplanes and rockets in fact, but the existence of turbulence has been tacitly accepted and characterized, not explained, using fundamental physics.

Now, experiments in three countries seem to be accepted, from Russia, the US, and France, all pointing out that molecular signatures for the onset of turbulence are observed. What this means is that nitrogen and argon differ in the way they become turbulent, and that the difference is ascribable to their molecular or atomic structure. Once such proof has been shown to the Scientific Committee, and after seven years of annual presentations to the World Laboratory in CERN to a group of skeptical engineers, a budget has been assigned to this new project. The title of the project is the first public acknowledgment that there exists a molecular theory of turbulence, partly developed at the Center for Fluid Dynamics, a tiny, loose collaboration of Filipino physicists and students who migrated from Diliman to Los Baños then back to Greenbelt, Makati. A referee for some of the papers has declared that, if the theory is true, turbulence research will be turned upside down.

This year, I organized physicists from Russia and the US to repeat experiments performed in Novosibirsk, at the Siberian Academy of Science, and a private laboratory in Cape Cod, Massachusetts. The latter was organized by an American physicist who retired at an early age after a well-rewarded life in industry. The controversial approach, in all honesty, could only be performed by people who are willing to risk their reputation. As an independently wealthy physicist, this American, Luke Hinkle, does not have to risk tenure. In truth, the majority of physicists only do safe or trendy physics. Very few physicists would buck fashion if they are building careers, but as it turns out, the molecular theory of turbulence has gone from the fringe of hydrodynamics to the mainstream of physics.

I see the following succession of events: duplication and even sales of American-built advanced apparatus to confirm our early, initially doubted results, the growth of a body of European graduate students in a new field of turbulence research at the molecular level, and, ultimately, the control of turbulence using new tools of research we have proposed. In the near future I will describe new dramatic results we have just begun to report in Europe. At that point, the applications of our research may attract governmental and corporate support. And Filipino physicists will have pioneered this new area of research. The itinerant Center for Fluid Dynamics will move back from CERN to the Philippines. Are we ready to respond to the challenge?

November 4, 2004

Alvin Culaba and
Raymond Tan

Soft Solutions to Hard Problems

In the information age, it is only fitting that significant research effort is being put into using intelligent computational techniques to address the world's growing environmental problems. The use of mathematical or computer models to simulate the interactions of technology with the environment sometimes yields counterintuitive solutions not immediately evident from a layman's perspective. For example, Kreith's recent article in the *Journal of Energy Resources Technology* dispels many of the myths of the much-touted hydrogen economy envisioned by today's energy specialists as the foundation of energy infrastructure for transportation in the future. Using straightforward life cycle-based thermodynamic arguments, the paper concludes that many of the proposed fuel hydrogen production and delivery systems, particularly those that rely indirectly on fossil fuel feedstocks, are actually less efficient—and consequently more environmentally damaging—than conventional systems in use today. Many similar environmental issues can be addressed through the use of mathematical modeling.

Closer to home, our research at the Sustainable Technology Laboratory in De La Salle University–Manila over the past decade has led to increased specialization in two specific areas—life cycle assessment (LCA) and pinch analysis. LCA methodology allows for a thorough accounting of environmental impacts associated with technologies, including indirect ones. For example, cellphones are not often regarded as generating air pollutants. A closer look reveals that the power used to charge cellphone batteries is often derived from a mix of electricity generated by different plants, including some which use coal, oil or gas as fuel. Thus, from an LCA perspective, cellphones generate greenhouse gases and other pollutants through their allocated share of power plant emissions. The actual procedure for computing this share, and accounting for it over the entire life of the phone, is somewhat more complicated. Considerable effort is being put into developing and perfecting such computational techniques. Techniques derived from operations research (OR) and artificial intelligence (AI) have been tested on different problem domains in recent years. At De La Salle, we have made use of expert systems, possibility theory, rough and fuzzy sets, Monte Carlo, and linear programming in LCA. Applications have included assessment and comparison of alternative fuels for road vehicles, and the development of pollution prevention strategies for the production of

Alvin Culaba: Center for Engineering and Sustainable Development Research and Mechanical Engineering Department, De La Salle University-Manila, Philippines.
culabaa@dlsu.edu.ph
Raymond Tan: Chemical Engineering Department of De La Salle University-Manila, Philippines.
tanr_a@dlsu.edu.ph

bulk goods (such as pulp and paper) and manufactured products (such as semiconductor devices). Similar novel approaches are being developed by our counterparts elsewhere in the world. A good example is a recent North Atlantic Treaty Organization-funded joint project by the University of Portsmouth in the UK and Moscow State University that used a mathematical model to develop optimal and economically viable pollution prevention measures for the production of bricks. We are now mobilizing a collaborative effort to develop more efficient solution algorithms for the same design problem. Potential approaches identified thus far include genetic algorithms, fuzzy integer programming, and inductive reasoning based on rough sets.

Mathematical models are intended to provide humans with simplified versions of reality to facilitate understanding of its behavior. It stands to reason that some resolution is lost in the process of translating real systems into a set of mathematical expressions or a computer code. The words of Zadeh, who developed the concept of fuzzy sets in the 1960s, summarize this dilemma:

"As the complexity of a system increases, our ability to make precise and yet significant statements about its behavior diminishes until a threshold is reached beyond which precision and significance (or relevance) become almost mutually exclusive characteristics." (McNeill, Daniel and Paul Freiberger. *Fuzzy Logic*. New York, NY: Touchstone, 1993, 43.)

Many of the tools we use are thus designed to deal with uncertainties inherent in the modeling process. In particular, we have been partial to the use of fuzzy numbers and fuzzy arithmetic, which offer some advantages with respect to computational techniques compared to conventional probabilistic approaches.

Pinch analysis is a highly specialized field that makes use of various mathematical techniques for efficient use and recycling of resources (such as energy or water) in manufacturing facilities. In the late 1970s and 1980s, much of the work in this field focused on energy conservation through heat recovery networks. The work of El-Halwagi since the late 1980s and the process integration group in the University of Manchester Institute of Science and Technology in the early 1990s led to extensions of these techniques to conservation of industrial solvents and water. Current basic research in pinch analysis is focused predominantly on developing simple, computationally efficient procedures for finding optimal water budgets or "targets" for plants, as well as developing feasible recycle networks to meet such budgets. Two schools of thought have emerged. The first advocates the use of simple graphical displays with visual and intuitive appeal. An example is the water cascade technique developed a few years ago in Universiti Teknologi Malaysia (UTM). The other school of thought prefers an approach based on mathematical programming "superstructures" which allows a whole array of algorithms to be brought to bear on problems with different features. While criticized as a "black box" approach that yields little in the way of insight in developing these solutions, they are able to solve problems that would prove intractable using graphical methods. In De La Salle, we subscribe to this second school of thought. We have also successfully demonstrated the first ever (that we know of) combination of pinch analysis with LCA concepts, thereby paving the way for the potential full integration of these two fields in the near future. That work was done as part of the master's thesis of Valarie Ku-Pineda, and will be published in the near future in the *Journal of Cleaner Production.*

We have also put considerable effort into dissemination of these ideas to industry, government agencies, our colleagues from the academe and, of course, our students. Dedicated LCA courses for engineering have been taught at the graduate level in De La Salle since the late 1990s. Some related concepts have also been integrated into a number of undergraduate courses as well as graduate programs in science and business. When LCA was offered at the masteral level earlier this year, a project-based approach emphasizing the principle of "learning by doing" was used with very encouraging results. At the moment a one-semester course on pinch analysis is also being offered in the master's program using a similar approach. Seminars and workshops are other venues for disseminating these concepts to a wider audience. Through the Center for Engineering and Sustainable Development Research (CESDR) De La Salle has hosted a number of these events. In the first quarter of 2005, the seminars on LCA and pinch analysis featured respected experts from the National Institute for Advanced Industrial Science and Technology (AIST) in Japan and UTM in Malaysia. The seminar participants included practicing engineers and managers from industry as well as officials of different government agencies and NGOs. In spite of these mild successes, much remains to be done; we are currently trying to develop innovative concepts for the delivery of these ideas, for example, through the Internet or through properly designed distance-learning materials.

As it happens, this area of research is also one that does not necessarily require extensive financial resources. This is a significant bonus for researchers and students working in the cash-strapped scientific environment of the Philippines. At the Sustainable Technology Laboratory, we have managed to produce internationally competitive work with only a handful of PCs, some freeware, and an ever-growing pool of talented research students. One major area of study by the research group is the assessment of alternative fuels and energy sources.

June 30, 2005

Raymond Tan

A Brief History of Pinch Technology

In 2005, a colleague from Malaysia gave me an odd Christmas gift—a photocopy of an obscure doctoral thesis on industrial energy conservation using heat exchange networks by Charles Hohmann dating back to 1971. This document, with its typewritten text and painstakingly hand-drawn figures, was a treasure of sorts, a lost document in the field of pinch analysis and process integration. Although the value of the methodology described in the work is quite evident in hindsight, especially considering today's inflated energy costs, Hohmann's work predated the oil crises of the 1970s and went largely unappreciated by his peers. Apparently, he eventually abandoned the field and moved on to other interests. It wasn't until the work of Linnhoff in the late 1970s and early 1980s that pinch analysis hit the mainstream. Linnhoff himself eventually left academia to co-found Linnhoff March, which is now an international engineering firm that runs a thriving global business built on services and products that use pinch analysis. One of the lessons here is that in R&D, luck and timing are critical. Hohmann's story is echoed by the experience of one of my colleagues in De La Salle, who did his master's thesis under one of Linnhoff's academic "descendants" in 1990. He, too, abandoned pinch analysis for bioprocessing, although today he still lectures on the topic to our undergraduate and graduate students.

What exactly is pinch analysis? The term refers to a family of computing techniques that aid in designing system networks or cascades. Proper design of such networks is essential for efficient use of resources, such as fuel, water, time, money, or even human resources. In the early days before the emergence of the ubiquitous PC, pinch analysis made use of graphical displays that used angles and distances as surrogates for various physical quantities. A skilled practitioner can turn a pencil, ruler, and several sheets of graphing paper into a crude analog computer that can solve pretty complex design problems. However, such feats are nowadays more notable for their intuitive appeal or mathematical elegance, rather than their necessity. I, for one, find it simpler to formulate mathematical models, and then write programs to solve them. This preference represents a different school of thought which offers the advantage of being able to solve a wider array of design problems, and in that sense is the "better" way, but I'm sure pinch advocates can make equally convincing arguments otherwise. Several months ago, Robin

Chemical Engineering Department, De La Salle University-Manila, Philippines.
tanr_a@dlsu.edu.ph

Smith of the University of Manchester Institute of Science and Technology gave a lecture in Kuala Lumpur on the possible future of this area of research—the fusion of the pinch analysis and mathematical modeling schools of thought into a single unified framework. It remains to be seen whether his predictions will pan out; I encourage any interested researchers to give it a try. In the meantime, for my lectures, I use the term "process integration" rather than "pinch analysis," as I find it more encompassing of various approaches. In contrast, one can argue that pinch analysis refers merely to a subset of techniques under the more general umbrella of process integration.

After early applications in energy conservation matured and reached commercial level, there emerged a whole gamut of other applications of pinch analysis to other areas. It is significant that many of the new applications were developed through analogies (on this topic, I always like to quote Richard Dawkins, the renowned evolutionary biologist from Oxford, who once said that the judicious use of analogies is essential for the advancement of science). In the late 1980s, Mahmoud El-Halwagi (and his less famous PhD adviser, Manousiothakis) used pinch analysis to determine how process plants can efficiently use "mass separating agents" such as industrial solvents. About half a decade later, Robin Smith (along with his less famous student, Y.P. Wang) extended the principle of pinch analysis to industrial water conservation. To this day, El-Halwagi (now with Texas A&M) and Smith are recognized internationally as two of the gurus in the field of pinch analysis and process integration. More recent extensions of pinch analysis include work on industrial supply chains by Uday Shenoy, now retired from the Indian Institute of Technology, and exotic applications such as financial planning by Toshko Zhelev at the University of Limerick in Ireland. One of my own recent contributions in this area, done jointly with the aforementioned Malaysian colleague, discusses how pinch analysis can be used in national or regional energy planning with a view to meeting carbon dioxide emission limits. That paper will soon appear in the journal *Energy*.

Having seen how the field has evolved over the past four decades, we then ask the question: what's next? New computing methods, perhaps? I have described in a previous column how my students and I have demonstrated the use of swarm intelligence to solve process integration problems. Would Petri nets or cellular automata work as well? What about new applications? For example, could pinch analysis be used to sort out storage difficulties encountered in biofuel supply chains? Could some bright electrical engineer figure out how to use pinch analysis for economic power dispatch problems? To me, the possibilities seem endless. At the same time, I find that this field offers significant opportunities for sufficiently talented and dedicated researchers to do internationally competitive work, even with the financial constraints faced by the typical Philippine university.

I recently recruited a promising PhD student to look into process integration as a possible dissertation topic. To get her started, I handed her a pile of old papers that represent a cross section of the history of this field. I'm hoping that a historical perspective will be what it takes to spark the speculative steps which will be the starting point of a significant body of work on process integration. Whether this will be the case remains to be seen. But as the experience of Charles Hohmann demonstrates, in research, luck is as much a factor as talent. And so researchers must be willing to gamble on new ideas—because that's where the future really begins.

April 19, 2007

Raymond Tan

Fuzzy World

"As the complexity of a system increases, our ability to make precise and yet significant statements about its behavior diminishes until a threshold is reached beyond which precision and significance (or relevance) become almost mutually exclusive characteristics."—Lotfi Zadeh, 1973

In 1965, Lotfi Zadeh published his seminal paper on fuzzy set theory. It took years of gestation before the rest of the world started to grasp the practical implications of this mathematical theory. Today, articles on fuzzy set theory and applications are routinely published in various scientific journals in fields that include engineering, decision theory, artificial intelligence, and mathematics. There is even one journal—*Fuzzy Sets and Systems*—which has focused exclusively on this topic since its initial publication in 1978. At the same time, a large number of modern everyday devices make use of intelligent control based on fuzzy sets. For example, the air conditioner in my car has a fuzzy controller; sometimes it seems to me that it is too "smart" for my own convenience.

What exactly is a fuzzy set? Basically, it is a set for which partial membership is possible. In contrast, conventional sets ("crisp" sets, in fuzzy terminology) allow only for complete membership or nonmembership. Consider, for example, the contrasting cases of the sets of "people in their 20s" (a crisp set, since a person is either in his or her 20s, or not) and "young people" (a fuzzy set, whose borders are not so distinct). This simple distinction gives rise to a wide array of possibilities. Rules of inference can be formulated using fuzzy logic to develop control algorithms for highly nonlinear, multiple input/output systems, as for the smart air conditioner I mentioned previously. Such devices may be difficult to regulate using classical control theory. Similarly, fuzzy rules can be used as a basis for expert systems—computer programs with specialized, domain-specific knowledge comparable to human experts and capable of functioning as "virtual consultants." Other applications include fuzzy optimization and possibility theory (as opposed to the more well-known probability theory).

Some of my research deals with the use of optimization models to design clean, efficient processes or industrial ecosystems. This work typically involves formulating mathematical models to optimize certain quantities (e.g., minimize the use of water resources) subject to constraints formulated as equations or inequalities that describe the properties of the system. These mathematical models are simplified representations of reality. They can then be manipulated using computers in order to give insights on how real systems should be built in order to reduce environmental impacts of industrial

Chemical Engineering Department, De La Salle University-Manila, Philippines.
tanr_a@dlsu.edu.ph

activities. It stands to reason that the validity of these insights hinges on how faithfully the models capture reality.

It is often the case that the model-building process requires a crisp mathematical formulation of real systems for which precise data is unavailable. Crisp mathematical constraints, for example, a restriction of energy usage in an industrial process to 20 megawatts or less, are often counterintuitive. In this case, 20 megawatts is perfectly acceptable, while 20.1 megawatts is completely unacceptable. This sudden transition from acceptable to unacceptable is a core feature of many mathematical models—just like the proverbial straw that breaks the camel's back. There is a certain intuitive appeal in allowing for a gradual transition from an acceptable to unacceptable solutions—and this is possible using fuzzy optimization models. Such models yield more robust and realistic solutions.

In a related problem, it is sometimes necessary to compare different technologies in order to select the environmentally preferable option—and in many cases such comparisons must be done with incomplete or imprecise data. For example, it can be quite difficult to predict the exact fuel economy of cars two or three decades from now. Such imprecise data can be characterized using a branch of fuzzy sets called possibility theory. Fuzzy numbers with "possibility distributions" can be used to model imprecise quantities, especially when probability theory is inappropriate. Furthermore, it is possible to quantify "degrees of preference" when comparing two or more different quantities.

Space restrictions prevent me from discussing theoretical aspects and practical implications of fuzzy mathematics in this article. However, it is possible to draw some useful lessons on the nature of scientific research from the history of fuzzy sets. It provides a useful counterpoint to the common notion that research must be fully utilitarian, or that the research community must work exclusively on urgent issues of practical nature. This way of thinking may work in the short term, but in the end it will be counterproductive for science and technology. After all, Zadeh did not set out to develop a mathematical theory for control systems, optimization, or decision theory in his 1965 paper. These applications came much later. What he did was develop the theoretical foundations that eventually led to all sorts of practical applications in the decades that followed, and which continues to find new and surprising uses to the present day.

July 19, 2007

Raymond Tan

Collective Minds

Michael Crichton's 2002 novel, *Prey*, tells the story of a high-tech plant in the desert being overrun by sentient, self-replicating swarms of runaway nanoparticles. Technophobes and would-be venture capitalists can rest assured that the events described in the book are, of course, fictional; there is, however, a grain of truth in the notion that apparently purposeful behavior of groups of simple entities, or agents, can emerge out of apparently simple dynamic rules. This is surprisingly true even when the groups are made up of simple individuals—hence the term "swarm intelligence."

In 1995, James Kennedy and Russel Eberhart published their seminal work on swarm intelligence (incidentally, their work provides an interesting lesson in cross-disciplinary collaboration; Kennedy is a specialist in social behavior of animals, and Eberhart is an electrical engineer). They described the use of a new computing technique, which they called particle swarm optimization (PSO), to develop empirical mathematical models known as neural networks (these, in turn, are patterned after the organization of neurons in brains). The technique was developed based on early attempts to simulate life-like movement of virtual animals in virtual space. The end result was an algorithm wherein "agents" or particles fly through multidimensional virtual space in search of a "best location" corresponding to the optimal solution to a problem; the flight of the particles is continually adjusted based on simple memory and information-sharing rules within the swarm. (The website *http://www.particleswarm.info* provides plenty of information for both novices and specialists). PSO turned out to be a very robust, effective algorithm for various problems, and in the past decade it has been used with varying degrees of success for applications that include optimal scheduling, engineering design, diagnosis, data mining, and forecasting.

PSO is only one of several algorithms that mimic the behavior of social animals. Other techniques based on ants, frogs, and bees have been developed, and they all fall under the general heading of swarm intelligence. Swarm intelligence, in turn, falls under a more general class of algorithms, known as metaheurisitics, which includes evolutionary and genetic algorithms as well as simulated annealing. These computing techniques combine adaptive behavior with random influences in order to facilitate the search for optimal solutions to problems, and to minimize the risk of getting trapped into deceptive local solutions.

Chemical Engineering Department, De La Salle University-Manila, Philippines.
tanr_a@dlsu.edu.ph

My own experience with PSO dates back to 2003, when I had the naive notion that it would be a magic bullet of sorts to solve difficult design problems in my field of process integration. The first program I wrote was a complete failure. Fortunately, I have since had increased success with the method for designing clean, environment-friendly industrial processes—for example, choosing the best combination of pollution management techniques to implement in an industrial plant, or designing the best scheme for recycling water through a network of pipes to reduce waste generation. Two of my graduate students, Hul Seingheng and Kristine Col-long, spent about a year each working on variants of the latter problem. The work has been the subject of both collaboration and friendly competition with other researchers in Malaysia, Cambodia, United Kingdom, Japan, and the United States.

The application of swarm intelligence to solve environmental problems is a fast-moving field. The first demonstrated use for designing industrial water conservation schemes was published in mid-2006 by a research group in China—two or three months ahead of the work of my own team, much to our dismay. The potential for development seems limited only by what can be imagined. Other applications that we are working on at the Center for Engineering and Sustainable Development Research (CESDR) in De La Salle University-Manila include designing eco-industrial networks, planning renewable energy systems, and predicting the properties of biodiesel from its chemical makeup. Other potential applications that I think should be explored include medical diagnosis and traffic routing to alleviate urban congestion. At the same time, there are many research opportunities in developing enhanced variants of PSO, for example, by developing hybrid techniques incorporating features borrowed from other algorithms. Our own experience in De La Salle has demonstrated the effectiveness of utilizing random mutation events and "population wipe-outs" for some engineering applications. Eventually, it may be possible to develop a uniquely homegrown algorithm, one inspired by direct observation of an as yet unnoticed natural phenomenon.

Looking at the whole concept of swarm intelligence from an entirely different angle, we might ask the question: Can we draw any lessons from these virtual swarms? In fact, I recently read a PSO paper by Mendes and Neves with the provocative title, "What Makes a Successful Society?" which hints at how swarm dynamics can give us insights on how real agents (e.g., people) behave under different conditions and on how these components, in turn, determine the ability of groups of individuals to progress. One of the fundamental lessons is that the group is smarter than the sum of its parts—apparently there really is such a thing as collective intelligence, at least in the world of the virtual swarms. The particles that make up a typical swarm are very simple entities, and the apparently purposeful behavior exhibited by the swarms emerges out of group dynamics through information exchange. On the other hand, it is quite clear, both from the swarm literature and from my own experience, that one of the main reasons swarms stagnate short of their ultimate goal is the loss of diversity. In our work in De La Salle, we have tried different strategies to preserve swarm diversity in our programs—using algorithmic operators with colorful names such as "mutation" and "turbulence." I am thus strongly inclined to believe that in any real-life enterprise, it is just as essential to always maintain a healthy level of diversity of thought, in order to maintain a solid foundation for future progress.

November 30, 2006

Part 11

Geobiology, Geology, and Paleontology

Ronie Calugay

Are There Martians among Us?

Nope, this isn't Hollywood or tabloid stuff. Truth is, for years, scientists have been intensely debating and scrutinizing evidence that life forms from Mars may have flourished on Earth, way, way back before the dinosaurs or the pyramids! Scientists have been breeding and meticulously studying these alleged Martians for almost thirty years! In fact, zillions are bred in US, Europe, and Asian facilities and some may have escaped from their breeding vessels! I swear this is not a hoax. Read on and you may realize that numerous species of these life forms may be gaining in numbers fast ... right in your neighborhood!

What on Earth Are They?

Eleven thousand years ago, a piece of rock ejected from the surface of Mars landed in Antarctica. According to some scientists, the 4.5-billion-year-old Martian meteorite tagged as ALH84001, must have shuttled life forms from the red planet. Their evidence? Magnetic crystals aligned in chains embedded in the meteorite with striking resemblance to the composition and arrangement of magnetic crystals found within a group of bacteria thriving here on Earth called magnetotactic bacteria.

Starstruck!

What are magnetotactic bacteria? An accidental observation on a routine, ordinary day, by an unknown and inquisitive graduate student of microbiology, led to a completely unexpected discovery of a unique group of microbes. While most of the young men of America were fighting in Vietnam and out in the streets getting high and yelling flower power in the early '70s, Richard P. Blakemore was busy peering through the microscope at bacteria from mud samples in a lab at the University of Massachusetts in Amherst. Rich wanted to isolate the microbes but he noticed something else ... a strange kind of movement like that of a flock of birds or a school of fish. He observed a group of bacteria moving in unison! The cells swam in the same direction with high speed when a magnet was brought near the microscope. After repeated tests, he finally confirmed that the cells were, in fact, magnetically responsive. Rich's discovery evoked the interest

Matsunaga Laboratory, Department of Life Sciences and Biotechnology, Tokyo University of Agriculture and Technology, Tokyo, Japan.
ronix5@yahoo.com

of microbiologists, evolutionists, biochemists, physicists, geochemists, paleontologists, animal behaviorists, biotechnologists, and astrobiologists worldwide! Rich shot to stardom in the scientific community, from serendipity and beyond!

Scientists so far have isolated magnetotactic bacteria from riverbeds, lakes, estuarine, and marine waters. They found out that the magnetotactic behavior of the bacteria is due to the presence of nano-sized membrane-bound magnetic crystals of magnetite (Fe_3O_4) or greigite (F_3S_4) aligned in chains perpendicular to the cell axis. These highly organized structures are termed magnetosomes. For geochemists and paleontologists, they are referred to as magnetofossils. The geomagnetic field lines are inclined downward in the Northern Hemisphere, upward in the Southern Hemisphere, and horizontal in the equator. Correspondingly, magnetotactic bacteria thriving in these geologic locations migrate according to the inclination of the magnetic field lines. Hence, it is proposed that the magnetic particles function as navigating compass needles for the bacterium to migrate through the Earth's geomagnetic field lines.

Deep Impact

In 1996, NASA geologist David McKay and his group shook the scientific community by reporting the striking morphological and chemical similarities of magnetofossils on Earth and in ALH84001 strongly indicating that life existed in ancient Mars. Scientists opposing McKay's theory demonstrated that the magnetic crystals in ALH84001 may have been formed by nonbiological reactions. The recent evidence gathered by NASA's rovers of the possible past existence of water, and hence, life in Mars, supports McKay's claim. McKay's report has ushered a whole new school of thought that meteorites can transfer life between planets in space. If this postulation is true, a mind-blowing question pops up: where in the vast universe did we come from?

References

1. Ronie J. Calugay, Hideaki Miyashita, Yoshiko Okamura, and Tadashi Matsunaga, Siderophore production by the magnetic bacterium Magnetospirillum magneticum AMB-1, *FEMS Microbiol. Lett.* 218 (2003) 371-75.
2. Ronie J. Calugay, Yoshiko Okamura, Aris Tri Wahyudi, Haruko Takeyama, and Tadashi Matsunaga, Siderophore production of a periplasmic transport binding protein kinase gene defective mutant of Magnetospirillum magneticum AMB-1, Biochem. *Biophys. Res. Commun.* 323 (2004) 852-57.
3. Ronie J. Calugay, Haruko Takeyama, Daikichi Mukoyama, Yorikane Fukuda, Takeyuki Suzuki, Kaneo Kanoh and Tadashi Matsunaga, Catechol siderophore excretion by the magnetotactic bacterium Magnetospirillum magneticum AMB-1. *J. Biosci. Bioeng.* 101 (2006) 445-47.

4. Takeyuki Suzuki, Yoshiko Okamura, Ronie J. Calugay, Haruko Takeyama and Tadashi Matsunaga, Global gene expression analysis of iron-inducible genes in Magnetospirillum sp. strain AMB-1. *J. Bacteriol.* 188(2006) 2275-79.
5. Hiroko Yokouchi, Yayoi Fukuoka, Daikichi Mukoyama, Ronie Calugay, Haruko Takeyama and Tadashi Matsunaga, Whole Metagenome Amplification of Microbial Community Associated with Scleractinian Coral by Multiple Displacement Amplification Using _29 Polymerase. *Environ. Microbiol.* (In press).

August 17, 2006

Ronie Calugay

Cool Boomerangs!

Boomerangs are fun and amazing. How could a spinning hunk of wood come back full circle to its starting point? For the flat-world thinkers in ancient times, it must have been magic! But for the modern man, throwing a boomerang is just an amazing application of some complex laws of physics.

According to anthropologists, boomerangs evolved from the hand-to-hand combat and hunting weapons used by Australian aborigines. The present-day boomerang is a specially crafted toy and when used to hit a target, it would stop itself from returning to its thrower. But what if there's a cool kind of boomerang, a kind not purchased from a toyshop, but emitted from your body and hits and delivers a target back to you? You might think that only a superhero like Spiderman with his sticky webs can have an amazing trait like that. Pure fiction, right? But move over Spiderman! Fact is, most microbes can produce and release molecules which hit and deliver back specific targets!

Ironing Out a Problem

Evolutionists proposed that before most life forms began, the primitive Earth existed under a condition where there was no oxygen. This favored the existence of iron in a soluble state which could be easily taken up by the bacteria already flourishing at that time. Conditions gradually evolved into an oxygen-rich environment converting the iron into an insoluble state. The ancient bacteria were then confronted with the problem of iron unavailability.

Microbes have no brains but they play no dice when their survival is on the line. Iron is so important in their metabolism that they have evolved sophisticated and highly efficient mechanisms for ensuring that the optimum supply of iron is acquired. Most microbes deal with the problem of iron unavailability by producing and releasing to their environment molecules designed to bind with high specificity and affinity to insoluble iron. These special kinds of invisible boomerangs are termed *siderophores* (from the Greek: "iron carriers"). Released siderophores return with their captured iron to bacterial cells via specific receptors on the surface of bacteria. In *Star Wars*, spaceships cannot enter the Deathstar unless they have the proper identification and authorization to land on the docks. That's how it works with siderophores, the Deathstar

Matsunaga Laboratory, Department of Life Sciences and Biotechnology, Tokyo University of Agriculture and Technology, Tokyo, Japan.
ronix5@yahoo.com

is the bacterial cell, the spaceships are the siderophores containing their cargo, and the receptors are the landing docks. This system ensures that the right cargo is delivered and no stealthy enemy attack succeeds. Once the iron-bound siderophore has crossed the surface, it is transported by specific proteins deeper inside the cell. The iron is eventually released and used up for a number of crucial biochemical reactions.

An Ironfist for Iron

If bacteria have a set of commandments to obey, "*Thou shalt not covet thy neighbor's iron*" is one commandment they have time and again broken. Scores of scientific evidence indicate that in the microbial world, iron thievery is rampant! It all boils down to survival. In Nature, different microorganisms coexist in one spot and competition for iron is tough. *E. coli* actually has seven sets of genes for siderophore iron transport systems, some of which are specific for alien siderophores! That's how the Deathstar lures and cheats cargo from spaceships not its own. For the sake of survival, that's how covetous bacteria can get!

In another classic case, once humans are infected, the body has the intrinsic mechanism of withholding iron by locking it away in special proteins. In response, the infecting bacteria release siderophores to rob iron from its hiding places. But not so fast! Hosts may have a mechanism to chase after these robbers! The breakthrough was reported by Flo et al. of the Institute for Systems Biology in Seattle, Washington in 2004. They reported that some host cells may strike back by releasing a protein called *lipocalin* or *siderocalin* which captures and strips iron from bacterial siderophores!

Pumping Iron

A special group of bacteria called magnetotactic bacteria takes up large amounts of iron, almost two hundred times more than what *E. coli* requires, to keep in shape and maintain the production of magnets. It seems like iron madness and gluttony. Iron assimilation of this proportion entails highly efficient and meticulous metering out of iron uptake. Just like *E. coli* and other bacteria, magnetotactic bacteria have several sets of iron uptake systems and a shut-off mechanism through a protein called ferric uptake regulator (Fur) which acts as a traffic enforcer regulating the influx of iron to avoid the toxic effects of too much iron.

Scientists propose that the magnets produced by magnetotactic bacteria may serve as magnetic compass needles for the bacteria to migrate using the Earth's geomagnetic field lines. However, no sufficient convincing evidence for this theory has been reported.

It's Ironic

It seemed crystal clear that most bacteria switch on an array of genes encoding for proteins involved in the production of siderophores and their corresponding receptors and transporters when exposed to low iron conditions. A controversy among microbiologists worldwide erupted in 1986 when the discoverer of magnetotactic bacteria himself, Richard P. Blakemore, published his results, claiming that siderophores are produced when magnetotactic bacteria encounter high and not low iron conditions. This ironic pattern of siderophore production was not settled until Professor Tadashi Matsunaga and his group at the Tokyo University of Agriculture and Technology demonstrated that when magnetotactic bacteria are grown in a high-iron medium, most of the iron within the first ten minutes is taken up and most of the free iron inside the cells is used up for magnet formation. The remaining iron in and outside the cell then reaches low levels that trigger siderophore production.

At last, the whole picture came together. With high iron supply, magnetotactic bacteria express genes to take up large amounts of iron to form magnetic particles, thus starving themselves of the element. At the onset of this iron starvation, enter the siderophores. What's left of the iron supply are the insoluble forms since they are difficult to assimilate without binding to special boomerangs like siderophores.

Nerves of Steel

For some of you readers, this article may just be a passing chuckle. But siderophores may play very crucial roles in the years ahead in the treatment of some of the most debilitating and paralyzing illnesses we humans suffer and endure. Popular biomedical applications of siderophores include the treatment of thalassemia, an iron-overload disorder. The treatment entails an infusion of a purified fungal siderophore called desferrioxamine (DFO) which works like a wireless vacuum cleaner collecting excess iron in the body and eventually flushing it out with the feces and urine. Studies suggest that in HIV-positive thalassemia major patients, the rate of progression to AIDS was significantly faster in patients with lower doses of DFO. Experts proposed that the toxic molecules produced by too much iron may trigger HIV to replicate and the addition of siderophores may prevent this phenomenon.

Siderophores were also investigated for their ability to control cancer cell proliferation. Bound iron in proteins and free toxic metals in human breast cancer cell lines were depleted by treatment with DFO which induced cell death by inhibiting cell division. A similar effect was observed in other cancer cell lines.

How can one assure that a drug can attack a specific pathogen with precision? To eliminate the effects of hit-and-miss targeting in drug delivery systems, scientists have come up with the brilliant idea of precise targeting by using an operation-penetration scheme called the *Trojan Horse Principle*. The strategy entails purifying or synthesizing an analogue of the pathogen's native siderophore and attaching it with an antibiotic. Since the pathogen has a natural high affinity landing dock for its siderophore, it is duped into accommodating a deadly present!

Iron Will

Siderophores are produced by most fungi and bacteria living in soil, rivers, lakes, coastal areas, and in the open ocean. Some plants also produce siderophores! This means there are still a lot of potentially useful siderophores waiting out there to be discovered. Siderophore applications have not reached a hundred percent efficiency but rest assured experts are painstakingly studying and perfecting strategies to make sure these cool boomerangs return to us quality life.

October 26, 2006

Marietta de Leon

Dinosaurs—And Why We Didn't Have Them in the Philippines

I believe that no other group of animals has captured the fascination of people more than the dinosaurs: gigantic creatures that once roamed the Earth millions of years ago. At one point in our lives we encountered the names *Tyrannosaurus, Brontosaurus, Stegosaurus* and other "saurus"–sounding names that we now refer to as dinosaurs. Although they are long gone from the face of this planet, their influence can still be seen and felt and are plenty.

The 1993 movie, *Jurassic Park*, and its sequels (*The Lost World: Jurassic Park* in 1997 and *Jurassic Park III* in 2001) had combined earnings of $767.319 million. These award-winning movies take moviegoers to an island where scientists found a way to revive dinosaurs. Superb visual effects backed by scientific facts made the dinosaurs life-like and maybe the closest to seeing the real thing.

Dinosaur cartoon characters are a big hit to kids around the world these days. TV shows like *Barney and Friends* glue children to television screens, making them sing and dance with their favorite purple dinosaur. And who can ever forget Walt Disney's animated movie The *Land Before Time* in the late 80s? Audiences were introduced to the adventures of Littlefoot, the baby *Brontosaurus*; Cera, the bossy *Triceratops*; Spike, the timid *Stegosaurus*; Ducky, the baby-talking *Hadrosaur*; and Petrie, the flying *Pterodactyl.* These led to the influx of various products in the market carrying a dinosaur theme. Items such as mugs, pillows, coloring books, stuffed toys, clothes, etc., either are shaped like dinosaurs or have dinosaur images in them. Dinosaurs have become popular endorsers so that the products bearing them not only become salable, resulting in big profits for the businessmen, but also further endear dinosaurs to children. Ask any kid and for sure, he can identify more dinosaurs than most adults can.

Facts about Dinosaurs

The word "dinosaur" comes from the Greek word that means "terrible-lizard." Scientists know, however, (and you will see later on) that not all dinosaurs are terrible and certainly they are not lizards. These reptiles lived during the Mesozoic era, which started about 225 million years ago, and lasted for about 160 million years. Dinosaurs

---

National Institute of Geological Sciences, University of the Philippines, Diliman, Quezon City, 1101, Philippines.
marietta.de_leon@up.edu.ph, marietta.deleon@gmail.com, or tootsie_deleon@yahoo.com

differed from present-day reptiles like alligators, crocodiles, and lizards in that their hips, hind legs, and ankles were specialized which allowed the legs to move directly under the body, rather than extending out from the side of the body. This feature enabled the dinosaurs to grow very big and sometimes stand erect on two legs. Skeletal features also separate dinosaurs from other ancient reptiles, such as the marine plesiosaurs and the flying pterosaurs, such as *Pteranodon* and *Pterodactyl* (no, they are not dinosaurs).

Dinosaurs are divided into two groups: the saurischians or lizard-hipped dinosaurs, and the ornithischians or bird-hipped dinosaurs. The famous meat-eaters like the *Tyrannosaurus rex* and *Velociraptors* (the two main dinosaur villains in *Jurassic Park*) as well as the giant plant-eaters such as *Diplodocus* and *Apatosaurus* (often mistakenly called *Brontosaurus*, but children know this already!) belong to the saurischians, while the horned (*Triceratops*), armored (*Stegosaurus*) and "duck-billed" dinosaurs (*Parasaurolophus*) belong to the ornithischians.

We all know that dinosaurs are big. But exactly how big are they? The largest dinosaur with a complete set of bones is the *Brachiosaurus* (meaning "arm lizard"). It is about 23 meters in length (about two large school buses!) and 12 meters in height (about four-stories high!). On the other hand, an adult *T. rex*, one of the largest predators that ever lived, can reach lengths of about 13 meters and weigh over five tons. Not all dinosaurs are big, however. The smallest dinosaur, *Compsognathus* (meaning "pretty jaw"), is just slightly larger than a chicken.

Except for whales, elephants and giraffes, no other modern-day animal is close to the size of the largest dinosaurs. Paleontologists, scientists who study fossils (remains of animals long extinct), think that the Earth, during the time of the dinosaurs, is much different from the world today, and that the climate and food supply must have been favorable for growing to large sizes.

Because of their huge sizes and strange features, some people think of dinosaurs as terrifying creatures and even monsters. But recent discoveries suggest that several dinosaurs are social animals with some migrating in large groups or herds. Suspected nesting sites of dinosaurs, including those of meat-eaters, have been found with badly crushed eggshells and skeletons of baby dinosaurs, together with bones of other animals (possibly their prey). This suggests that some babies stayed in their nests after hatching and were probably fed by parents, a feeding method similar to those of modern birds.

Where Have They Gone?

Imagine if dinosaurs were still alive today. What do you think the Earth will look like? How will man interact with these creatures? Well, you only have to watch dinosaur movies like *Jurassic Park* in order to have an idea. But in any case, dinosaurs are long extinct so you don't have to worry. But what happened to them? Why did they die out?

To start with, the Mesozoic era is divided into three periods, each marked by mass extinction at the end. Different groups of dinosaurs lived in each period. The age of the dinosaurs began with the Triassic period (245-208 million years ago). During this time, dinosaurs were relatively small, and most were quite similar. Pterosaurs also first appeared during this period. *Coelophysis* and *Plateosaurus* flourished in the Triassic. This period is followed by the Jurassic (208-145 million years ago) where the Earth was dominated by the largest animals that ever lived which include the *Diplodocus, Brachiosaurus,* and *Stegosaurus*. The ancestors of the first birds also lived during this time. The Cretaceous (145-65 million years ago) marked the end of the Age of Dinosaurs. The first flowering plants developed in this period, accompanied by the abundance of insects. *Tyrannosaurus rex* ruled this period, along with the famous horned dinosaur *Torosaurus* (meaning "bull lizard"). As you can see *T. rex* couldn't have possibly preyed on *Stegosaurus* because they lived during different periods.

The massive extinction that occurred at the end of the Cretaceous killed not only dinosaurs but other animals as well, on land and even in the seas. One of the most famous theories used to explain the death of the dinosaurs is that a large heavenly body (meteorite or asteroid) hit the Earth 65 million years ago. This impact led to a sharp drop in the Earth's temperature, which killed the dinosaurs, among others. Other theories include large-scale volcanic eruptions, epidemics, and changes in the Earth's orbit or magnetic field. Incidentally, man and other mammals flourished after the death of the dinosaurs which makes us wonder whether God did that on purpose, assuming that dinosaurs did become extinct.

From Dinosaurs to Birds

Ask most paleontologists and they will tell you that birds probably did originate from dinosaurs in the same way that man may have evolved from apes. They can even tell you at least twenty different characteristics that are shared by a certain group of dinosaurs and the first birds. But how did they come to this crazy idea? It all began with the discovery of *Archaeopteryx* in Germany during the early 1860s. *Archaeopteryx* had all the characteristics of an early reptile, including sharp, pointed teeth. What sets this creature apart from the other reptiles during its time is that it is covered by what appears to be feathers. Since then many other supposedly "missing links" were discovered all over the world, further strengthening the idea that the feathery creatures of today did come from the giant reptiles from the past. Now you will never look at chicken the same way again.

Why weren't there dinosaurs in the Philippines?

In 1987, paleontologists from the National Institute of Geological Sciences at the University of the Philippines discovered bones of *Stegodon* sp. in Tres Hermanas, Antipolo, Rizal. Although the name *Stegodon* sounds like that of a dinosaur, it is not. It is an extinct elephant that once lived in the Philippines thousands of years ago. There are also other fossils found in the country but most are microscopic marine organisms.

So why weren't there dinosaurs discovered in the Philippines? Scientists believe that there were no dinosaurs in the Philippines because at the time dinosaurs were alive there was no "Philippines" yet. The "Philippines" was under the deep blue sea, its mass was formed only during the Tertiary period (65-1.64 million years ago), a few millions of years after the last dinosaurs lived. You see, some of the more than seven thousand islands of the Philippines came from different parts of the world and converged in this part of Asia a long time ago.

But I wouldn't be surprised if one day someone finds dinosaur bones in the Philippines. Take New Zealand as an example. Scientists first thought it would be impossible to find dinosaurs in New Zealand for a number of reasons. It took an old woman with no formal paleontological training to prove that the conventional knowledge about where dinosaurs lived in the past is wrong. Joan Wiffen, the "Dragon Lady" as she is now called, placed New Zealand in the world map when Mesozoic reptiles were found there. "I was too ignorant to know that dinosaurs officially never existed in New Zealand," she said. Paleontologists initially thought that the islands in New Zealand were too small and isolated and that the country has a very turbulent geologic history for dinosaurs to survive: conditions not unlike those of the Philippines.

So if there were, indeed, dinosaurs in the Philippines, where are they likely to be found? Palawan and Mindoro are the most logical places to begin the search. This is because these islands were once part of mainland China, with rocks older than the dinosaurs themselves. Like the "dinosaur hunters" of the past, all it takes is a lot of patience, a little luck, and an idea on the geology of the area you are looking at in order to find dinosaur fossils. So, it may be just a matter of time before someone discovers a form which would later be named *Tyrannosaurus philippinensis*!

Printed in two parts on July 21 and July 28, 2005

Marietta de Leon

Fossil—A Thing of the Past

Hey, have you ever wondered where your friend's favorite t-shirt, watch or leather bag brand, Fossil, came from? Or when you stopped for gas, have you noticed the "Fossil Fuels" sign at some gasoline stations? Well, the word fossil actually came from the Latin *fossilis* which means something dug up. Not a pretty picture, huh? The paleontologists or the experts studying fossils (and experts in fossil hunting and digging, too), say that any remains or traces that could serve as an evidence of life of an animal or a plant that once lived, can be called a fossil. They are very important inclusions in any type of rock—may it be sedimentary, igneous and in very rare cases, even in metamorphic rocks. Now, you may throw me this question, "How about the remains of Cro-Magnon man, cave paintings, potteries, and stone tools?" Can't these things be considered fossils? The answer is yes. Cave paintings, potteries, and tools were made by man and their mere presence indicates man's earlier existence. The mold of a carabao and its calf under the lahar of the 1991 Mt. Pinatubo eruption is preserved, or if you bury the shell of the *tahong* you ate for dinner last night, they will be modified and could become true fossils after thousands or millions of years. One qualification for something to be called a fossil is that it must give us some hint of the size, shape, form, and structure of the life form that once lived. The organism, however, doesn't have to be extinct.

Generally, there are two types of fossils. First are those formed by the body or by some parts of the organism like the footprints or the burrows left by ancient worms and crabs. The second are those which include the body itself—be it only the teeth or scales. Furthermore, they can be altered or unaltered. One perfect example of an unaltered fossil is the frozen mammoth or extinct hairy elephant found in Siberia and Alaska. The amazing thing is you can still find its blood fresh, their eyes, skin and flesh intact, and most of all, even tell what it ate on its last dinner as if it only died yesterday.

On the other hand, altered remains are less spectacular than the mammoth fossil. This group includes the bones of the first mammal discovered in the Philippines, the *Stegodon mindanensis.* This fossil found in Tanay, Rizal in 1887 is considered to be the ancestor of elephants. Actually, there are two species of elephants, *Elephas beyeri* and *Elephas* cf. *namadicus,* and one species of rhinoceros, *Rhinoceros philippinensis,* that once resided in our archipelago. There are also four species of stegodons: *Stegodon* cf. *trigonocephalus, Stegodon luzonensis, Stegodon* cf. *sinensis* and *Stegodon mindanensis.*

National Institute of Geological Sciences, University of the Philippines, Diliman, Quezon City, 1101, Philippines.
marietta.de_leon@up.edu.ph, marietta.deleon@gmail.com, or tootsie_deleon@yahoo.com

These fossils have survived the tests of time—physical and chemical weathering. The latter could be the filling up of pore spaces by minerals and/or the replacement of the original components of the bone (or wood) by minerals from circulating solutions. The petrified wood (or turned-into-stone wood) looks just like ordinary wood but it is heavier and more compact and its texture more polished. The fibers and cellulose of the original tree were gradually replaced by silica or calcite, retaining its form but not its composition.

Another type of fossil is the impression produced by organisms. Under the right conditions, impressions can serve as molds that may be filled by sediments which would later harden. If they are split, the impression (mold) and the cast (figure produced) can give the features (size, shape, markings, etc.) of the organism. On the other hand, leaves and delicate animals (e.g., jellyfish and worms) can be preserved when they leave a thin carbon film.

Another Philippine fossil example is the first and oldest coral called *Gshelia* sp. found in a clast near the base of the Late Tertiary Punso Conglomerate found by Easton and Melendres in 1963. This coral is approximately 290 million years (my) old!

Since 1959, several discoveries were done in the Philippines starting in the Puerto Galera area in northern Mindoro. Most of these remains are very tiny marine protozoans, ranging from only 1 mm to 5 cm. Examples are the deep marine foraminifera (*Orbitolina and Globorotalia velascoensis*—65.5 my) and algae (*Distichoplax biserialis*— 65.5 my). In 1966, another significant discovery is the 250-million-year-old rice grain-like foraminifera called fusulinids (*Neoschwagerina* and *Paraschwagerina*) by Andal. In 1968, Hayami also described the Jurassic or ~160-million-year-old mollusks, while Shuto worked on gastropods obtained from Iloilo.

It is good to note that only a fraction of the total number of organisms that lived in the past is written on the fossil record. Some buried plants and animals were completely destroyed by scavengers or decomposed by bacteria. The sediments where they were originally laid down could be disturbed and eroded again hindering preservation. That's the reason why rapid burial is needed. Another necessary condition is for the organism to possess hard parts.

Uses of Fossils

So, why do we have to dig up old bones and traces of animals and plants? Is it just one weird hobby? Yes, fossils have a very crucial role in the environment and in our economy. Don't you know that the diesel, gasoline, and liquified petroleum gas (LPG) that we use every day as well as coal used to power some factories like cement factories come from fossils? The organisms that formed into these fuels died a natural death and were buried deeply by tons of sediments that have accumulated for millions of years. Whereas the diesel, gasoline, and LPG come mostly from single-celled microorganisms that once lived in the oceans and lakes and other bodies of water, coal comes from woody plant materials. By the perfect combination of long periods of time, plus the extremely high temperatures and pressures in the earth, these organisms were "cooked" to produce crude oil which was later stored within the spaces of rocks. The Philippines and the world as a whole get 60 percent of its energy from petroleum and natural gas. We have the Malampaya deep water gas-to-power project, the biggest natural gas project in the world today. The project produces about 3.4 trillion cubic feet of natural gas which can supply 2,700 megawatts of power to Luzon for twenty years. Since January 2002, it has been providing 30 percent of the country's energy requirements.

Coal, on the other hand, is the second largest source of energy, contributing about 23 percent of the world's total energy demand in 1995. When woody plants decompose in ancient swamps and deltas and get accumulated for millions of years, they form layers upon layers of sediment and decaying plant materials. These materials experience elevated pressures and temperatures which would cause the changes within these beds to form coal deposits. Approximately, an accumulation of decaying plants as high as a tall building like the Shangri-la Plaza can produce just a meter thick of coal deposit. One coal mine here in the Philippines can be found in Semirara Island off Mindoro island-province.

Fossils of plants and animals are also essential tools in identifying old environments and interpreting paleoclimates. By looking at the manner their living representatives or relatives behave and react to conditions of today, we are able to understand the environment under which they existed. This is an idea introduced to us by the Father of Modern Geology, James Hutton. According to him, "The present is the key to the past." Following this thinking, we would know that fishes are aquatic organisms, so that if we see bones of fishes embedded in a rock, that particular rock must have formed underwater. If that rock is already at the mountaintop now, some processes, like lowering of the sea level or continental uplift, could have happened for it to be exposed in its present location. Another important issue is about the climate. Are we approaching another ice age or another period of global warming? By studying fossils like foraminifera, we can produce a "saw-tooth" graph that can give us patterns of the climate for millions or thousands of years and help us predict future climate changes. Migration patterns of plants and animals living in the past can also tell us where the polar ice extended to at a particular time. Fossils can also give us clues on how plants and animals reacted under extreme cold or warm episodes in order to survive. Also, the appearance and extinction patterns of different species, like the reptiles, can also help us study lineage and better understand what caused some living forms to be erased from the fossil record. Were there catastrophic events like volcanic eruptions, asteroid impacts, tsunamis or landslides which caused their extinction?

Did you know that these fossils are "buried clocks" which can assist geologists give an age and establish the order of deposition of layers of rocks? Yes, fossils, indeed, can tell us the time when the rocks that bear them were formed. In oil companies, for instance, core samples (sediments containing fossils) are analyzed to know if they have already reached the particular rock containing petroleum. Certain fossils called index fossils characterize a certain age range. Fossils can also help correlate rocks with the same age but from separate locations.

REFERENCES

Andal, P.P., Esguerra, D., Hashimoto, W., Reyes, B., and Sato, T. 1968. The Jurassic Mansalay Formation, Southern Mindoro, Philippines. *Geology and Paleontology of Southeast Asia*, vol. 4, 179-97.

De Leon, M.M. and Ocampo, J.A.A. Philippine Paleontology in the Second Half of the 20th Century and Its Direction in the 21st Century. (In preparation)

Schoell, W.U., Militante-Matias, P.J., and Peleo, A.M., 1991. Stegodon Fossil Remains from the Plio-Pleistocene Laguna Formation, Antipolo, Rizal, Philippines. *Journal of the Geological Society of the Philippines*, vol. 46, no. 3-4, 50.

Shrock, R.R. and Twenhold, W.H. 1953. *Principles of Invertebrate Paleontology*. McGraw-Hill Book Company, Inc. The Maple Press Company, York, PA. 2nd Ed.

Tarbuck, E.J. and Lutgens, F.K. 1984. *The Earth: An Introduction to Physical Geology*. Charles Merrill Publishing Company. A Bell and Howell Company, Columbus, Ohio. 1st Ed. 146-48.

http://www.malampaya.com

http://www.txtmania.com.trivia/only.php

Printed in two parts on July 7 and July 14, 2005

Leopoldo de Silva Jr.

The Geology of the *Panghilod,* the Traditional Filipino Stone Tool

Even before the introduction of loofahs and other body scrubs in today's beauty bars, there has always been the lowly *panghilod*. To the young and uninitiated, a panghilod is a piece of rock or stone used in scrubbing out dirt and grime from the skin during bathing. The type of rock used for this purpose varies, but the panghilod generally possesses three important attributes: small enough to be hand-held; a naturally polished or smoothened surface; and flat or oblate in shape.

These are the same characteristics exhibited by pebbles found on stream beds, the natural source of panghilod. Pieces of rocks found on stream beds were derived from rock exposed upstream. Weathering causes these large bodies of rocks to break down into smaller pieces. They are then removed (erosion) and transported by river water. In the process of transport, the pieces of rock are tumbled and abraded until the surface is smoothened and the shape flattened. Rocks with similar characteristics can also be observed along rocky shorelines where constant wave action causes smoothening of the surface without having to transport them long distances. In either case, the natural processes responsible in making a panghilod involve hundreds even thousands of years. Should you choose to use the panghilod anytime in the future, remember that you are literally holding a piece of Earth's history in your hands.

But what type of rock is more commonly used as panghilod? To answer this question, we need a little understanding of petrology (study of rocks) and geology of the Philippines. There are three basic rock types: igneous, metamorphic, and sedimentary. An igneous rock results from the solidification of molten material known as magma. The solidification process involves the formation of solid phases known as minerals. If the rate of cooling or solidification is rapid, minerals are not given time to grow, resulting in a very fine-grained texture (mineral grains are so small that they cannot be seen). This texture is typical of volcanic rocks. Volcanic rocks form from lava (magma extruded to the surface). If magma, however, solidifies beneath the surface of the Earth, the rate of cooling is relatively slow, allowing the formation of larger crystals or a coarse-grained texture. The resulting igneous rock, called a plutonic rock, is composed of large interlocking grains of minerals. What about sedimentary rocks? Sedimentary

---

National Institute of Geological Sciences, University of the Philippines, Diliman, Quezon City, 1101, Philippines.
desilva_son@yahoo.com

rocks are literally recycled rocks, formed from previously existing rocks. Rocks exposed to the surface are mechanically or chemically broken down to pieces, transported and then deposited. These fragments of the original rocks are called sediments, an example of which is the panghilod. Compaction resulting from burial and cementation (pore spaces filled with cementing material) transforms the loose sediments into a solid rock known as a sedimentary rock. Rocks buried deep beneath the surface are subject to extreme temperature, pressure, and chemical conditions. These may cause the rocks to undergo changes in the solid state (no melting involved), forming a rock type known as a metamorphic rock.

The choice of rock type to use as a panghilod depends on mineralogy (type of minerals present in the rock) and availability. Sandstone, for some people, is an ideal panghilod. This type of sedimentary rock is composed of compacted and cemented sand-sized sediments. The surface of the rock tends to be grainy, quite similar to sandpaper. The thought of rubbing sandpaper against the skin is, of course, not very pleasant; it is, however, very effective in getting rid of dead skin (exfoliation). The disadvantage of using sandstone as a scrubbing agent is the tendency of this rock to be composed of very hard minerals like quartz. The mineral quartz is so hard that it can scratch steel. Processes involved in the formation of sedimentary rocks (e.g., weathering, erosion, and transport) tend to favor the enrichment of very stable and hard minerals. Therefore, sandstones are not the rock of choice for people claiming to have soft "baby" skin. They, however, are ideal for getting rid of tough and stubborn skin such as calluses.

Metamorphic rocks are usually associated with very old terranes similar to what we find in Palawan and Mindoro. Most of the Philippine islands are relatively young (geologically speaking), less than 30 million years old. Thus, metamorphic rocks are rarely used as panghilod. The most abundant and widely distributed rock type in the Philippines is volcanic rocks. For those who are familiar with the theory of Plate Tectonics, the Philippines is an example of an Island Arc: an arcuate group of islands formed through submarine volcanism. Philippine rivers, the natural source of panghilod, are littered with volcanic rocks. In fact, most of the panghilod*s* I have seen so far are volcanic rocks. I have yet to hear somebody boasting of an imported panghilod. We simply use what is readily available.

To this day, many swear to the effectiveness of the panghilod as a stone implement for bathing. Although I have not yet consulted an anthropologist, I suspect that the use of the panghilod predates the Spanish period. In Africa, paleontologists believe that the first stone tools used by the ancestors of modern humans were stones similar to the panghilod. Of course, they think that these river pebbles were used as a means of defense against predators (convenient things to throw). This just adds merit to the use of the panghilod. You're clean and you're safe. Scrub away...

November 24, 2005

Part 12
Population Issues

Mercedes
Concepcion

Why Population Growth Is Still an Issue in the Philippines

According to the 2000 Census, the enumerated population of the Philippines reached 76,504,077, making it the second largest population in Southeast Asia and the thirteenth in the world. An average 1.58 million Filipinos were added yearly during the 1990s. Death rates have fallen slowly in the recent past, while birth rates have remained relatively high. Consequently, the corresponding average intercensal growth rate in the eighties was 2.35 percent and 2.34 percent in the nineties. If this rate were to continue in the future, the number of Filipinos will double in less than thirty years.

The long history of high birth rates and, to a lesser extent, the more recent declines in infant and child mortality have resulted in a young population. According to the 2000 census, at least 37 percent of Filipinos are below age fifteen. In contrast, 24 percent of the Republic of Korea's population and 15 percent of the Japanese population are under fifteen years. On the other hand, Bangladesh and Pakistan have 39 and 42 percent of their populations, respectively, in the 0-14 age group.

The contraction in the proportion of those aged 0-4 (13.6 percent) and the almost identical proportion in those aged 5-9 (13.7 percent) suggest a major demographic shift in the population age structure. Some 3.8 percent of the population enumerated in the 2000 census is aged sixty-five or older, leaving 59.2 percent in the working ages of 15-64. If each person of working age is economically active, the dependency ratio of 69 implies that every 100 productive people would be supporting about sixty-three young dependents and around six elderly dependents.

The youthful age structure imposes an economic burden because a significantly greater share of development resources must cover the immediate requirements of the young. For example, education commanded 33.8 percent of the 2003 national budget, up from 31.2 percent in 2002. The large percentage of children generates a built-in momentum for future population growth. Even if fertility falls to the "replacement" level of a little over two children per woman, births would still exceed deaths and population would continue to grow for at least fifty years.

Fertility—Data from the 2003 National Demographic and Health Survey (NDHS) indicate that the average number of children born to a woman by the time she reaches

---

UP Population Institute, University of the Philippines, Diliman, Quezon City, 1101, Philippines.
concepcion_mercedes@yahoo.com

age fifty (total fertility rate) had dropped from 6.0 children in 1970 to 3.5 in 2002. Most of the reduction has taken place among women aged 25-34. Moving the current fertility level to the 2.1 replacement level will demand a relatively similar magnitude of fertility reduction over a nearly equal duration with a lower starting-off point.

One of the challenges facing the country during the next twenty years is the expected growth in the number of women of reproductive age, 15-49, from 19.6 million in 2000 to an expected 28.4 million in 2020. If the Philippine government is to provide at least the same level of reproductive health services to women in 2020 that it provides now, its budget for these services will definitely have to rise correspondingly in the future.

The increasing proportion of high-risk births and the unmet demand for family planning services suggest a need for attaching high priority to reproductive health. The 2003 NDHS reported that nearly three in four (72.9 percent) women gave birth between the ages of 20 and 34. The rest (27.1 percent) bore their children in the high-risk age groups below 20 and above 34.

Mortality—Deaths of infants under age one are still relatively high, averaging around 29 per 1,000 live births. In contrast, infant mortality rates in the United States, Canada and Western Europe usually range from five to seven deaths per 1,000 births. About 4 percent of all Filipino children die before reaching their fifth birthday. The chief causes of death among infants and young children are communicable diseases and diarrhea with consequent dehydration and malnutrition. Improved diets, better sanitary practices, and health care could reduce death rates further.

Notable improvements in general health conditions and survival prospects of Filipinos were noted during the period 1950-2000. In 1950, the expectation of life at birth was around 43 years. Currently, it is in the neighborhood of 68 years for both sexes. This steady increase since post-World War II reflects medical and health improvements. In 2000, the UN estimated the corresponding life expectancy in Thailand to be 70.2 years, while that of the Republic of Korea was 74.9 years.

Swift growth in numbers hampers improvements in health and in the delivery of health services in several ways. First, in the face of economic stagnation, it is difficult to maintain, and more so, upgrade services for the growing number of people. Second, elevated birth rates generate a sizable fraction of young children in the population—the group with the highest illness and mortality rates in developing countries and hence, with the utmost need for health services. Third, too many births as well as those that are closely spaced are linked with high rates of maternal and child mortality.

Internal migration—The Philippine population is still distinctively rural. But the country has been urbanizing at a fast pace, fueled by a constant stream of migrants from the countryside and the high rate of natural increase in the cities themselves. Population movements during the 1960s signaled a definite shift from a frontier-ward to an urban-ward orientation, while development policies shifted to industrialization as exemplified by import-substitution schemes. Urban dwellers, only 27 percent of the 1950 population, now make up about 48 percent of the population.

For many Filipinos, the possibility of a better life in the cities is an irresistible magnet. The fortunate few who find jobs in industry, business or government earn incomes many times greater than they would in rural areas. Cities also claim the largest share of modern social services—schools, health facilities and entertainment. Despite the squalor, the housing shortage, pollution, congestion, and physical danger, urban centers still provide a faint hope for a brighter tomorrow.

Population and development—Philippine social and economic development is being hampered by fast population growth. Most visibly, it hinders efforts to elevate overall living standards, feed people adequately, maintain the resource base essential for future development, meet the demand for employment, education and health services, and cope with exploding cities.

A high rate of economic growth does not necessarily mean widespread improvement in living standards, especially where national incomes are distributed unequally. On the other hand, when the population is growing faster than the economy, improvements in living standards are highly unlikely.

The Philippines has witnessed decades of economic stagnation and declining living standards. The Gross National Product (GNP) per capita in 1960 was reported to be $165; in 2002, it was $1,020. In 1950, per capita Gross Domestic Product (GDP) in the Philippines, in purchasing power parity (PPP) terms, was midway between that of the richer city states and that of China. By this measure, some fifteen years later, it slipped behind Taiwan and the Republic of Korea and was similar to Thailand. By 2000, China and for a period, Indonesia, had overtaken the Philippines. The growth record of the Philippines was comparable to that of its neighbors in the 1960s and 1970s. The really pronounced divergence occurred in the 1980s, when its neighbors averaged about 6 percent annual growth, in contrast to the Philippines' 1 percent. The differences were larger still in per capita terms since by the 1980s, Philippine population growth rates were significantly higher. The differences in growth rates narrowed in the 1990s, but for most of the decade, the Philippines continued to fall further behind her ASEAN neighbors.

The imbalance between available jobs and workers is one of the most fundamental development issues facing the country. The National Statistics Office estimates that in 2000, 11.2 percent of the workforce was unemployed. The current estimate (April 2004) is 13.7 percent. In earlier decades, unemployment rates had been markedly lower with the highest unemployment rate during the 1970s, 6.3 percent, recorded in 1972.

The working age population expanded from 55 percent in 1980 to 59.2 percent in 2000. This implies that the large number of babies born during the high-fertility '60s and '70s swelled the size of the working group. In effect, the population problem of the 1960s transformed itself into the employment problem of the 1980s and 1990s. Over 800,000 new entrants are being added to the working population annually. The capital stock must continually increase merely to maintain current productivity and work opportunity for the mounting number of workers. Unless this occurs, each worker will be producing less, thus leading to lower or stagnating incomes. When wages fall in relation to expenditures, leading to income inequalities, it is again the poor who suffer.

Unemployment among young and educated workers is a common phenomenon. In the 1970s and early 1980s, most unemployed persons were in the 15-24 age group and had been educated in the primary or secondary level. But an increasing proportion of the unemployed now possess a tertiary education. The percentage of total unemployed with at least some college education increased from 17.4 percent in 1980 to 33 percent in 2000. Even more telling was the proportion with a college degree, which rose from 8.5 percent in 1980 to 14.8 percent in 2000. This trend could reflect the "choosy youth" phenomenon or the overseas worker phenomenon, which tends to raise the reservation wage. Nevertheless, it implies a serious waste of resources.

Ironically, improvements in educational attainments have only aggravated the labor situation. Among Filipinos, education is seen as the path to a better job and escape from the tedium and poverty of traditional rural life. Obtaining a modern job is not assured, but the rewards are large for those who do. Those who fail are forced into informal jobs in the countryside and in the urban areas. The frustrated aspirations of innumerable young people who fail to land jobs after leaving school may eventually escalate social and political demands.

For incomes to rise, investment must grow faster than the labor force. But this would entail a similarly rising demand for spending for education, health, energy, public works and so forth, just to keep things going. For an already cash strapped economy, it should be difficult to expand educational spending if population multiplies too rapidly.

Environment—A recent assessment has revealed major environmental problems, including deforestation, fisheries depletion, land and water system degradation and urban pollution. Such problems have costs, which include adverse impacts on the health of the population and on the growth potential of the economy. Rapid population growth and changes in economic structure are associated with the direct and indirect impact of government policies. The same inward-looking protectionist policies that partly explained the lack of sustained economic growth in the Philippines compared to her Asian neighbors in the last thirty years, are the same policies that aggravated environmental problems. For one, the net effect of economic policies has been to perpetuate poverty in the rural areas, and this, in turn, encouraged population movement to crowded cities and ecologically fragile uplands.

Forests are being depleted at a disturbing rate due to uncontrolled logging, forest fires, encroachment of swidden agriculturists onto forests, pests and diseases, and other causes. In 1991, the Department of Environment and Natural Resources (DENR) estimated that between 1980 and 1989, a total of 252,649 hectares of forests were destroyed. In 1972, the country had about 10.4 million hectares of natural forest covering 34 percent of the total land area. By 2000, only 5.8 million hectares remained, or about 19.4 percent of the total land area.

As the Philippines transforms itself into a world of modern medicine, education, wage employment, and scientific farming, fertility will certainly drop to replacement level. For the present, however, a multitude of cultural and economic factors combine to keep Philippine fertility higher than that of most of her ASEAN neighbors.

Determinants of high fertility—Historically, desired family size and, consequently, actual fertility have fallen as development progresses. Based on the experience of the developed world and some developing countries like the Republic of Korea, Singapore and Hong Kong, it seemed that industrialization, a large proportion of city dwellers, and relatively large income per capita were indispensable for fertility decline. Now, it has been demonstrated that fertility can decline markedly in countries such as China, Indonesia, Thailand, Sri Lanka, and even Vietnam, where economies are not yet highly industrialized, where per capita incomes are low, and where the populations are still largely rural. Fertility declines since the 1980s have been associated less with per capita income levels than with improvements in life expectancy and in women's education, and a relatively equitable distribution of income and access to health services, including family planning.

Conclusion—The Philippines has undergone relatively rapid population growth compared with many of its neighbors. The outcomes of this growth have been unevenly felt across the archipelago as a consequence of significant spatial movements. The swift growth of population has also contributed to the persistently high poverty levels and to a rising number of people with incomes below the poverty threshold. Fast population growth has adversely affected the environment, particularly the rural resource base. Although health status indicators for the population as a whole have shown a modest improvement since 1950, continuing high levels of fertility have had negative consequences on the health of mothers and children.

Public policy is intended to promote improved standards of living and to bring about the conditions whereby all citizens can realize their full potential. The slow pace of development cannot be blamed on unchecked population growth alone. Nor will moderating population growth by itself ensure economic progress. But retarding population growth can complement other development efforts in enhancing human welfare and improving the quality of life. Continued development requires equal commitment to a judicious population program as a vital part of the leadership's effort to improve the lives of all Filipinos in the wake of high infant and child mortality, poverty and limited economic opportunities, insecurity and restricted access to education. Still, large families and the resultant rapid population growth are prejudicial to the country's social and economic development effort, thus negatively affecting most families.

Along with this dilemma at the individual level, there is a vicious cycle at the public level. Ultimately, poverty and limited opportunities produce a break between public and private interest, between the need to decelerate population growth rates to advance development and the desire for large families. But rapid population growth rates render it increasingly difficult for the Philippines to keep poverty down and improve people's lives.

Printed in two parts on February 10 and February 17, 2005

Ernesto M. Pernia

Population, Economy, and Poverty

Is it any wonder why we have fallen behind our Asian neighbors and are likely to be left behind by the rest of developing Asia? We're still debating such rudimentary matters as the population issue and fiscal deficits, while our neighbors have moved on to focus on more contemporary economic concerns, such as global competitiveness, investment climate, and productivity growth. The lack of a resolution of the population issue in the Philippines can be attributed to its soft *State* and hard *Church*. Our government has been unable to adopt an unambiguous and coherent population policy because it is more concerned with the opposition from the Catholic Church hierarchy rather than listening to the voice of the people.

The link between population growth and economic development was the subject of intense research from the 1960s to the 1980s. A common view was that rapid population growth—of 2 percent or higher per year then prevailing in many developing countries—was more likely to hinder than foster economic development. This negative effect operates via reduced child care and human capital investment, lower household savings for private and public investments, and constraints on allocative efficiency, entrepreneurship, and innovation. Rapid population growth results in available capital being thinly spread among many workers, as well as in fiscal and environmental externalities.

Many developing countries have imbibed these lessons well, with good results, and since have moved on—but not the Philippines. Other countries in East and Southeast Asia have experienced sharp reductions in poverty as a consequence of rapid and sustained economic growth, attributable to sound economic strategy that included strong population policy. These countries have benefited from a "demographic bonus" resulting from marked increases in the share of workers (population ages 15-64) relative to young dependents (ages 0-14), while the Philippines continues to bear a "demographic onus"—a large share of young dependents relative to workers (and savers). Thus, in the 1990s, research and the debate on the population issue in the developing world began to taper off, except in the Philippines.

Population growth in the Philippines diminished slowly from 3.0 percent yearly in the early 1970s to 2.4 percent in the 1990s and dropping slightly to 2.1 percent currently. Thailand's and Indonesia's population growth rates, which were about the same as the

Economics Department, University of the Philippines, Diliman, Quezon City, 1101, Philippines.
empernia@up.edu.ph

Philippines' in the early 1970s, have dropped to 1.4 percent and 1.5 percent, respectively. At the same time, while poverty incidence in the Philippines remains high at 33 percent, Thailand's poverty incidence has fallen to 9.8 percent and Indonesia's to 18.2 percent.

These comparisons are quite telling about the links between governance, population policy, and poverty. Thailand is reputed to be the best among the three countries on all three counts, suggesting that good population policy combined with good governance results in rapid economic growth and poverty reduction. On the other hand, the experience of Indonesia, where governance and corruption ratings are worse than those of the Philippines, implies that good population policy by itself can contribute to significant poverty reduction.

Moreover, contrary to claims that significant population growth slowdown can happen only in countries at high income levels, Indonesia, with a lower per capita income and lower literacy rate, was able to reduce population growth faster than the Philippines. A similar experience has been shown by Bangladesh, Sri Lanka, and India's Kerala state.

Macro data are corroborated by household data, indicating consistently over time the close association between poverty incidence and family size. Data for 2000 show that poverty incidence rises monotonically from 9.8 percent for a family size of one to 57.3 percent for a family size of 9+. And the larger the family size, the more likely it is to become destitute owing to exogenous shocks (e.g., typhoons, droughts, and price increases).

The country's current total fertility rate (TFR) is 3.5 children per family. But this average TFR masks the wide variance across asset groups: 5.9 children for the poorest quintile (or 20 percent), 3.5 for the middle quintile, and 2.0 for the richest quintile. Desired fertility declines monotonically from bottom to top asset class: 3.8 children for poorest quintile, 2.6 for the middle, and 1.7 for the richest quintile. The wide gap between actual and desired fertility (2.1 children) among poor households suggests that family size adversely impacts on their living standards. Behind this gap is a large unmet need for family planning services: 27 percent for poorest versus 15 percent for middle class and 12 percent for richest. And the poor depend on the government for these services which it has failed to provide.

It is imperative for the national government to adopt an unequivocal and coherent population policy for three reasons: (a) negative externalities, (b) imperfect information, and (c) poverty reduction.

Negative externalities (costs imposed by some people on others) are often associated with the environmental effects of rapid population growth, such as congestion, environmental degradation, and resource depletion. Moreover, in poor countries like the Philippines, population growth also results in fiscal externalities: the more poor people, the higher the taxes that the nonpoor must pay just to maintain the quality of education, health services, and infrastructure.

Information about and access to family planning services are inadequate. Low-income or less educated couples are often ill-informed about the health risks to both mothers and children of many and closely spaced births. And even those who are sufficiently informed about the advantages of family planning may not know how to operationalize the information they have or often do not have access to suitable services.

The large gap between wanted and actual fertility, the high unmet need for contraception, and the low contraceptive use among the poor constitute cogent justifications for the government's provision of effective family planning services. Population policy is a necessary complement of a poverty reduction strategy.

Population policy must be initiated by the national leadership and must be national in scope. First, local government leaders typically get their cues from the national authority. In fact, controlling population growth at the local level is incentive-incompatible with internal revenue allotments, which increase with population size, as well as with politicians' electoral chances. No wonder why at present only a handful of local government unit (LGU) executives take the population issue seriously.

Second, there are negative spillovers since population is mobile across LGU boundaries. Hence, a town or province with successful population management, good economic performance, and adequate infrastructure and social services would find itself swamped with migrants from poorly performing areas. And, third, varying fiscal resources and technical capabilities among LGUs are likely to work against a consistent and effective implementation of population policy.

There are three key objectives and instruments of an effective population policy. First is to reduce unwanted fertility (or to meet unmet needs for contraception) through a strong national family planning program, i.e., one that allows a choice among both traditional and modern methods of contraception. Family planning services, consisting of good information and effective contraceptive means, should be made readily available to low-income couples who want such services.

Second, raising the quality of basic education, reducing infant mortality, fostering women's empowerment, and increasing employment opportunities for women would result in a smaller desired family size and reinforce the decline in fertility.

Third, women's empowerment and job opportunities would also lead to later childbearing and wider birth spacing that slow population momentum. Slowing population momentum, like the first and second objectives, also requires fully responsive and effective family planning programs.

In sum, rapid population growth remains a critical national concern for our country. A sound population policy must be part of good governance to promote faster economic growth, lower income and wealth inequality, and hasten poverty reduction. The Church and State must arrive at an entente on this critical issue—an understanding on the need for a national population policy—as has long happened in other countries. The national government must take the lead in formulating, financing, and implementing a population policy so that LGUs would carry out effective programs. This is what the majority of Filipino women across all socioeconomic classes want, and Filipinos in general have affirmed the importance of addressing the population issue, as surveys have consistently shown time and again.

February 24, 2005

Pro-poor Programs or Pro-poor Growth?

Ernesto M. Pernia

Malacañang recently announced that the government would fund substantial programs for the poor, such as the provision of subsidized rice, instant noodles, and medicines, among others. Funding would come from the putative ₱35-billion "budgetary savings"—the difference between the programmed and actual budget deficit in 2005. This announcement has attracted comments from some quarters but arguably not enough for an informed public debate on an issue as crucial as poverty alleviation involving a nontrivial amount of fiscal expense. Should the government be doing pro-poor programs or focus instead on raising economic growth? This question is not unimportant especially for a country that is under a serious fiscal strain and where poverty is chronic and widespread.

One of the most robust results of empirical research in development economics is that rapid and sustained economic growth leads to significant and long-term poverty reduction and that direct pro-poor programs are inefficient and unsustainable. The much-heralded dynamic development-cum-poverty reduction of our East and Southeast Asian neighbors is living testimony to this. For example, South Korea, Taiwan, and Malaysia have been able in two decades to bring down their poverty to levels close to those of Western countries. Similarly, Thailand and Indonesia, which had poverty rates of above 40 percent—comparable to that in the Philippines—in the early 1980s, have reduced those rates to less than 10 percent and below 20 percent, respectively, by the early 2000s, while the Philippines' poverty incidence remains at around 33 percent.

Several observers and analysts have characterized the economic development of our Asian neighbors as "pro-poor growth." So, what is pro-poor growth? Simply put, it is growth that enables the poor to actively participate in and significantly benefit from economic activity. This is a departure from "trickle down" which was the dominant development concept in the 1950s and 1960s. Trickle down means that the benefits of economic growth go to the rich first and the poor gain only indirectly through a vertical flow. Poverty could diminish but only slowly at best, as has happened, unfortunately, in our country.

In contrast to trickle-down development, pro-poor growth requires a strategy that is biased in favor of the poor so that the poor benefit proportionally more than the rich, thereby allowing faster poverty reduction. Key to the definition of pro-poor growth is the

Economics Department, University of the Philippines, Diliman, Quezon City, 1101, Philippines.
empernia@up.edu.ph

joint consideration of growth and the distribution of such growth, i.e., a concern about not only how fast the national pie expands but also how the increment to the national pie is distributed. In other words, pro-poor growth requires that the proportional income growth of the poor exceeds the overall average income growth. For instance, if the Philippines' GDP per capita growth rate in 2005 were 2.7 percent (overall GDP growth of 5 percent minus population growth of 2.3 percent), average GDP per capita growth rate for the poor should appreciably exceed 2.7 percent, and such excess sustained, for the country's poverty incidence to go down significantly over time. Analysis of historical data, however, reveals that the poorest households' income rises by less than 60 percent of the overall per capita income increase. This explains our country's mediocre poverty reduction record.

In general, for a given increase in average incomes, poverty falls fastest if growth results in lesser inequality, meaning that the poor are benefiting proportionally more than the non-poor. The opposite situation obtains if growth results in greater inequality, i.e., the distribution of the increment to the pie favors the non-poor. A neutral outcome is when growth keeps inequality unchanged.

What policies promote pro-poor growth? A growth strategy that makes efficient use of labor—the poor's principal asset—and makes the required investments in education and health, is good for both growth and distribution. Sound macroeconomic management that emphasizes fiscal prudence and good tax administration results in manageable deficits and public debt, and facilitates physical and social investments that benefit the poor besides ensuring long-term growth. Economic openness, underpinned by a realistic exchange rate, promotes exports and foreign direct investment, creates employment, lowers prices of consumer goods, and facilitates the adoption of advanced technology for the economy to move up the global value chains. Private sector development, stimulated by a favorable investment climate, generates jobs, raises productivity, and reduces the strain on the public sector, thus allowing it to concentrate on the provision of public goods and services, including social safety nets. Efficient financial intermediation lowers the cost of capital, eases access to credit, and spurs investment and employment growth. Labor market deregulation facilitates labor mobility and the efficient use of production factors, contributing to output growth.

Pro-poor growth also entails the removal of institutional and policy-induced biases against the poor. Discrimination on grounds of class, gender, ethnicity, and religion hurts the poor more than the rich; the same is true of artificial barriers to entry into certain trades and occupations, or into the formal sector in general. Big-city-oriented industrial location policies and public infrastructure spending for urban areas tend to be biased against smaller towns and rural areas where there are large concentrations of poor and disadvantaged people. Similarly, there are micro policies that work against the poor, such as monopoly enjoyed by some firms that result in high prices, subsidized public utilities (e.g., low water fees) and state universities (low student fees) that benefit primarily the nonpoor, and housing policy (rent control) that limits housing supply. Removal of these biases is likely to enhance market efficiency besides promoting social equity.

The foregoing policies can be reinforced by direct pro-poor policies. These include, among others, adequate public spending for basic education, health and family planning services, easier access to micro-credit, promotion of small and medium enterprises, and targeted infrastructure investments in bypassed or underserved rural areas. Human and physical capital investments favoring the poor will improve their productivity and contribution to the economy.

The above are examples of policies that would promote efficient and equitable growth. Earnest implementation of such policies would result in mean incomes of the poor rising faster than overall average incomes and sustained poverty reduction. Our government already has some of these policies on the drawing board but falls short of consistent and sustained implementation. Thus, rather than get its attention and resources distracted by unsustainable pro-poor programs, it would be better for the government to relentlessly pursue a policy reform agenda that fosters pro-poor growth.

Fish or fishing skills? We all know from the old adage that the poor will be better off with the latter.

March 9, 2006

Arsenio Balisacan

High Population Growth Threatens Food Security

More people suffer from hunger in the Philippines than most of its neighbors in the East Asian region. According to the UN Economic and Social Commission for Asia and the Pacific (UNESCAP), the Philippines has the largest proportion (23 percent) of the population below the minimum level of dietary energy consumption from the late 1990s to the early 2000s among such countries as China (9 percent), Vietnam (18 percent), and Indonesia (6 percent).

Equally troubling is the fact that progress on poverty reduction has been particularly sluggish in the Philippines. Notably, China and Vietnam have managed to reduce absolute poverty at remarkably rapid rates. From a level of about 30 percent of the population in the early 1990s (based on a poverty line of $1 a day per person), China has reduced its poverty level by almost half; it is now about 16 percent only. Similarly, Vietnam, whose absolute poverty level was about 15 percent in the early 1990s, managed to bring that figure down to close to two percent in less than a decade. A fantastic feat!

Is the rapid population growth in the country a key factor contributing to the dismal state of our population's overall welfare and food security situation? Definitely, as we will demonstrate below. We note that the population growth rate in the Philippines is one of the fastest in the Asia-Pacific region.

Food security and poverty reduction are foremost among the development objectives of developing countries, including the Philippines. And these are not just a matter of making sure food supply is available. The population must be able to have access to food as well. The fate of the agriculture sector plays a pivotal role in this endeavor. As it is true for most developing countries, the development of the agriculture sector is crucial, not only because it contributes significantly to national income and employment, but also because a substantial proportion of the poor—three out of every four—is dependent on this sector. Even poverty in urban areas is partly an indirect effect of poverty on agriculture because extreme deprivation or lack of livelihood opportunities in rural areas induces rural-urban migration.

But the performance of the agriculture sector has been quite weak. The weak performance is evident in the comparatively low growth of agricultural output. Among the major Asian economies, the Philippines had the lowest average agricultural growth

Southeast Asian Regional Center for Graduate Study and Research in Agriculture, Philippines
ambalisacan@agri.searca.org

rate during the past two decades, averaging only one percent a year in the 1980s and 1.6 percent in the 1990s.

Improvement in efficiency—that is, increases in total factor productivity (TFP) growth—holds the key to successful cases of rural development. A study showed that from 1980 to 2000, the Philippines was the worst performer among selected Southeast Asian countries; its annual TFP growth rate was 0.2 percent, while that of Thailand was 1.2 percent and Indonesia's was 1.5 percent.

These low growth rates translate into high food prices, low farm incomes, and demand for high nominal wages. It should be stressed that farmers do not benefit from high food prices since the large majority of them are net buyers of food.

With such lackluster performance from the agriculture sector, it is therefore not surprising that poverty levels continue to be high in the country, especially in rural areas. Although agriculture has recently been generating impressive growth rates relative to those registered in the 1980s and 1990s, sustainability is another cause for concern.

The predicted trap of rapid population growth and widespread hunger was considered inevitable with what was perceived as fixed resources and slow technological growth. However, other countries in the region were able to sidestep this predicament. For one, most of these countries were able to make investments in infrastructure and rural development. And besides, their population growth rates were slower than that of the Philippines.

By 2003, the National Statistics and Census Board (NSCB) estimated the Philippine population at 82.3 million. The most recent average annual growth rate of the population is 2.36 percent. At this rate, the country is expected to double its size in less than 30 years. On the other hand, most neighboring countries have been growing at less than 2 percent.

This scenario compels the Philippines to channel its limited budgetary resources to meeting the basic needs of its burgeoning population, instead of directing it to investment programs that could boost the productivity of the agriculture sector.

As it is, productive land is already scarce in the country, so the agriculture sector can only be ameliorated by land-augmenting investments such as irrigation development. The Philippines, however, severely lacks the investment resources required for agricultural growth and rural development.

Worse, the population pressure compels poor families to move to and cultivate fragile or marginal lands. Indiscriminate logging may not be the sole culprit of the recent devastation from floodwaters brought on by the series of tropical storms that hit the country. Human intrusion into forest areas and the uplands and implementing lowland agricultural practices on them also result in resource degradation. Furthermore, those who are forced to settle on these marginal lands face the possibility of landslides and flash floods during times of bad weather.

Though nothing can be done to control the forces of Nature, there are feasible efforts to be made to assist the hungry and the helpless. One area for serious consideration is addressing the rapid population growth. And what effect would a slower population growth have on the Philippine agricultural sector, and to poverty reduction efforts for that matter?

In an earlier study (Balisacan et al., 2003), we have looked into how population dynamics affect economic growth and, in the process, have also attempted to answer such concerns.

The study focused on comparisons between the Philippines and Thailand because, about thirty years ago, these two countries had more similarities than differences in socioeconomic terms, including food security situation. But looking at them now, the countries are on quite divergent growth paths, and have been so for the past couple of decades.

According to the study, if the Philippines had followed Thailand's population growth path from 1975 to 2000, Gross Domestic Product (GDP) per capita for the Philippines could have been higher by 22 percent, reaching $1,210. The relationship between economic growth and poverty reduction has been established by empirical studies. So going further, simulating a slower population growth for the Philippines results in a lower incidence of poverty, translating into some 3.6 million more Filipinos who could have been lifted out of poverty in 2000.

A slower growing population, particularly a smaller youth population, would have meant substantial savings from the provision of basic education and health services. These savings could have been channeled to improving the rural sector, specifically agricultural development. As previously mentioned, about two-thirds of the poor are found in the rural areas where most of them are engaged in agriculture-related activities. In the study, the estimated savings from basic education for the 1990s were ₱128 billion (or roughly ₱12.8 billion per year), and from basic health, ₱52 billion (or ₱5.2 billion per year). In the simulations, allocating these for agricultural investments would have increased incomes and reduced poverty incidence in the agricultural sector by more than half!

What needs to be appreciated is that enormous amounts of resources are required to provide the basic needs of a rapidly growing young population. Since fiscal resources are very scarce (and government is almost always in serious deficits), government spending for a rapidly growing population crowds out investments required for raising productivity, opening up employment opportunities, and sustaining gains in human development.

By no enormous exertion of the imagination, it may be gleaned that arresting the rapid growth of the Philippine population could very well address concerns regarding food security as well as significantly reduce poverty and food insecurity in rural areas. The status quo is not acceptable, and the facts reflect it.

March 17, 2005

Edgardo Gomez

Are There Enough Fish for the People?

Time was when fish was truly the poor man's affordable protein. This situation has changed drastically in the recent past and there are clear indications that the situation is worsening rather than improving. This is the consequence of a vigorous population growth, the continued deterioration of natural aquatic environments, and overfishing in many parts of the world, especially in the Philippines. And it is not natural disasters like tsunamis and typhoons that are the principal causes of the degradation of our natural ecosystems but man, who has evolved from just one of many species on this planet to become a technological, "geological agent," capable of altering the natural characteristics of our planet.

But after having mentioned tsunamis, let me call your attention to the advice of one of the world's renowned fisheries scientists, Dr. Daniel Pauly, formerly with the World Fish Center (known to many Filipinos as ICLARM or the International Center for Living Aquatic Resource Management) and who is now a faculty member of the University of British Columbia. I was recently in Newcastle, England for a meeting about a new project on coral reef restoration in which I am involved. Having arrived there a couple of days early from an earlier meeting in Rome on a parallel project, I decided to spend one afternoon in the library of the University of Newcastle upon Tyne, since our local libraries are so lacking in scientific journals. While in that modern, fully equipped library, I picked up the current issue of the weekly science journal *Nature* (February 3, 2005) and read a letter of Dr. Pauly to the editor. He was reacting to an article in the journal about the recent tsunami that has claimed more than a quarter of a million lives. That article entitled "Scientists seek action to fix Asia's ravaged ecosystems" highlighted the frenzy among donor-countries to get started immediately to reverse the devastation of the tsunami that did extensive damage to coastal ecosystems. In addition to the call on scientists to take measures that will contribute to the rehabilitation of "ravaged ecosystems," the article pointed to the need to rebuild the decimated fishing fleets in the countries most directly affected by the natural disaster. This is the normal mindset of the investor-countries that fail to see the broader issues about the present condition of marine and coastal ecosystems.

Dr. Pauly's letter was captioned "Rebuilding fisheries will add to Asia's problems: Overfishing has already caused depletion and conflict. Instead, train people for new

The Marine Science Institute, University of the Philippines, Diliman, Quezon City, 1101, Philippines.
edgomez@upmsi.ph

jobs." In effect, he answered the question of the title of this article. His letter ends with the exhortation to amend the old adage that went: "If you give a man a fish, he will eat for a day. If you teach him how to fish, he will eat for the rest of his life." He pointed out that since this no longer holds, it should now be amended to teaching man "to repair bikes, sewing machines and water pumps." Basically his message is the need to train the displaced fishers to do land-based jobs instead of letting them go back to sea to fish and contribute to the further decline of fisheries stocks. There just are not enough fish left for all the mouths to feed!

His message is at least ten years old. In 1995, Population Action International, a US-based NGO, produced a poster with the title "Catching the Limit: Population and the Decline of Fisheries," which outlined some of the pessimistic statistics in fisheries. The global fish catch peaked in 1989 at 89 million metric tons (mmt) and hovered around 85 mmt. The United Nations Food and Agriculture Organization (FAO), then already estimated that nearly 70 percent of the world's conventional fish species were already "fully exploited, overexploited, or rebuilding from past overfishing—in other words, fished up to or beyond natural limits." In many parts of the world, fishers were catching less fish than they once did, a situation one sees in virtually all coastal villages in the Philippines.

What is alarming is that with the human population continually growing, there is less fish available on a per capita basis, and this is most true in our country. Because of the rising demand for fish and fishery products globally, the price of fish is becoming prohibitive to more and more poor people (if not to the middle class), for whom fish is the major source of animal protein. What this means for future generations of Filipinos is malnutrition, resulting in decreased brain capacity, particularly among the poor who are thus condemned to servitude and menial tasks. As more than one writer has pointed out, we seem to be unwittingly raising more and more Filipinos who are most suited to wiping the bottoms of foreigners, whether at home or abroad. The poster I mentioned points out that in 1990, our per capita consumption of fish for food was 36.5 kg, a level that was significantly higher than fish consumption in many developed countries. However, it is projected that by the middle of this century, the fish consumption may drop to 21 kg, 17 kg, and 14 kg per capita, respectively, depending on the low, medium, or high projection scenarios of population growth in the country.

Since my work is related to coral reefs, one may ask: what is the connection between reefs and fish? In many of our coastal areas, the relation is direct. Healthy reefs can support 15 to 30 metric tons of fish per square kilometer, while bombed reefs would support only two to five metric tons. Who bombs the reefs? The greedy and the desperate. What is the relation to the population? Since there is a growing demand for fish, the greedy fishers often use illegal and destructive methods for fishing, thus degrading the coral reef ecosystem. The desperate fishers become guilty of what Dr. Pauly refers to as Malthusian overfishing. With so many people to feed, many coastal fishers become desperate to have fish for the next meal that they often bomb or dynamite the reefs they have access to since traditional methods are often inadequate for catching enough fish for a family for the day.

Others may say that we can solve the problem with aquaculture. While it is true that aquaculture is showing strong growth, it is also certain that it is slowly degrading natural ecosystems, thus aggravating the plight of poor coastal fishers. The public has been made aware many times about spectacular mass fish kills in coastal towns like Bolinao and Binmaley in Pangasinan. But one has to witness a municipal beach carpeted with rotting *bangus* to truly appreciate what can happen with aquaculture. Culturing fish is not a quick fix to a serious problem that is on the rise. It should contribute to the solution but too often it adds to the problem.

It is high time that government officials and church leaders alike acknowledge the fact that this planet has a limited carrying capacity, not just individual ecosystems. Even with the best social systems, the problem of reaching the limits of the natural carrying capacity is not going to go away unless we take serious and immediate steps to curb our runaway population growth.

March 10, 2005

Ben Malayang III

Population and Environment: Why They Matter Together

When "population" pops up in ordinary conversations in the Philippines, the topic usually centers on numbers (how many we are now—about 83.7 million as of mid-2004). Or we talk of the rate we are increasing over time (presently 2.0 percent each year without counting migration—referred to as Rate of Natural Increase). Some would note that this is higher than other Catholic countries like ours, like Columbia (1.7 percent), France (0.4 percent), Italy (-0.1 percent) and Spain (0.1 percent). Or the conversation can go into the more controversial aspect of how we can control our population, such as to whether we should promote artificial contraception or not. Some would point out that the prevalence of the use of artificial contraception in the Philippines (33 percent among married women) is much lower than in other Catholic countries like, again, Columbia (64 percent), France (69 percent), Italy (39 percent) and Spain (53 percent). We make the comparisons because we see that these other countries (and most countries with smaller populations relative to the size of their land and seas from which they derive their basic wealth) are generally much more prosperous than us. But beyond these, we frequently talk of "population" as something that really does not concern *ourselves*. We are concerned about our children and how many we need to feed, clothe, and send to schools and to doctors for check-ups, but "population" is something for government to worry about. Many would say that the government should worry about our population, but not for us to do so ourselves. It really does not matter to us personally.

Also, when "environment" pops up in our conversations, we dwell on only the usual topics like air and water pollution, garbage and wastes, deforestation, declining fisheries, and the protection of certain prominent species like the Philippine eagle (*P. jefferyi*), the tamaraw (*B. mindorensis*), turtles (*Cheloniidae* and *Dermocheliidae*), whales and other sea mammals like the sperm whale (*P. macrocephalus*), bottlenose dolphin (*T. truncatus*) and *dugong* (*D. dugon*), and the tarsier (*Tarsius syrichta)*. Then, too, beyond talking about it, most of us hardly take environment as our personal worry. We say that the government must do something about it because we worry about how a bad environment will increase our doctors' bills, lower the market value of our homes, or allow toxic pollutants to seep into our foods, the air we breathe, and our drinking water. But the "environment" itself is something we hardly fret about most days in our lives.

Philippine Sustainable Development Network, Philippines.
beniiim@yahoo.com

Not only do we not think of the two as matters that concern us personally, we also hardly link them as *interwoven causes* for *our* worry. Yet, they are, based on what is known in today's study of ecology:

1. *Population exerts a generally net negative pressure on the environment.* Humans are on the top tier of the food chain (referred to as trophic structure). This means that they derive their subsistence from an array of plants and animals in the lower levels of the chain, which cascades into ever increasing amounts and numbers as they go down the chain. It works like this: each plant requires an array of nutrients from the soil where it grows and the large number of plants that a plant-eater (herbivore) will need in order to survive will also require a large amount of soil nutrients that is only a multiple of the amount required by one plant. Eaters of the plant-eaters (carnivores) will correspondingly require an even higher multiple of the amount of soil nutrients to support the amount of plants that the number of herbivores would need to support one carnivore. And if in higher trophic levels there is one other organism (say, an omnivore, or an eater of both herbivores and carnivores), which also needs to survive, then the pressures on lower carnivores, herbivores, and ultimately on the pool of nutrients in the soil, will multiply at a rate that increases ever higher (or geometrically) as one goes down the food chain. Humans are on top of all these. We consume omnivores, carnivores, herbivores, and plants for food, housing, fuels, clothing, industries, commercial products, and medicines. A rapid increase in human populations exerts pressure on the food chain with an increasing probability that the pressure rises above the rate that Nature can replenish each affected element in the food chain (e.g., like if the demand for certain plant foods rises above the rate they can be grown). When the pressure continues to rise to a point beyond the ability of Nature to replenish the *lowest* elements in the food chain (e.g., soil nutrients), the food chain may collapse. Add to these the complex of factors in the trophic structures of our environment like pests, diseases, drought, flooding, and severe climate events, which can further multiply the pressures on any one element of the food chain, and hence, ultimately, on their ability to support the other elements in the chain. This can be true for food chains on land, in seas, or in those spanning both land and seas.

A collapse of the trophic structure of the environment can be, of course, mitigated. That is, if each human is able to put in more into the structure than he/she derives from it. Meaning, a collapse can be prevented if each human that consumes plants, animals, and nutrients is able to replenish what he/she consumes at a rate higher than his/her consumption. Indeed, there are many ways this is done today. The genetic traits of plants and animals are being modified so that we are able to produce more of them than before. The quality and regularity of our water supply are being improved and stabilized so that more plants, animals (and ourselves) can have better quality of life. And industries are being redesigned so that we can produce more of the things we need with lesser amounts of natural resources used to produce them.

But one problem remains: each human represents from birth to date a *certainty* of consumption but only a *probability* of production. That is, every human being is *sure* to need something to survive and to live a life but is *not sure* if he/she would be able to put back into Nature more than what he/she takes from it. And if our population is both large and rising fast (to a point when it becomes harder to fully feed, clothe, ensure the health, and educate our children), the probability increases that more of us would be net takers from Nature rather than net givers. Consequently, assuming all things equal (say, we don't send our people abroad), pressures on our environment will rise and trophic collapse can become even more probable. This matters to us. And it is a cause for our concern.

2. *The symptoms of net negative human pressures on the environment are escalating.* Data from global surveys are showing that in places with higher populations per unit area of accessible land and marine economic zones, the quality of the environment is generally deteriorating. The quality of lands and soils, forests, biodiversity, freshwater resources, coasts and seas, air and water bodies, are observably on the decline in Africa, Latin America and the Caribbean, the Asia-Pacific and West Asia. In contrast, the quality of lands and soils in North America is slightly improving, and air quality and the quality of water bodies, forests, biodiversity, freshwater resources, coasts and seas, and of lands and soils is relatively stable in North America and the polar regions. In the Philippines, the denser regions outside Metro Manila—e.g., Regions 1 (third densest), Region 2 (first densest), Region 7 (second densest) and Region 6 (fourth densest)—have among the highest percentage of degraded lands among all regions (Region 1 is first, Region 3 is second, Region 6 is fourth, and Region 7 is sixth). The denser regions also generally have the lower percentage of remaining forests—e.g., Region 7 (second densest and the lowest percentage of forests left), Region 1 (third densest and fourth lowest in forests left), and Region 5 (fifth densest and second lowest in percentage of forests left). Also alarming are the regions with otherwise low resident population densities but have among the lowest percentage of forests left (e.g., Regions 1, 9, 10, 11, and Caraga). This is because their forests are cut to supply higher demand elsewhere (e.g., where population densities are highest like in Metro Manila, Cebu, and other urbanizing centers in the archipelago).

Human pressures on coastal and marine resources have been increasing as well. This is particularly so in areas with high densities of coastal populations. It is well-known that less than 25 percent of our corals (which host the most of our nearshore fisheries) remain in good condition. Illegal and destructive fishing like the use of poisons and coral-destroying methods have been on the rise thereby reducing our ability to catch fish in our municipal waters in less costly ways. Estimates point to our country losing as much as perhaps ₱130 billion a year in fisheries due to illegal and destructive fishing and pollution. Catch per unit effort (CPUE) of fishing has been estimated to have declined by as much as 90 percent since the 1950s.

There are many other indications of increased human pressures on our environment. For example, our mangrove cover has declined from over 500,000 hectares in 1908 to less than 150,000 hectares today; our soils are getting more salty in many areas, or more acidic in others, or exposed to inorganic pollutants from agriculture and mine tailings. And the air we breathe in many of our cities is ever increasingly laden with dust and other dangerous pollutants (like carbon dioxide, sulfur oxides, and nitrogen oxides).

Overall, the estimated value of environmental degradation from four economic activities in the country—activities meant to meet the rising demands for materials and services by our burgeoning population, including (1) agriculture, fishery, and forestry; (2) manufacturing; (3) mining; and (4) land transportation)—has risen 137.8 percent in six years, from ₱3.7 billion in 1992 to ₱8.8 billion in 1998. Of the four, manufacturing seems to show the biggest increase from ₱2.2 billion in 1992 to ₱7 billion in 1998. Air pollution from manufacturing rose 92.3 percent from 789,020 metric tons (MT) in 1992 to 1.5 million metric tons (MMT) in 1998. Water pollution from manufacturing rose 23 percent from 2,094 MT of biological oxygen demand (or BOD, a measure of the rate of oxygen depletion in the water associated with microbial activities to decompose organic wastes in it, used as an indication of the amount of wastes in the water) in 1992 to 2,579 MT BOD in 1998. The value of degradation estimated from land transportation represents the emissions we are producing from heavily congested streets and highways which tend to accumulate as vehicles rev up in slow traffic. These are linked to our large and fast-growing population which is making it harder for the government to sustain spending for additional infrastructure.

3. *A weakened environment generally weakens the ability of human populations to meet their needs.* Human needs, for the most part, are procured from ultimately natural sources. Foods, fuels, fiber, most clothing materials, construction materials, medicines, and many other products used by humans come from the flora, fauna, and other biological entities (or their extracts), minerals and processes (like water, carbon, and nutrient cycles) in Nature. Synthetic products like nylon, rubber, pesticides, inorganic fertilizers, and molluscicides are laboratory mimics of information from Nature. When the regular functions of Nature are disrupted (say, by pollution or by nutrient loss in soils), its ability to produce (or to support the production of) goods and services needed by humans would be disrupted as well. This makes it harder for humans to derive additional products and services from Nature. For instance, it has been shown that, in general, in our country where forest cover is low, water recharge tends to be also low. Thus, water availability can decline and this seems presently indicated by the price of delivered water in our urban areas which has been mostly rising in the past ten to fifteen years. Or, when large mining areas are improperly abandoned (after having been mined out), their sites remain as wastelands degraded by pollution from tailings and cyanide and mercury from processing ores. Consequently, they can no longer support agriculture or biodiversity

or become sites for healthy human settlements, which are equally critical needs of the population. When forests decline, their capacity to hold water, or to act as wind and pest breaks to protect agriculture and human communities, could also decline to such a point that catastrophic events (like typhoons, heavy rainfall or pest infestations) can become even more catastrophic (a phenomenon I earlier referred to as "enviromagnification"). When human populations increase, the land they would need for siting their homes, schools, roads, hospitals, and other necessary infrastructure (like seaports and airports, parking sites and recreational areas) also increases, thus requiring large areas of land and water that otherwise could have supported agriculture, the production of trees, the protection of biodiversity, or water processing. Today, 22.9 percent of our total land area is built-up (or used for these purposes). Some 33 percent of our lands are devoted to farming, leaving only 44.3 percent for other equally important environmental production like forests, freshwater reserves, pastures and lands for wildlife. The general trend is that degraded environments offer less opportunities for the production of important ecological services, or if the services were to be maintained under these conditions, they would be costlier than before. This, too, matters to us.

Conclusion. Population and environment are not *apart* from how we are able to pursue a quality life. They are *a part* of it. We can deny their impacts on our personal lives and choose to disregard the evidence of science; that is our choice. But it does not make the impacts untrue if, as most scientific studies suggest, they are, in fact, true. And if only because they *might* be true, let us at least think about them as two issues that *do* matter to us personally.

(The data used in this article came from many sources. The author acknowledges them. Limited publication space, however, precludes their being listed here. The list can be provided upon request.)

March 3, 2005

Author Index

The Editors

GISELA P. PADILLA-CONCEPCION has a PhD in Chemistry and is a professor at the Marine Science Institute, College of Science, University of the Philippines, Diliman. She is the Chair of the Office of Special Initiatives for the Advancement of the Sciences, Dean's Office, College of Science, UP Diliman. She is also a member of the National Academy of Science and Technology, Philippines and the recipient of the 2008 University of the Philippines Alumni Association (UPAA) Outstanding Award for Science and Technology.

EDUARDO A. PADLAN has a PhD in Biophysics and was a research scientist at the (US) National Institutes of Health, until his retirement in 2000. He is currently an adjunct professor at the Marine Science Institute, College of Science, University of the Philippines, Diliman. He is a corresponding member of the National Academy of Science and Technology, Philippines. He is also the Editor-in-Chief of *Philippine Science Letters* (*www.philsciletters.org*), a peer-reviewed, online journal that showcases science done in the Philippines and science done by Filipinos elsewhere.

CAESAR A. SALOMA has a PhD in Physics, and is a professor at the National Institute of Physics, College of Science, University of the Philippines, Diliman. He is presently the Dean of the College of Science and is a member of the National Academy of Science and Technology, Philippines. He is the recipient of the 2004 Galileo Galilei Award of the International Commission for Optics and the 2008 ASEAN Outstanding Scientist and Technologist Award. He also received the Metrobank Foundation Outstanding Teacher Award (Higher Education categoy) in 2007.